Practical Solutions to Integrated Oil and Gas Reservoir Analysis

Practical Solutions to Integrated Oil and Gas Reservoir Analysis

Geophysical and Geological Perspectives

Enwenode Onajite

ELSEVIER

Elsevier
Radarweg 29, PO Box 211, 1000 AE Amsterdam, Netherlands
The Boulevard, Langford Lane, Kidlington, Oxford OX5 1GB, United Kingdom
50 Hampshire Street, 5th Floor, Cambridge, MA 02139, United States

Library of Congress Cataloging-in-Publication Data
A catalog record for this book is available from the Library of Congress

British Library Cataloguing-in-Publication Data
A catalogue record for this book is available from the British Library

ISBN: 978-0-12-805464-2

Publisher: Candice Janco
Acquisition Editor: Amy Shapiro
Editorial Project Manager: Tasha Frank
Production Project Manager: Mohanapriyan Rajendran
Cover Designer: Matthew Limbert

Cover: Contains special section on pore pressure prediction with case application GOM.

Typeset by SPi Global, India

CONTENTS

SECTION 3 PORE PRESSURE PREDICTION

FOREWORD

As an oil and gas reservoir matures it becomes more complex. People working an oil and gas field breaks down into small groups of geoscientists and reservoir engineering specialists, intent on solving their own problems. Groups become more focused and tend to talk less with those outside their specialty. So, it becomes harder for professionals to keep up with other fields. Not only is their time limited, but the increased complexity of other fields makes it harder to understand them in a reasonable amount of time. This makes cooperation and communication that much harder.

This is a serious problem in the petroleum industry. A lack of communication and understanding between geologists, geophysicists, reservoir engineers, drilling and exploitation professionals can be very costly. In extreme cases it costs human life.

Drilling and production personnel need the input that geologists and geophysicists can give, particularly in the offshore where over-pressured formations pose a real drilling risk. But those geo-professionals are on the whole, ignorant of the methods and needs of those who drill and produce the wells. Drillers on the other hand do not, on the whole, appreciate the limitations of the information coming from these other fields and cannot adequately quantify the data's reliability. A potentially dangerous state of affairs exists. The problem of communication is not restricted to professionals, but is also noticeable with academic researchers, students and lecturers in the geoscience disciplines. There is just not enough time to delve into all the technologies available.

In this book Enwenode Onajite seeks to foster communication and understanding between geoscience groups in the interest of oil field safety and good practice. Without sufficient communication and understanding, drilling accidents will continue to happen putting great amounts of money, the environment and most significantly, human lives at risk.

In his first book, Seismic Data Analysis Techniques in Hydrocarbon Analysis, Mr Onajite laid out the field of geophysics in a simple to understand form. In the last several pages he touched on the uses of seismic data to evaluate the fluids filling rocks and identifying potential accumulations of gas and oil.

In this book, he pushes further into the science and art of basin analysis (both conventional and unconventional reservoirs), seismic interpretation and quantitative seismic interpretation (reservoir geophysics) with an eye to eventually linking it to pre-drilling planning and detailed case studies from the Gulf of Mexico and Bossier and Haynesville Shales (unconventional reservoir).

Although he provides background information on each topic, Onajite adopts the novel approach of letting working professionals educate us through practical questions/challenges encountered by professionals in oil and gas reservoir development. The answers are provided by an intercontinental selection of knowledgeable professionals from a number of working oil and gas development companies.

Using the voices of actual professionals gives a far better, more applicable picture of the uses and misuses of information provided by the geosciences. It provides a shortcut to understanding the intricacies of the different technical fields since only real questions directly from users are presented. Theory may enter in to the discussion, but it is always related to actual exploration, drilling and exploitation problems. A detailed table of contents allows the reader to quickly find the section and questions that are of interest to them. This makes the book easy to use by professionals in all the roles identified in the book as well as other interested parties.

Since cooperation, understanding and communication between specialists in our industry is vital, a thorough reading of this innovative book is time well spent for any professional involved in exploration and production. Also, this process of increased communication can, through use of this book, yield positive effects with students and instructors as well. It should prove especially valuable to student and applied researchers.

Robert W. Avakian
Geophysicist Instructor, Oklahoma State University,
Institute of Technology, Okmulgee, OK, United States

ACKNOWLEDGMENTS

I am most grateful to the many people who have helped and inspired me in the writing and development of this book.

With a heart full of appreciation, I say thank you to Rev. Chris Oyakhilome, PhD, for building a strong Godly core in me and for instilling in me the right mindset to succeed in any endeavor in life. I am indeed grateful for the words you said to me during those youth conferences you graciously organized for us: "Refuse to let your mind go to sleep; find a need and reach out to meet that need. Be a solution provider and make a difference where you are." These words of yours have found expression in my life. Thank you for having been an inspiration to me.

Thank you to the late Dr. Myles Munroe. He taught me to "seek to become a person of value" and that you become a person of value when you discover your unique gift and make your significant contribution to the world.

Thank you to Bayo Ojulari, MD, Shell Nigeria Exploration and Production Company, and Nedo Osayande, former GM Corporate Social Responsibility, Shell Nigeria, for granting me the opportunity to be exposed to the oil and gas industry.

Thank you to Dr. Kingsley Ojoh, Executive Director, Geosciences and Reservoir Total E&P Nigeria Limited, Dr. James Edet, MD Research and Development Total E&P Nigeria Limited, Vincent Nnadi, EGM Corporate Social Responsibility Total E&P Nigeria Limited, for their encouragement.

Thank you to Osagie Lee Izekor for all the support you have given me, which I deeply appreciate.

Thank you to Louisa Hutchins, Elsevier Limited, Oxford, United Kingdom, for believing in my idea for this book and for sending the manuscript proposal to five global reviewers. I am indeed grateful.

Thank you to Frank Tasha, Elsevier Limited, United States, for your support throughout the manuscript development and submission process.

Thank you to Robert W. Avakian, geophysicist, instructor, Oklahoma State University, Institute of Technology, for the excellent job you did in editing the manuscript and for your contributions to the book. You brought excellence into this book.

Thank you to Susan Smith Nash, PhD, Director of Innovation & Emerging Science and Technology/AAPG, University of Oklahoma, United States, David Greeley, geologist at British Petroleum America and Rm Iman Argakoesoemah, senior petroleum exploration geologist, for their contributions to Chapter 1 (Basin Analysis and Unconventional Reservoirs).

Thank you to Son Le, senior geophysicist/lead seismic interpreter at Schlumberger Asia Pacific region; Peter Rowbotham, geophysical advisor at Apache Corporation, Aberdeen, United Kingdom; and Andrew Forbes McIntosh, geophysical technical lead at BP Houston, Texas, for their contributions to Chapter 2 (Qualitative Seismic Interpretation).

Thank you to David J.B. Dushman, Borehole seismic specialist at Chevron, Houston, Texas; Zaheer Ali Shah, technical coach (Geosciences) at Petronas

Carigali, Kuala Lumpur, Malaysia; and Sergio Filippo Rodriguez Spadavecchia, DCS Borehole Geophysics En Schlumberger Maturin, Venezuela for their contributions to Chapter 3 (VSP and Time-to-Depth Conversion).

Thank you to Shiyu Xu, geophysical advisor at ExxonMobil, Kingwood, Texas; Erick Alvarez, senior reservoir geophysicist at Shell Norway; Diego Timoteo, exploration geologist at Sechura Oil and Gas Peru; Brian Schulte, geophysical advisor at Repsol Calgary, Canada; Ariel Kautyian, principal senior geophysicist at Total Austral; Jim Applegate, CEO and managing member at Seismic Utensils LLC Greater Denver Area, Colorado; Sagnik Dasgupta, senior rock physicist at Schlumberger Houston, Texas; and Rob Simm, senior geophysical advisor at Cairn Energy, for their contributions to the section on reservoir characterization.

Thank you to Mark Tingay, senior geomechanics specialist at Chevron Perth, Australia; Selim Simon Shaker, consulting geoscientist at Geopressure Analysis Services, Houston, Texas; Alejandro Garcia, geophysicist and geomechanics at Total E&P UK; Tim Rolph, pore pressure analyst at BP, Houston, Texas; and Cory Moore, senior real-time pore pressure analyst at Ikon Science, Houston, Texas, for their contributions to the section on pore pressure prediction.

Also, a big thank you to Jincai Zhang and James Wieseneck, Shell Exploration and Production Company Americas, for their contributions to Chapter 9 (Pore Pressure Prediction with Case Application: Gulf of Mexico and Bossier and Haynesville Shales (unconventional reservoir)).

Finally, I am grateful to the giver of life, the creator of the universe, God almighty, from whom the inspiration and ability to create this book comes.

For the complete list of acknowledgments of the over 350 professionals in the oil and gas industry globally that contributed to the development of this book, please check after references.

SECTION

Basin Analysis and Qualitative Seismic Interpretation

1

CHAPTER

BASIN ANALYSIS: OVERVIEW AND USES

1

CHAPTER OUTLINE

Basin analysis is vitally important in oil and gas exploration. It incorporates the processes used to determine where there is a high likelihood of encountering commercial quantities of recoverable hydrocarbons, and thus is integral to prospect generation. In the oil and gas industry, the petroleum geologist determines the possible presence and extent of hydrocarbons and hydrocarbon-bearing rocks in a basin, and basin analysis is often conducted through use of reflection seismology and data from well-logging (Fig. 1.1).

USES OF BASIN ANALYSIS

Basin analysis is used in more than simply generating prospects. Geoscientists have expanded basin analysis by incorporating new technologies and techniques and it can now be used:

- For identifying sweet spots through projecting enrichment along migration pathways.
- For determining the best fluids to use in drilling and completions.
- For developing a reservoir model, and for estimating recoverable reserves.
- In identifying and avoiding geo-hazards.

Practical Solutions to Integrated Oil and Gas Reservoir Analysis. http://dx.doi.org/10.1016/B978-0-12-805464-2.00001-9

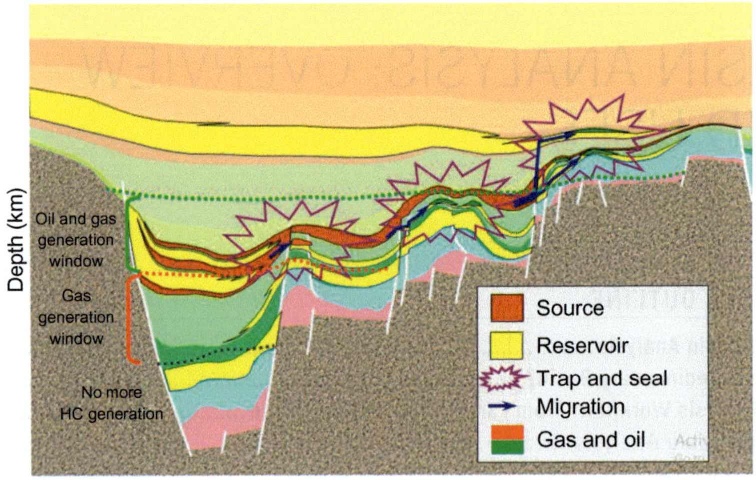

FIG. 1.1

Cross-section of sedimentary basin where petroleum system analysis is carried out. From AAPG slide show.

Courtesy: ExxonMobil and F. Schroeder.

WHAT IS A SEDIMENTARY BASIN?

Basin analysis covers sedimentary basins, which are areas on the earth's surface where sediments have been deposited. In general, basins are quite extensive and can cover tens of thousands of square kilometers. Sedimentary basins can be very deep and also can be shallow. Deep basins are often marked by high pressures and temperatures at depth (Fig. 1.2).

Sedimentary basins form because of tectonic activity in the earth's crust. Crustal subsidence occurs over time, which results in punctuated episodes of sediment accumulation, deformation, and structural activity. There are several mechanisms that result in sedimentary basins, and they include the following:

Rift basins: These sedimentary basins emerge due to the action of sea-floor spreading and pull-apart forces. They are very important for oil and gas exploration because they tend to have had excellent heat flow, which has allowed the generation of oil and ongoing thermodynamic/thermochemical processes which encourage the movement of oil along migration pathways (fractures, faults, connected pore spaces, etc.) (Fig. 1.3).

Compressional-convergent plate basins: These occur where there is structural movement, which creates normal faults with significant throw, and the subsequent shedding of sediments from uplifts and topographic highs (Fig. 1.4). It is important to keep in mind that tectonic activity can occur in many different pulses within a basin, and any pulse may not cover the entire extent of the basin.

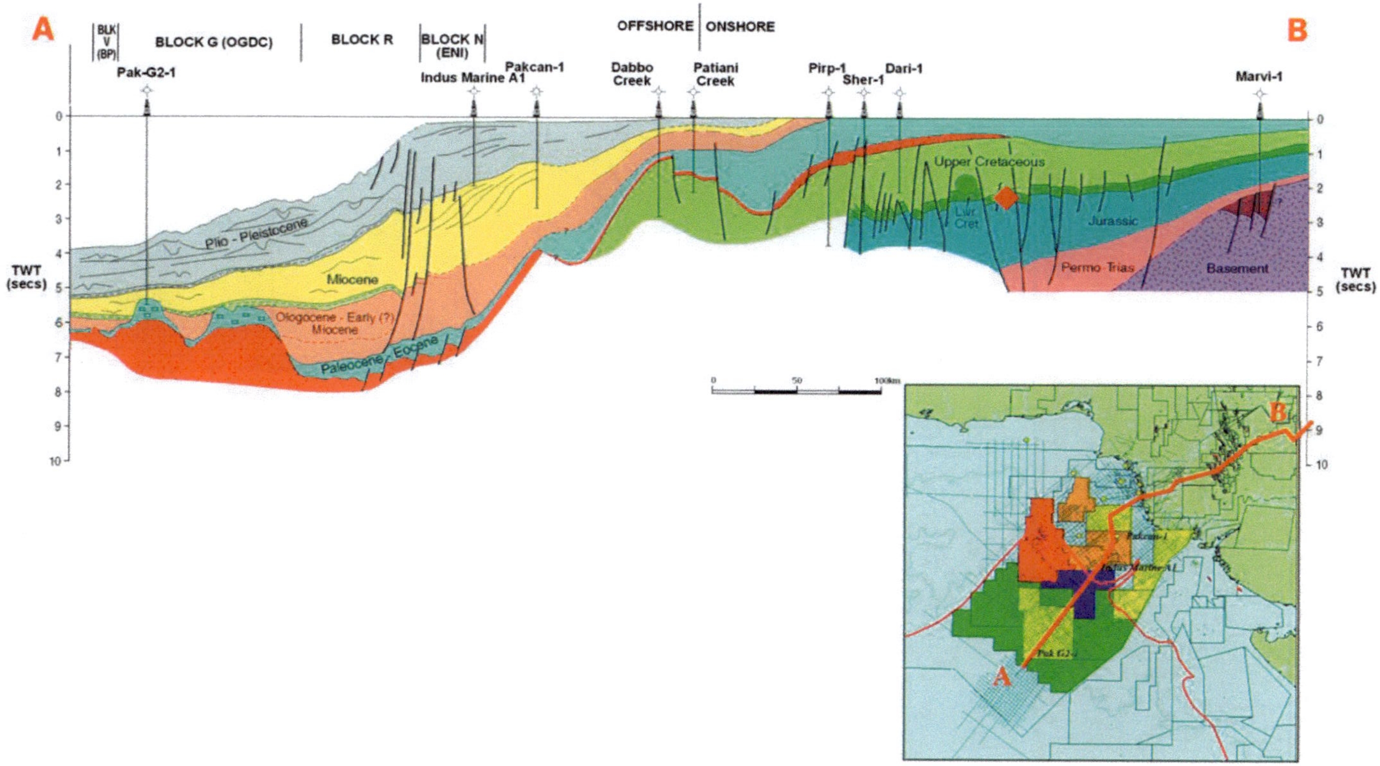

FIG. 1.2

Show Geology and hydrocarbon potential of the offshore Indus Basin, Pakistan.

Source: Carmichael, S.M., Akhter, S., Bennett, J.K., Fatimi, M.A., Hosein, K., Jones, R.W., Longacre, M.B., Osborne, M.J., Tozer, R.S.J., 2009. Geology and hydrocarbon potential of the offshore Indus Basin, Pakistan. Petroleum Geoscience 15, 107–116. http://dx.doi.org/10.1144/1354-079309-826.

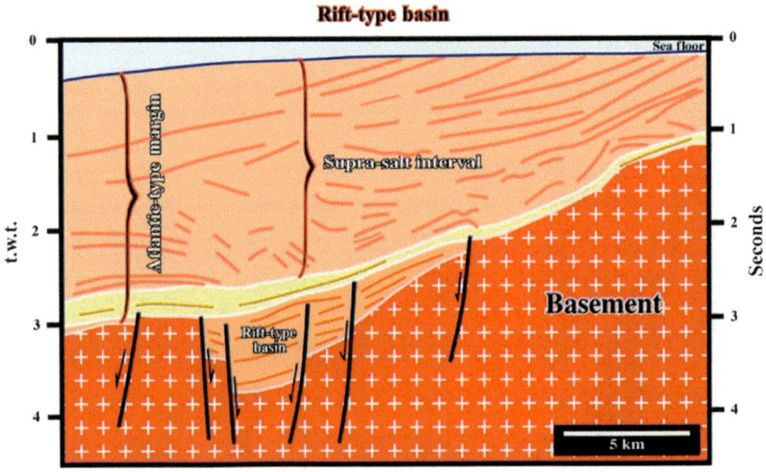

FIG. 1.3

Rift basin.

From http://homepage.ufp.pt/biblioteca/RoleOfSaltTectonicsInPetroleumSystemsAngolaGulfOfMexico/Pages/Page1.htm.

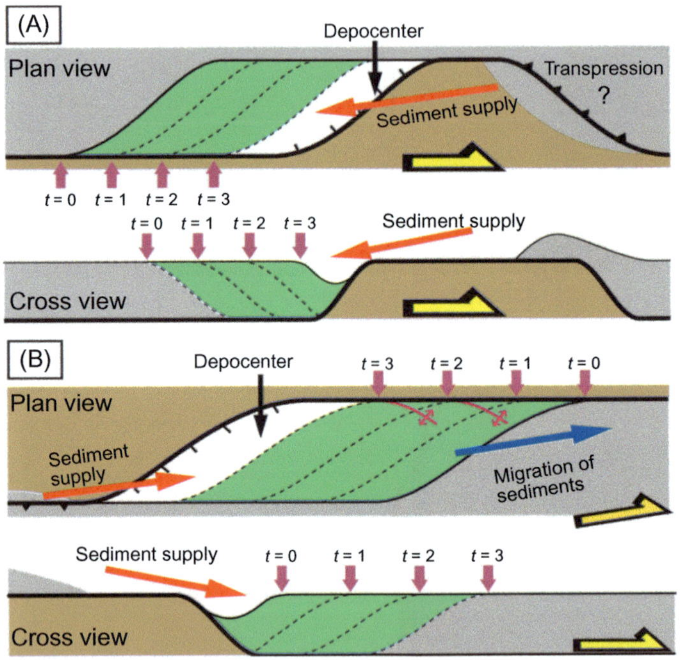

FIG. 1.4

Conceptual models for depocenter migration and axial sediment supply in fault-bend basins. (A) Progressive right-lateral migration of paired bends on the foot-wall generates compressional uplift and extensional depression on the hanging-wall. Sediments are always supplied from the same direction along the long-axis of the basin. (B) Depocenter fixes along the releasing bend result from the right-lateral migration of sediments deposited on the foot-wall. A transpressional component would be required to generate the sediment source, and en échelon folds may form along the master faults. Both models generate deposits with axial sediments whose thicknesses are greater than the burial depths.

Reproduced with permission from Atsushi Noda.

BASIN ANALYSIS WORKFLOW: FROM LARGE-SCALE TO MINI-SCALE

Large-Scale Analysis

This is essentially reconnaissance work, which can be done on both frontier basins and those that have been developed. It is often a good idea to return to a basin and conduct a reevaluation of the basin, especially if new data are available, and if new processing methods can result in a new way of viewing existing seismic, drilling, completion, and production information.

Step 1: Define the extent of the basin. The basin framework can be determined by the most important regional structural features. It is important to keep in mind the tectonic framework as well, and to locate the basin within plate movement.

Step 2: Identify the main structural features and the depocenters. Depending on the extent to which the basin has been developed, it may be possible to incorporate well data. If not, initial basin analysis generally requires the use of satellite imagery, seismic, gravity, magnetic, and information from outcrops. Outcrop information is often correlated with seismic data in order to create a preliminary subsurface basin model that includes major structural features as well as an idea of sediment depths. This is a good stage to start studying the sediments very closely in order to propose depositional environments and depositional models. By proposing working hypotheses with respect to the depositional environments of each of the formations and basin thermal history, it is possible to start to determine which ones could be good source rocks, or reservoir rocks.

Step 3: Incorporate petroleum system information. It is often possible to conduct tests on cores or to calculate total organic content (TOC) using the Passey Method. Favorable levels of TOC and types of kerogen can indicate where there is the possibility of finding economically viable recoverable reserves of hydrocarbons. The next step would be to continue with geochemical methods and determine maturation by using vitrinite reflectance or other tests. Further, obtaining thermal information can provide information useful in deciding whether the thermal history was adequate for maturation and the generation of hydrocarbons.

Step 4: Determine the structural and depositional history, at basin level. Determining the timing of major structural events is extremely important for determining the timing of hydrocarbon generation, and then the expulsion and migration through faults, fracture networks, and porous connected sediments. This is also a good time to identify possible structural traps, by defining reservoir rock, seals, and migration pathways into the possible trap.

Once the large-scale analysis has been completed, it is very important to start identifying subbasins, subbasin level depocenters and local structural movements. The medium-scale analysis is important in narrowing down the scope and starting to rank the sediments and subbasins in terms of relative prospectivity.

Medium-Scale Analysis

Deposition that results from continued episodes of tectonic activity (mainly subsidence or uplift) and the compartmentalization of the basin into subbasins and mini-basins is important for starting to pinpoint likely areas for oil and gas.

Step 1: Identify subbasins and mini-basins

Step 2: Create cross sections that reflect the stratigraphic architecture, that incorporate available information and data. First, use 2D, 3D, and 4D seismic to identify faults, thicknesses of sediments, and the changes through time.

Step 3: Include any field location information in the study. Select the most prospective subbasins and start to map major depositional events. An example is the Cretaceous onlap of the Gulf of Mexico (Fig. 1.5). Keep in mind that separating out the sand-prone sediment system will aid in age-specific studies that incorporate the time sequence and movement over the deposits.

Step 4: Begin to map hydrocarbon types, as well as reservoir fluids. Then place these within a framework of structure, so that it is clear where the major faults and structural features exist, as well as fracture networks at as low as nanoscale.

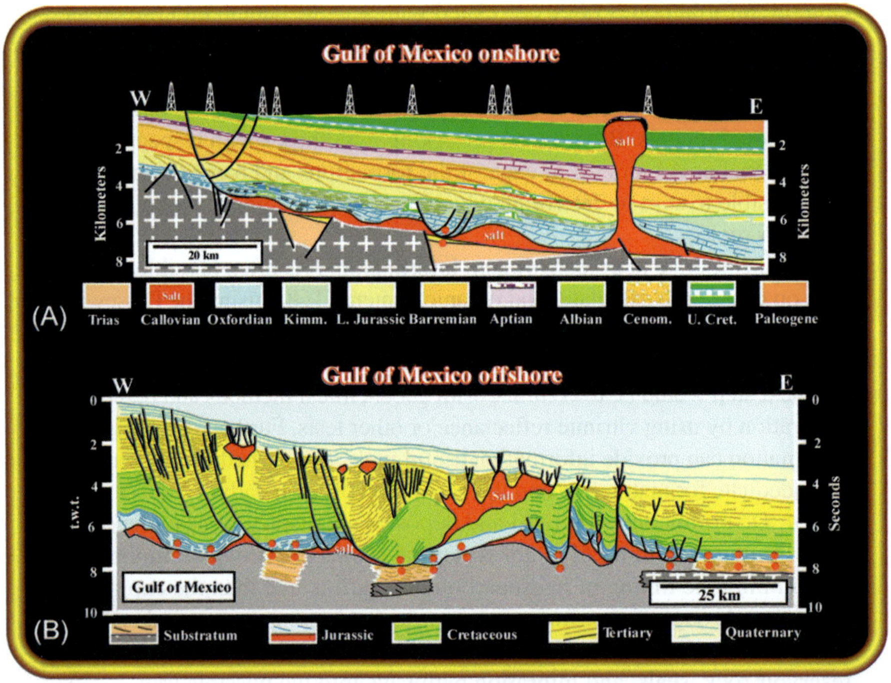

FIG. 1.5

Regional seismic lines showing characteristic style of salt tectonics in (A) onshore proximal part of Gulf of Mexico and (B) offshore distal part of Gulf of Mexico.

Data from Fonck, J.M., Cramez, C., Jackson, M.P.A. The 5th International Conference of Brazilian Geophysical Society.

Where possible, combine this information with thermal flows in order to gain an idea of maturation, expulsion, migration, and then the possible diagenesis patterns due to hydrothermal fluids. Diagenetic alteration is critical because alteration can both enhance and destroy porosity (sparrycalcity). Chert can clog pore space, but dolomitization can result in enhanced intercrystalline porosity. Possible seals and traps (structural and stratigraphic) can be identified, and the relative likelihood of commercial deposits can be assessed.

It is useful to keep in mind that this is the point in the analysis when it is possible to start creating a time-focused (chronostratigraphic) model of the sediments. Sequence boundaries (unconformities or correlative conformities) correlate to depositional sequences. They can help one to map where the transgression surface occurred and also the point of maximum flooding. The resulting "genetic" depositional sequences are extremely helpful in being able to map the lateral extent and thicknesses of different formations.

Small-Scale Analysis

Many geoscientists have looked at the small-scale analysis stage as the one in which drillable prospects are identified. Yet, with new techniques of data mining, geochemical analysis, geomechanical modeling, and microseismic analysis, it is possible during this stage to identify sweet spots for acquisition or drilling, to plan fluids for drilling and completion, and to develop a reservoir model appropriate for enhanced oil recovery.

Structural movement and migration pathways

Movement along faults, the opening and closing of fractures and fracture networks, along with nanoscale structural behavior due to hydrocarbon generation (gas) and expulsion can be used for proposing migration pathways and pinpointing areas of comparative enrichment. Being able to map facies patterns can also identify relative permeability and porosity, and the ultimate storage capacity in the reservoir. Thus, initial and very preliminary reserve potentials can be estimated. Large-scale analysis of surface lineaments as correlated to different subsurface structural features can provide new insights.

Using a combination of well-log information, petrophysics, and microseismic data can also help develop mini-basin models that can pinpoint likely geohazards.

Hydrothermal fluid movement

By understanding where heated fluids spent the most time in the reservoir, and understanding their chemical composition (as well as the lithology of the reservoir rock), it is possible to project where there may be porosity enhancement via dissolution or secondary dolomitization. Likewise, it is possible to project where porosity was destroyed due to precipitation and/or mineral filling of pore spaces and fracture networks. New thermal sensors and data-mining techniques are helpful for finding relationships between vast data sets.

Small-scale depositional modeling with typing of grains/mineralogy, along with reservoir fluids

Being able to map facies patterns can help identify the lithologies and their chemical composition. This analytical process will assist in selecting drilling and completion fluids to avoid formation damage. It is possible to fine-tune this process with new geochemical tests and fingerprinting using isotopes.

Fine-scale depositional elements can help one determine chronostratigraphic relationships, and thus identify connected reservoirs, especially in clastics. Microfossils can also be very helpful in carbonates, particularly where there are correlations between the paleoecological environment and reservoir quality.

QUESTION 1

What data, other than seismic pre-drill data, can be examined (in general) to help constrain the hydrocarbon potential in poorly known basin? I have an analogue sub-basin with oil in this formation. I wonder what methods in particular could I use to better understand my plays, like aero-magnetic survey etc?

Sergey Onuchin

Sr Geologist, Rosneft Oil Company

ANSWERS PROVIDED BY INDUSTRY EXPERTS

Sukumar Roy Chowdhury

CEO at Tex Geophysical Private Limited

Sergey's question, "What data other than seismic can be examined (in general) and will help to constrain the hydrocarbon potential in poorly known basin?"

My comments

1. Seepage study.
2. Analogous basins with similar tectonics, structural styles and evolution history.
3. Satellite imagery.
4. Geochemical study - GORE.
5. Surface geology study.
6. Structural geology study & structural balancing.
7. Aero Gravity-Magnetic studies.
8. E-M studies.
9. If nearby wells are available - try to correlate.
10. Do a regional sparse 2D seismic to delineate basin boundaries and separate areas of interest.
11. Do some 2D seismic in selected areas of interest and integrate all the above data.
12. Drill one exploratory test in a well selected location (there is oil where there is oil).

U Samanta

Geologist

Sergey, To know the presence of source rock, paleogeographic studies of each formation must be carried out. Paleoclimate, tectonics, sedimentation pattern must be favorable for source rock deposition. Maturation modeling must be carried out to find out hydrocarbon kitchen in the basin.

Arun Das

Experienced Petroleum Geologist

Sergey, Poorly explored basins (or better known as Frontier basins) are a real challenge for hydrocarbon exploration. One should gather as much information as possible regarding geology (outcrop data if available, if not then photogeological/remote sensing data), surface geochemical survey data (if carried out) etc. These will help in narrowing down the area of interest for precise subsurface imaging (2D & 3D seismic surveys).

John Olaifa

Senior Geophysicist at Consultant Services at Seis Tech Inc., P.Geo, MSc

Sergey, From the work I have done in exploration, you need to know about the depth of your prospective targets relative to the surrounding basin. Has hydrocarbon been found at such depths before? What is the structure in your area of interest, compared to other basins that have hydrocarbon? You need to know the age of your formations compared to the other basins that have hydrocarbons, etc.

Dhiman Mondal

Processing Geophysicist at TGS

Sergey, What a challenge! You can look for any previous work done on that basin such as reconnaissance survey (I believe there is no place on earth that does not have a reconnaissance survey). I presume there have been some age old Gravity and Magnetic or possibly airborne surveys done on that basin. Perhaps you can get lucky with reports or even data. If you get some 2D seismic data then you have a solid foundation from which to validate your arguments to constrain the hydrocarbon prospect with high confidence. I also presume, the basin has previous geological information (models, outcrop studies, old seismological studies indicating faults etc) which you can find by searching for research reports, government reports and geological surveys. I would put emphasis on finding information in the following ascending order:

1. Geology – outcrop data could be the best argument in this case. If you can get some more data regarding stratigraphy, rock ages in the sequence from that basin or surrounding basins this is a PLUS
2. Geophysical data: 2D or if lucky 3D Seismic
3. Gravity and Magnetic - don't underestimate them, they were the key methods in early surveys and were very successful regarding finding some of the important structural reservoirs.
4. Any experimental EM surveys

David Robichaux Jr

Mudlogger at Stratagraph, Inc.

SERGEY, One of the most under types of surveys that has proved effective in some areas is the geo-microbial survey. This survey was originated in Russia in the late 1930's and has proven to be a good indicator through the years. Through the years with computer enhancement these surveys has become more and more accurate. In the 1960's thousands of pre-drill tests were done in the USA by Phillips Petroleum and other companies. Testing done before drilling showed a prediction rate of 70% or more. This is an easy test to do, taking soil samples from just a foot or two below the surface. The test was also shown to be predictive in water environments.

QUESTION 2

If all elements of a Petroleum System are in-place, can one estimate the possibility of "TRAP BREACHING" after hydrocarbons migration from source rock, prior to drilling an exploratory well?

Atif Hussain

Development geologist at Pakistan petroleum limited

ANSWERS PROVIDED BY INDUSTRY EXPERTS

Rien Nederlof

Director At Geology And Energy Analysis "GAEA"

Hussain, One case I can remember is that gas chimneys on seismic may indicate that the top seal is leaking. But sometimes the chimney arises from gas charge trying to fill a trap more than the structural spill point allows. Hence a gas chimney may not exclude the existence of a gas (and oil) accumulation. Experiential data from many analogs has shown, by statistical analysis that the important top seal qualities are: lithology, thickness, faulting and fracturing and whether recent movements have taken place.

E.B. Setyo

Petroleum Geologists

Hussain, I agree with Rien Nederlof's comments above. I worked in a basin with deltaic environments where the top seal is not that perfect, thus gas chimneys were often seen above traps. In one way it works for us: the gas leakage allows heavier hydrocarbons (oils) to be left in the trap.

Julien Bourdet

Senior Lecturer, Center for Energy Geoscience, University of Western Australia

Hussain, you should read Gartrell et al. (2006) in AAPG bulletin (volume 90, number 12). There is also a group at CSIRO in Perth that developed seismic attribute recognition for leakage zone; you can contact laurent.langhi@csiro.au for that matter. Otherwise, by combining structural analysis and fluid inclusions analysis (for OWC depth variation) of fields already drilled, it is possible to identify which families of faults are likely to leak.

Gary Chapman

Mine General Manager

Hussain, It sounds like the "seal" element of your prospect has the highest risk. If the seal is shale, a fairly quick assessment might be to make some isochore maps e.g. thickness of seal, to a seismic marker a few hundred and/or thousand feet shallower. If the map shows an isochron thin above your target, you have an "early" structure with somewhat less likelihood of structural breach. But if no thinning is observed there is a significantly higher risk of seal since a late structure might have folded more lithified (brittle) seal rocks causing fracture and leakage. A large risk with late structures is migration timing. If there are fields in the basin where your target reservoir (or age or quality-similar reservoir) is present and sealed, that's probably the best way to assess risk, pre-drill.

Atif Hussain

Development geologist at Pakistan petroleum limited

Chapman, Is it possible that basin exhumation can result in seal breaching and cause the hydrocarbons to escape? If so, how can you judge such risking in the preliminary stage of your work, prior to drilling a well?

Gary Chapman

Mine General Manager

Hussain, Absolutely, basin uplift can cause trap breaching. In some basins, a large percentage of previously trapped oil and gas has been displaced by uplift and erosion. On the Arabian Plate, from east to west an increasingly thicker section has been removed by the Hercynian orogeny, starting in Carboniferous, ending in Triassic. Hercynian uplift had the regional effect of tilting the plate up to the west, removing up to several kilometers of previous deposits in the west (western Saudi and western Yemen). Usually the unconformities that result from exhumation can be seen seismically as angular unconformities, but sometimes it's not so obvious. Good paleo or palynomorphology can help if you have it. Your question is very similar to the question we had at Anadarko when we were exploring the panhandle area of Jordan. The question there was, did Hercynian uplift occur before, or after the basal Silurian source rock had subsided through the oil window? I'm not sure we ever really understood the situation, even after drilling an exploration well.

QUESTION 3

I have two reservoirs N Sands and M Sands. Both lie at the same depth and juxtapose each other. The N Sands are producing gas on the up-thrown side while the M Sands are producing oil from the downthrown side of a fault. What can be the possible reasons that both reservoirs are producing different hydrocarbon while they are lying at the same depth and also juxtaposing each other? (Fig. 1.6).

Farah Khan

Geologist at United Energy Pakistan Limited

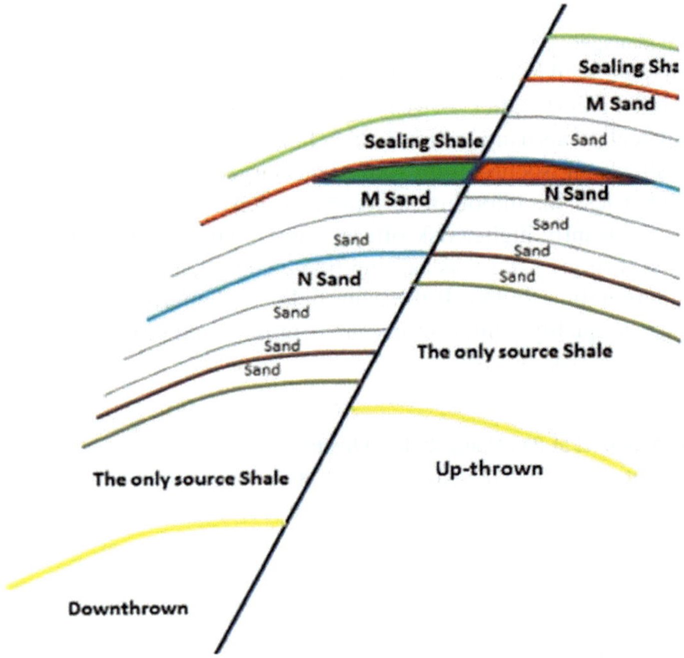

FIG. 1.6

N and M sands.

Reproduced with permission from Farah Khan.

ANSWERS PROVIDED BY INDUSTRY EXPERTS

Syam Sundar Yalamarty

Petroleum Geology Consultant

Farah, These two reservoirs could be of two different stratigraphic horizons fed from different sources having different burial histories and juxtaposed (post migration) across a sealing fault. Detailed tectono-stratigraphic evolution of the area should lead to better understanding that can also help further exploration and development of these reservoirs.

Grant Hanson

Consultant at DISYS

Farah, Obviously the fault that separates them is a sealing fault. Consequently they are really two different reservoirs.

Tyler J. Hargrove

Operations Geologist at Arrington Oil & Gas Operation LLC

Farah, did you happen to look at the rock itself? Also, up-dip of the M Sand is there gas production or dry hole/tight DST or flow back? And down-dip of the gas zone is there water or is it equally tight? I would say hypothetically that the situation with the

M and N Sands sounds like a selective permeability issue, where tighter permeability in the N Sand only allowed gas migration into the zone, while the M Sand might have a greater permeability which allowed oil migration into the reservoir. The fault itself, depending on the type of rock juxtaposed to one another, and the type of fault, can either be a barrier or a conduit for fluid flow both laterally and vertically. This could also have resulted both in conjunction or separately in the production scenario we see in the M and N Sands.

Gonzalo Villacorta

Geólogo Senior, Consultant E&P

I agree with Tyler J. Hargrove. **Farah**, could you check your reservoir gradient pressures on both sands to verify if they are really two reservoirs or that they only have different lithology and capillary pressure properties. I have found on sidewall cores, 1.8 cm (one point eight) in diameter, at Comodoro Rivadavia, San Jorge Basin, one half of the cross section shows heavy oil while the other half shows very light oil. Also, check the depositional environment of the sands remembering that surfaces of accretion act as permeability barriers. Review papers about the waste zone, al AAPG bulletin. 720661398182

Douglas Wilson

Oil & Energy Professional

Farah, This happens often in structurally complex reservoirs. Remember the hydrocarbons did not necessarily form where they are currently. An interesting question for you to look at is where were the hydrocarbons formed? And how did they get to their current location? You might locate the sources and potentially more hydrocarbons. Faults seal via a variety on mechanisms, often just clay smear which is a function of shale being moved along the fault plane. Milne Point field in Alaska is a very typical example of a highly compartmented reservoir.

Adeosho Ayodeji Saheed

Reservoir Geophysicist

Farah, in addition to a combination of all the answers given, you can also check the petrophysical properties on each individual sand body to ensure permeability and porosity are equal in both or not (which leads to an inference of a change in depositional environment). If these are equal and these sands are of the same section of a depositional environment, then a probable explanation goes thus: hydrocarbon charge started in the downthrown sand section as the fault was being developed, and then at maximum charge, associated gas escaped through a not fully sealed clay smear along fault planes into the upper sand section (recall the Klinkenberg effect - gas molecules will slide through the tiniest connected porosities). Then we can call for an SGR analysis to check the fault integrity to know if the fault is fully sealing.

Thomas Klopf

Senior Geophysicist at KUFPEC

Farah, You could look at the M Sand on the up-thrown side of the fault to see if there are any indications of oil. If none, it might be that the oil source has migrated into the trap from the left side of your cross-section drawing. Likewise, you can look at the N Sand on the down-thrown side of the fault. If there are no indications of gas, it is possible the gas source is migrating from the right side of your drawing. The fault gauge seal would then be separating the two reservoirs.

Dr Natt Arian

Managing Director at Arian Petroleum

Farah, The answer does not lie within the reservoir but is in the history of the basin, I mean trap/charge history. It seems that the fault is totally sealing in both vertical and horizontal directions allowing the accumulations to exist and to be separated at the same level. You may have two different migration pathways into each reservoir (different sides of the fault) and possibly from two different sources: an oil charge to one side of the fault and a gas charge to the other side, it's most likely that charges are different times. All you need is a 3D basin model to explain everything.

QUESTION 4

What should be the approach for drilling a side track / horizontal well in a naturally fractured reservoir having a history of mud loss even while drilling with water?

Suresh Mayal

Geologist at ONGC

ANSWER PROVIDED BY INDUSTRY EXPERT

Hans Schoendorfer

Consulting Petroleum Geoscientist at Crail Petroleum Ltd.

Suresh, We have done that in a naturally fractured carbonate reservoir with high viscosity heavy oil and just drilled ahead pumping in lots of seawater. It needs a great water source (ocean) of course and in our case the sea water is of similar composition to the formation water. We have lost up to 180m3 per hour of "drilling fluid". We subsequently wash the hole to remove any skin. In most cases this has worked. I admit these are a particular set of circumstances. Sometimes it takes a week of pumping on test before the oil cut starts to climb. But we have had wells flowing clean oil after an extended test. These wells are up to 5900m at TD and have up to 1500m of open reservoir, usually completed with slotted and blank liner with ACP's dividing zones or occasionally swell packers. We had first tried pumping lost circulation pills but found, in our case, that this did not completely seal the fractures and completions generally were less than ideal in terms of oil volumes.

QUESTION 5

What else can I decide from an isochrone map? Could I say that if the travel time taken is longer, then the depth of that place is deeper?

Fadzlin Hasani
Petrophysicist, PETRONAS

ANSWERS PROVIDED BY INDUSTRY EXPERTS

Angga Direzza

Sr. Geophysicist at PHE Nunukan Company, Indonesia

Fadzlin, With an assumption that the layer has same velocity regime and no reverse deformation structure, I would say yes. It can represent the thickness: thicker equals deeper.

Mazhar Ali

Geophysicist at Pakistan Petroleum Limited

Fadzlin, You can assume/correlate deeper depth with longer time to some extent but time factor depends on some other variables such as heterogeneous medium, velocity changes in the vertical as well as horizontal direction.

Elive Menyoli

Exploration Geophysicist at Marathon Oil Corporation

Fadzlin, What Angga and Mazhar, said is accurate. Let's assume those factors are taken care of. What about geologic implications of an isochrone map? An Isochrone map shows you "thicks" and "thins" of the basin. You can correlate them to depositional history of the basin and how the basin center shifts with time. The "thicks" may indicate that at the time of sediment deposition, there was an accommodation space in that area compared to other areas. The "thins" might imply less sand deposition, probable basin margin or you might expect more channelized system.

Rashed Abdullah, Geophysicist

PHD Researcher, University of Queensland

Fadzlin, I do agree with Elive Menyoli. In addition, you can think about this tectonically. Consider your isochrone with syn-rift growth strata or post compressional packages for you section.

Gerald Phillips

Senoir Geoscience Application/GIS Analyst, Statoil

Fadzlin, From 1975 to 1982, my career was focused on seismic exploration for oil in the Williston Basin with the old Gulf Oil Corporation. As a Gulf Oil exploration geophysicist, I identified and mapped out prospects that lead to discoveries from conventional reservoirs in the Paleozoic - typically the main pay was found in the Mission Canyon. I used isochorn mapping as the key exploration method to locate and validate the low relief structural closures. These maps were the hand contouring of isochron times from 2D seismic sections. The time picks were carefully measured between strong regional marker horizons (Greenhorn to Last Mississipian Salt or the Bakken to the Winnipeg). The isochon thin anomalies correspond to areas where structural uplift had occurred during depositional history. You could validate prospective highs by measuring isochron "thins". If you mapped out an apparent high - that was only a high based on reflection travel time - no thinning, it was likely an anomaly due to incorrectly removed static or near surface velocity variations. Looking for the closed thins and ignoring the total reflection time undulations was the secret to successful exploration in the WB from 1975 - 1985. The total travel time map over Little Knife was inconclusive as to whether or not critical structural closure existed. However, a broad extensive isochron "thin" closure existed over two townships in areal extent. That thin exactly outlined the structure Gulf discovered with the drill bit.

So: Isochron mapping equals understanding and insight into tectonic movement during historical deposition. This is true for both basin wide trends and local structural features.

QUESTION 6

Which technique would be better 1D or 2D/3D, to model maturity of source rock?

Faisal Qayyum

M.Sc Geologist Student

ANSWERS PROVIDED BY INDUSTRY EXPERTS

Alexey Romanov

Petroleum Geologist

Faisal, 1D is good for points, that is, wells. 2D usually utilizes cross-sections, or sometimes maps, aka "pseudo 3D". 3D on the other hand is part of a three dimensional volume of rocks. That is, part of a basin. Depending on your objectives and available package, choose one of the above.

Amdad Ali

Insite specialist – Data operator at Halliburton

Faisal, 1-D Modeling is also called maturity modeling. It is easy and quick technique to evaluation. So, 1-D modeling may be better. I agree with Romanov, it totally depends on your objectives.

Fabio Rivolta

GeoLog International B.V

Faisal, I think it depends on what your target is. 1D can be good for an easy and fast modeling about maturity, but with low information about expulsion and no modeling about migration. For 3D you need much more detailed information and that can be hard to develop. If you have not done yet check this link http://www.petromod.com/

Verena Lampart

Sale Supervisor at PGS Geophysical AS

Faisal, I follow Fabio, and recommend Petromod. It's excellent for 1D and 3D. You just should make sure you know what the software is doing/calculating with the data you feed it. Do you have enough data for a 3D model? Maybe you could start with 1D since it's easier and then go on.

Amit Singh

Geologist at ONGC

Faisal, For an MS thesis I believe you should go for 2D/3D basin modeling. But yes to begin with to understand the source rock you should always do the 1D modeling in different parts (depths of SR) of your area. You would need depth, SR, paleo-water depth, thermal history, etc. data for the area. For software, I recommend TemisFlow. Contact; temis@beicip.com or visit http://www.openflow.com

Faisal Qayyum

M.Sc Geologist student

Which software would you people recommend for 1D and 2D/3D?

Shane Wooding

Senior Geologist, Corporate Geology, PTTEP

Faisal, It depends on what data you have available in your study area. If you have tight 2D or 3D seismic plus some good well control then 3D modeling will work well for you and provide a detailed 3D analysis. However, if you are in a frontier basin with a coarser 2D seismic grid and very little or no well control then 1D (single point) or maybe 2D (transect) modeling is all you will be able to achieve. For 1D or 2D I prefer Basin Mod and, Trinity software for the full 3D.

Jaron Lelijveld

Senior petroleum systems Consultant at Schlumberger Information Solutions

Faisal, Given the limited information about your MSc. thesis, first you should define your objectives; estimation of generation, migration or other basin parameters such as pore pressure prediction, heat flow analysis or tectono stratigraphic history. Then, depending on the amount and type of data you have, go for a 1D study first and then combine multiple 2D sections which you can easily convert into a 2.5D study if you don't have 3D data.

PetroMod is the most complete and easiest to handle basin and petroleum systems modeling software on the market and, is able to work in compressional tectonic regimes. Upscale your field observations and apply them in the software and, in general I can recommend; start simple, complicate later.

Eloy Jerviz

Planning Manager, Petrolera Paria (Joint Venture PDVSA/Sinopec)

Faisal, First, as many have already mentioned, Petromod is an option that suites your needs. I have worked with it for some time now and I find it pretty easy to learn and use. However, you should always keep in mind that "software" is just a tool to get your calculations done. But what matters is what lies under it: knowledge. Whatever the input you will always get an output but, will it be the right output? It will depend on your understanding of the calculations and the quality of your data. Now that we are talking about data, I believe that vitrinite will represent the key issue in your research since it is not available for all the wells. Its use depends entirely on the ability of the technician and sample quality, and you will need it to calibrate your model. Secondly, for a MS thesis you might want to deepen your topic a little bit. Here is couple of ideas.

1) You could have a look of the pressure regimes in your study area, if overpressured you can try to study its relationship with HC maturity/expulsion.

2) You might also try to study some drilled prospects in the area and, with the help of your maturity/migration results, try to explain successes or failures from the point of view of "timing" and "charging". I have some papers about the topic that might help you. Email me to ejerviz.syky@sinopec.com and I will send them back to you.

Musbahu Abdussalaam

Project Intern at CGG

Faisal, Field experience is always an added advantage to whatever analysis you do with the computer so, a better option would be to combine field, observation, manual results/data analysis (quantitative as well as qualitative), then use the software analysis for better more accurate result/conclusion.

QUESTION 7

I am looking for analogs for Indian prospect. Can someone please pass on references or any information on maturing sub thrust plays?

Pranaya

Geologist

ANSWERS PROVIDED BY INDUSTRY EXPERTS

Darko Spahic

Senior PetroMod Engineer at Schlumberger

Pranaya, Thermal maturation of sub thrust plays is an extremely interesting but tricky issue. It is necessary to quantify and restore slip history of each block associated with the petroleum system - as for HW as for FW. A good case would be to restore the entire sequence i.e. towards the basin initiation (main extension phase) not only back to the basin termination i.e. main compressional stage. In that manner it is possible to track overall maturation paths, especially if blocks are covered (each separate block) by vitrinite data. One of the nicest papers that deals with maturation in compressional systems is from Parra in Venezuela. Parra et al. (2010), The Monagas Fold–Thrust Belt of Eastern Venezuela. Part I: Structural and thermal modeling, in Marine and Petroleum geology).

David Greeley

Geologist, BP America

Pranaya, If you haven't already done a Google/Bing search on U.S. sub thrust plays, you should look at the Utah-Wyoming Sub thrust which has been explored and producing for decades. Field names: Anschutz Ranch East, Anschutz Ranch, Pineview, Ryckman Creek, Painter Reservoir, Whitney Canyon, and a host of others. There are a ton of papers out there, some fairly recent.

Vicky Ruliansatri

Geophysicist at Medco E&P Indonesia

Are there also references about how to reduce the pitfalls of seismic expression in sub-thrust play?

Anthony Hardwick

Principal Geophysicist at TGS

Vicky, Clari-Fi now includes a full receiver de-ghosting solution. But some of the principles behind the method are described in these articles and abstracts. You can download from the TGS website or through Earthdoc

Hardwick, A., Travis, T., Stokes, S. and Hart, M., 2010. Lows and highs: using low frequencies and improved velocity tools to image complex ridges and basement highs in the Faroe-Shetland Basin. First break vol. 28 (9).

Woodburn, N., Hardwick A., Travis T., 2011 Enhanced Low Frequency Signal Processing for Sub-basalt Imaging. Extended abstracts, EAGE 73rd conference and exhibition.

Hardwick, A.J., Whittaker, J. & Woodburn, N., 2012. Combining Bandwidth Extension Techniques and Pre-stack Q Inversion for Enhanced Imaging of Intra-Basalt Plays. EAGE 74th Conference and Exhibition.

Hardwick, A., Woodburn, N. & Whittaker, J., 2012. Broadband Verses Conventional Marine Seismic: The Importance of Compensating for the Earth Filter. Workshop 11 - Broadband Seismic - Technical Value to Reservoir Characterization Workflows? EAGE 74th Conference and Exhibition.

Woodburn, N., Hardwick, A. & Travis, T., 2012. Enhanced low frequency signal processing for sub-basalt imaging on the Faroese Continental Shelf. 4th Faroe Islands Exploration Conference.

QUESTION 8

Does anyone have global figures on the overall chance of success when drilling stratigraphic traps compared to structural traps? I have a sneaky belief that most companies are overestimating the chance of success when drilling stratigraphic traps. This may be because risks like presence of top and base seal, lack of up-dip pinch-out and lack of lateral extent have not been factored in to the overall risking. It would be interesting to compare the calculated POS's (probability of success) or COS's (chance of success) for drilling stratigraphic traps with the final well results.

Jon Noad

Geological specialist, Husky Energy

ANSWERS PROVIDED BY INDUSTRY EXPERTS

Abdullah Alnibari

Experience Oil & Gas E&P in Middle East & North Africa

Jon, I think there are no precise figures for chances of success for drilling a stratigraphic trap. It all depends on the way you interpret your seismic data. For example a major oil company drilled a well in Egypt to test a reef build-up depicted from seismic, but instead they found intrusive body instead of a reef, so the chance of success is less than chances of failure. It all depends on how you define your parameters and especially the probability of trapping integrity and the probability of reservoir rock.

Scott McCleery

Excutive Vice President of Geoscience, Thunder Horse Energy, LLC

Jon, Risking is a risky business. Some companies may use only 5 risking factors, others may use 9 (or more). Evidence in support of any factor can vary quite a lot, so risking between plays and basins can be somewhat subjective. However, the elements one risks for structural traps are the same for stratigraphic traps. Arguably, some elements for stratigraphic traps are more difficult to demonstrate. In my experience, a risking team would tend to raise the level of risk on a stratigraphic play because of this factor, not lower it. As far as "global figures" goes, I don't have those statistics. However, C&C Reservoirs might have something because they generate global field analogs. Also, Nehring Associates compiles a lot of field statistics for the US. I don't know if either has considered COS for stratigraphic vs. structural traps. The problem is we usually only study successes, we don't spend as much time documenting failures (and we certainly don't like to publicize them).

Paul Post

Gelogist at Bureau of Ocean Energy Management

Jon, Hard figures on success are often difficult to find. My feeling, like yours, is that POS and/or COS are often assumed higher than drilling supports. I would add another reason for why pre-drill POS are higher than post-drill statistically documented: seismic attributes. The assumptions of a simple, two-component system with only reservoir fluid affecting amplitude can be wrong. In many cases marls and limestone were not considered, but were encountered instead of the targeted hydrocarbon filled reservoir. This was the recent the situation in a CSEM supported, seismically defined prospect offshore West Africa. In addition, unfortunately, hydrocarbon saturation is very difficult to predict from amplitude analysis, and a number of new field wildcats encounter non-commercial saturations of "fizz gas".

Rose & Associates have been doing studies of DHIs over the years for their consortium with occasional papers in the public domain; e.g., http://www.touchbriefings.com/pdf/1736/ACF1A23.pdf, http://www.docstoc.com/docs/123302119/PowerPoint-Presentation---DHI---Rose-_amp-Associates, http://www.searchanddiscovery.com/documents/2010/110139forrest/ndx_forrest.pdf.

Steve Adcock

Geophysical Consultant at dGB Earth Sciences

Jon, I have seldom seen this kind of data shared, or even broken out within a company's own analysis. Additionally, I have often seen stratigraphic traps "disguised" as structural traps during the interpretation process to make them more palatable to management, which could confuse these kinds of numbers even if you had them. However, regardless of the success rate, I do accept that old rule of thumb that 70% of the traps out there to find are stratigraphic (3-way or less structural closure). Having said all that, here's one of my absolute favorite articles on this subject and it may help with your analysis, particularly figures 9, 15, and 16: Exploration Play Analysis from a Sequence Stratigraphic Perspective, 2003 by Snedden, Sarg, and Ying. http://www.searchanddiscovery.com/documents/snedden/

Garry Kolafa

Owner, Exploration System International Ltd.

Jon, The way I look at it, a structural trap carries less uncertainty because of its' possible definition, but the overall exercise demands a process of elimination, where uncertainty is minimized. This is achieved by applying the correct techniques in initial surveys that will improve your knowledge about your target. Drilling in marine environments has had better success rates than on land due to the fact their budgets allow for added surveys which may be not conducted in a land survey due to cost. The success rates in the marine environment are roughly twice as high due to this, regardless whether the trap is stratigraphic or structural.

Jim Rockett

Owner at JimRockett Geo-Digitizing Resources, Tulsa, Oklahoma

Jon, For me stratigraphic traps for me always bring to mind questions and more questions. For example, how do you know the limits of a trap in all directions? The answer is: you don't know unless you drill. Provided the first well is successful, one can drill a step out any assumed correct direction from the first well to define the outer margins. By the time you get through drilling wells, step out by step out, the outer limits are defined. Of course there is no longer any need to drill around the first well since all locations are drilled. But at least you have proven in which direction not to go. In one case, the very first well we drilled was a stinker of about 10 bbls a day. We moved north across the road about one location and got a 500 bbl a day well. We moved to the east about one location and got another small well but it was better than the first. We moved to the south about one location and got a thousand bbl a day well. So you pay your money and take your chances. Therefore, I agree with Garry's comment doing extra surveys could make life easier, even on land.

UNCONVENTIONAL RESERVOIRS

Unconventional and Tight Gas Sands: What We Know Now

Unconventional reservoirs typically fall into four main categories. All are fairly imprecise and ambiguous but have become convenient labels for reservoirs that were, until the early 2000s, considered uneconomic and more or less unproducible by "conventional" means, (which meant vertical drilling and only one or two stages for hydraulic fracturing). Now, "unconventional" reservoirs have become the "new conventional," since at times the majority of wells drilled, at least in the continental United States, are horizontal wells 15,000 ft in horizontal length, with 40 or more stages. The original categories of unconventional reservoirs are described in the following sections.

Shale oil and gas

This category refers to hydrocarbons produced from shale reservoirs. Shale plays are called "shale" for convenience, but even though they contain clay minerals, they are not technically shales. Instead, they are simply extremely fine-grained clastics, often with clay minerals, subarenites, and carbonates, to the point that they are often referred to as mudstones, a rather imprecise denomination which also fails to capture what they are. So, for convenience, it is a good idea to confine oneself to the notion that they are fine-grained clastics typified by extremely low permeability, with extreme heterogeneity with dramatic lateral facies changes and variations (Fig. 1.7).

Production occurs through fracture networks, which are usually a combination of natural fractures and induced fractures which have been propped open by proppants, introduced during the completion phase.

Drilling and completion techniques involve placing horizontal wells at an ideal distance from each other to optimize the fracture connectivity and to make efficient use of the drilling infrastructure. In the case of the Bakken, there are often several productive zones in a single lease, which allows "stacked" plays with the potential for additional economically viable production (Fig. 1.8).

Examples of shale gas include the Marcellus, which is located primarily in Pennsylvania, Ohio, and West Virginia, and the Haynesville, which is found in the Gulf of Mexico region in the Southern United States. The Marcellus has been a good producer of dry gas, but the decline rates have been steep and pipelines, processing plants, and other infrastructure must be built. In the case of the Haynesville, it has been difficult to economically produce the gas at less than $8 per mmcf due to depth, high heat and pressure.

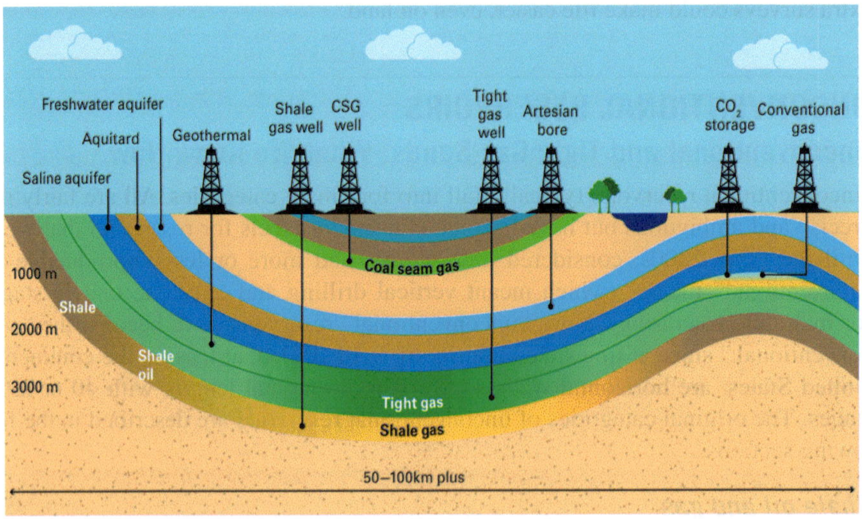

FIG. 1.7

Coal seam gas and tight gas.

From http://theconversation.com/coal-seam-gas-is-just-the-latest-round-in-an-underground-war-35164.

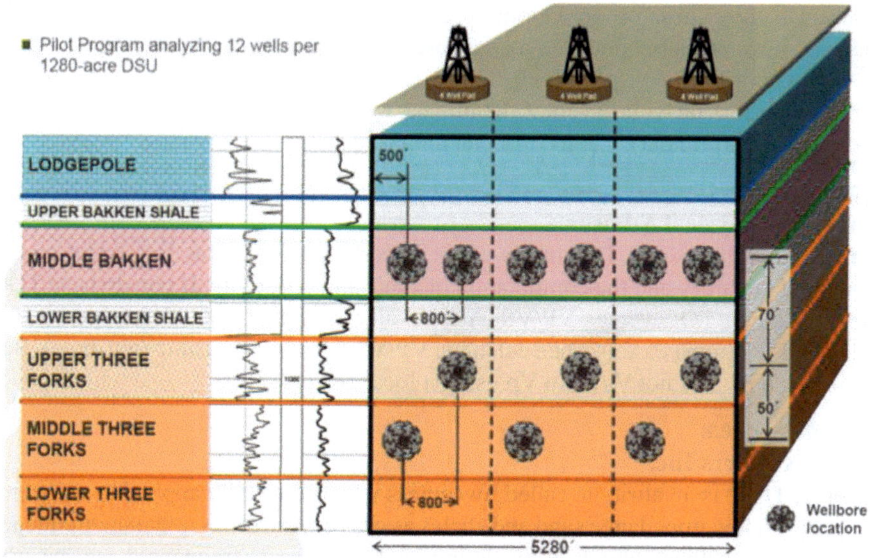

FIG. 1.8

Example of shale oil includes the Utica Shale, the Eagle Ford, the Woodford, and the Bakken.

From http://info.drillinginfo.com/5-top-bakken-shale/.

QUESTION 9

Can organic rich shale be detected on 2D seismic data (post-stack)?

Aqeel Ahmed

Assistant Geologist at Oil & Gas Development Company Ltd. Pakistan

ANSWERS PROVIDED BY INDUSTRY EXPERTS

Mohamed Salah Sedek

Geophysicist

Unconventional reservoirs always needs a regional scale seismic (even 2D), to be able to be identified correctly. Back to the basics, seismic amplitude represents the contrast in velocity and density at the interface between two formations. So in case of shale it will show that contrast even in post-stacked seismic. You can apply seismic inversion to the 2D seismic section (you need at least one well) and try to describe that contrast quantitatively.

Raju Arasu

Senior Geophysicist & Lead Triassic Paleozoic Prospect Evaluation Unit at Kuwait Oil Company

Ahmed, Do you see a sharp high amplitude event in the midst of low amplitude section in your data? Check in a set of dip lines, keeping basin geometry in mind. Is its phase opposite to a sand event's phase? If you get one, select all lines and obtain an amplitude attribute map. Does the shape spread over a considerable area basin-ward? If so, flatten a dip line close to this event and look at the geometry of the reflectors below. Do they show progradational characteristics? If these things are observed, then you have probably detected a condensed section, a candidate of TOC.

If you don't observe any such things, respond accordingly. Let us do another experiment. (So far you have not mentioned whether you are working in clastic or carbonate environment). Let us try in an unbiased away.

Jonny G. Pinto

Senior Geophysical Advisor

Amhed, first things first. Do you have a well nearby to see if it has TOC? TOC comes from the laboratory and the combination of sonic and resistivity. If you do have TOC, then cross plot TOC versus VP/VS. A low value of Vp/Vs should give you a high value of TOC (most cases). Then calculate VP/VS from your seismic (attribute). Best is pre-stack data, if not Vs from Vp using a local correlation.

Azer Mustaqeem

Petro-Explorers Inc.

Amhed, There is an attribute called Sweetness which is usually applied to differentiate between low impedance sand and shales as gas bearing sands attenuate frequency. The same Sweetness which is by definition Energy/sqrt (frequency) can be used within the amplitudes of shales to differentiate the areas of possible organic rich shales or maturity (as gas is adsorbed in the shales with higher organic content). Jonny is right that Vp/Vs from pre-stack AVO inversion will give you better answer. Good rock physics analysis, with TOC against elastic properties can be helpful at least in understanding the ranges of impedance.

Jonny G. Pinto

Senior Geophysical Advisor

Ahmed, Good call about sweetness. I have used it in the past in clastic environments looking for potential sands (low frequency/high reflection-strength or trace envelop). You will still need to calibrate the attribute to the well for possible TOC content. Remember, if the well has a TOC calculated from logs, it needs to be calibrated with core data. Azer, thanks for the tip about using sweetness in Shale having a high TOC.

Oil shale

Often referred to as kerogen oil, oil shale is basically undermature petroleum which must be heated in order to achieve pyrolysis and to liberate the hydrocarbons. It tends to be in very fine-grained sedimentary rock and the high level of organic matter makes it possible to use different processes to release it. Oil shale of lacustrine origin is often associated with continental margins in rift basins. The reservoirs tend to be marked by extreme heterogeneity. Oil shale deposits are found in North America in Utah and Wyoming (Green River formation), where they are generally produced using a process of crushing, heating, and chemically treating the oil shale, which can be done either in situ or in a processing facility (Fig. 1.9).

An in situ process developed by Shell involves heating the hydraulically fractured formation and then, once the synthesized oil has been generated, pumping it to the surface. Oil shale processes are also found in China, Brazil, and Estonia.

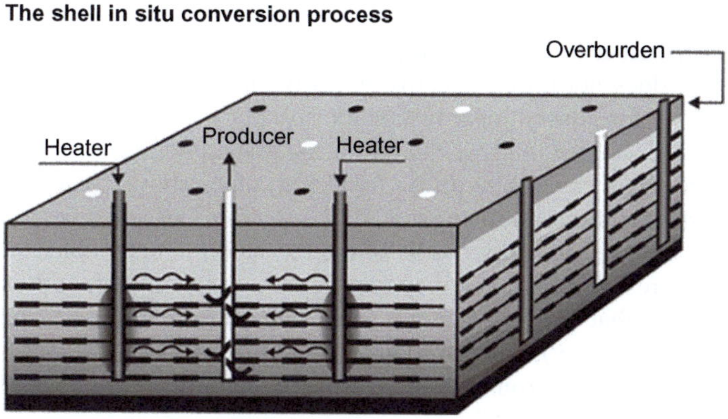

FIG. 1.9

The shell in situ conversion process.

Based on material provided by Shell Exploration and Production Company. http://ostseis.anl.gov/guide/oilshale/.

The advantages of creating synthetic oil from kerogen include national security and a secure supply if there are no nearby conventional oil and gas reserves. The disadvantages are numerous in that the process is expensive due to the need for energy to heat the rock and water used in the processing. Furthermore, the process tends to produce a number of contaminating byproducts.

QUESTION 10

I know that shale does not have any effective porosity, if this is the case how come shale is the main hydrocarbon source? How can the hydrocarbon escape from shaly formations?

Samer Hafez

Student at King Saud University

ANSWERS PROVIDED BY INDUSTRY EXPERTS

Anne-Kari Furre

Geophysicist at Statoil

Samer, Firstly, shale is not one uniform "thing". Mineral content, compaction history, etc. will have influenced the properties of the particular shale you're looking at. Shales can have varied porosity (and permeability). Check e.g compaction trends in the literature or any published data on this.

Secondly, it also depends on the time scale you're working with. Effective porosity is a term useful when producing a reservoir. A shale might very well be tight at that time scale. At the same time: nothing is tight over geological time; and when talking about source rocks, we're talking a long time.

Dan Krygowski

Senior Petrophysicist at the Discovery Group

Samer, Shale does not have "effective porosity" in the way that we calculate effective porosity for conventional reservoirs (usually there is a shale volume term (Vshale) in the equation, and we start with the assumption to begin with, that the effective porosity in shale is zero). As Anne-Kari notes above, "shale" can be both the hydrocarbon source and the hydrocarbon reservoir, but you have to look at the mineralogical and sedimentary details of the formation to determine exactly where the source and reservoir are (and they may be closely intermingled). It is important to remember all that youhave learned about conventional reservoirs (clastics and carbonates with porosity), but also to understand that many of the concepts that we use regularly in those environments have different definitions in the world of shales-organic mudstones-"crap-rock".

Thalbert E (Thal) McGinness

Independent Oil & Energy Professional

Samer Hafez, As noted by both Dan and Anne-Kari, shale is a catch-all term with a large range of definitions. Core analysis in the GoM will give you ~ 15% porosity in what for that area is seal. But it is above a 30+ % porosity, half a Darcy permeability sand. The pay zones in the various shale oil and gas plays currently under development will encounter some porosity and nano permeability but the rock will still produce if completed properly. That GoM shale has just been compacted down into 1 tenth or less of it is original thickness and has had most of the water and hydrocarbons squeezed out of it in the process.

Dung Trung Tran

Petroleum Engineer

Samer, The word "effective" refers to the interconnected pore network - which governs rock permeability. You can have shale with porosity ~10% but having micro-darcy permeability.

OP's two questions are not related to effective porosity, but more to geochemistry. The first one is due to the nature of shale which has higher organic matter (OM) which could be preserved with burial. OM can mature (change, breakdown) into oil/gas given time and temperature. For the second one, I prefer the "microfracking" hypothesis as the main cause. Note that the rock (shale) must be buried deep enough before oil/gas generation occurs (temperature >~50-60 deg C). As solid OM ("kerogens") matures and turns to HC liquid/gas; the pressure is built up tremendously inside the shale until it cracks the rock and "some" of the HC escapes to relieve the pressure. With time, the lighter HCs can migrate to adjacent (reservoir) rocks and even make to the surface where they are lost. But still, there are a lot of HCs which get trapped and never make that exciting trip. The implication of that is what we see today - the shale movement. With new technologies we can and are cracking the shale to get the remaining gas/oil out. There are many papers on kerogen porosity & micro fracture in shales that you can look up (i.e. onepetro).

Rm Iman Argakoesoemah

Senior Petroleum Exploration Geologist

Samer, In unconventional gas shale plays, porosity and permeability in shale are at "nanno-scale" as nannopores. This is based on observation on scanning electron photomicrographs showing nannoscale pore architecture in organic matter in shale. Larger pores are formed when kerogen is converted to hydrocarbons.

mamdouh A. Shebl, Ph. D.

Senior Petrophysicist at Chevron

Samer, Shale effective porosity is not the same thing as porosity in siliciclastic and carbonate reservoirs. It took the oil industry over 70 years to understand siliciclastic and carbonate reservoirs, and they are not fully understood yet. Understanding shale source rocks is less than 25 years old, and a lot of research needed to be done to completely understand that. On the other hand, shale gas and oil reservoirs are less than 10 years old, and very far from being completely understood. In the oil industry there is the tendency to use analogs of what we know to understand what we do not know. When it comes to shale that might be dangerous! Why? Just as an example take the topic of today: "effective porosity". It is a concept with values derived for conventional reservoirs controlled by among other things, total porosity, pore size distribution, connectivity between pores, permeability, fluid composition and diagenesis. In unconventional reservoirs, such as shale, most of these controlling factors act quite differently, and therefore the concept does not apply, or at best does not mean the same thing, particularly when total porosity does not mean only pore spaces, and matrix permeability does not act the same way as in a conventional reservoir. For porosity you have matrix porosity, porosity in the kerogen, and fracture related porosity. System permeability is controlled by natural and induced fractures in the shale. Clays and kerogen play great and not fully understood parts in matrix permeability and that is much different from conventional reservoirs. Another level of complication: the term "shale" is ill defined in the oil industry event today. In literature you see people describing clay-rich (above 50% clay) rock as shale, others describe organic-rich rock with clay as shale while, others describe unconventional rocks where clays control the behavior of the rock (with as little as about 20% clay) as shale. And others just uses "shale" as synonymous with "unconventional". As Dan mentioned, to better understand shale reservoirs and their properties, we need first to define and specify what we are talking about: namely "shale". Secondly, we need to develop new way of thinking about shale. We cannot use thinking based on conventional reservoirs and apply it to unconventional reservoirs. We need to develop unconventional thinking, (thinking "outside the box"), to match the unconventional reservoirs we are dealing with. Thirdly, consider the fact that unlike conventional reservoirs, there are great deals of vertical and horizontal heterogeneity within shales as well as between different shales. Therefore, using analogs, as we used to do in conventional reservoir, may be less effective and may be misleading. Each well, each locality, each shale reservoir is unique and should be treated as such. Then you can compare and contrast these wells to establish trends and develop better understanding of that particular shale and/ or shale reservoir.

Biswajit Thander

Geophysicist/ Seismic Interpreter at Gujarat State Petroleum Corporation Ltd

Samer, The source of hydrocarbon is shale as it contains a large percentage of organic matter. This organic matter needs to be preserved to get hydrocarbons. So, oxidation must be minimized in the depositional environment. When these organic rich materials (kerogen) matured they generate hydrocarbons. Temperature plays a vital role in generating hydrocarbons from kerogen. Generally, kerogen generates hydrocarbon at temperatures above 60 degree centigrade (oil window). If the temperature is less than 60 degree then, those organic rich materials form oil shale deposit. Hydrocarbon cannot be extracted from this oil shale by the application of conventional techniques, as it has very low effective porosity and is impervious. Unconventional methods like hydraulic fracturing are used. Hydraulic fracturing is used to extract hydrocarbons from highly impermeable rock like shale rock with the help of injecting large amounts of fracturing fluids at high pressures to create connections of fissures (gaps) in the shale rock and, as a result hydrocarbons get a pathway to travel to the well for extraction.

Steve Hobart

Sr. Pore Pressure Advisor

Samer, Addressing the question about how shale can be a hydrocarbon source and taking into account that shale is not understood to have "effective porosity" and is thus impermeable: "Impermeability" is a relative thing. Shales are indeed "impermeable" compared to the permeability of other reservoir rocks. However, their permeability is not zero. Over geologic time, fluids, especially water, can be transported through shales or expelled from compacting shales. Water and hydrocarbons are considered immiscible; however, the solubility of hydrocarbons in water is not absolutely zero, either. It depends on temperature, pressure and the existing load of solutes in the water, primarily salt. The movement or expulsion of hydrocarbon-bearing waters from shales into reservoir rocks is driven by compaction processes as well as, by the increase in pore pressure in the rock volume containing the complex organic materials which break down into smaller, and more numerous molecules. This explains how the hydrocarbons get from the shales and into the reservoir rocks.

Alain Brie

Give occasional petrophysics training courses at retired petrrophysicist expert

Samer, Shale is a rock made of clay and silt (fine grain quartz). Most shales are deposited in deep sea basins, in a very quiet environment, often together with organic material coming from remnants of living organism (plankton), algae, etc. The decomposition of the organic material in anaerobic conditions yields the kerogen which then transforms to oil and gas when temperature increases. The clay fraction is made of sheet alumino-silicates of very small sizes forming clay flakes. These have a layered structure with a sizable amount of water between the layers when they are first deposited. There is also some inter-granular porosity in between the flakes. Water is trapped inside the clay by very high capillary forces caused by the tiny pore sizes, resulting in very low permeability. As the shales are buried at increasing depth, the confining pressure increases and squeezes out the fluids during compaction. At that stage shales are ductile and do not fracture, the fluid movement is caused by pore pressure exceeding capillary pressure. Think of a sponge, as long as you don't squeeze it the water remains inside, but squeeze it and the water comes out. Some mineral transformation also occurs with depth and increasing temperature releasing more water (diagenesis of smectite to illite). In the process if kerogen maturation occurs, hydrocarbons can be squeezed out of the shale to a neighboring reservoir (migration). If no porous reservoir is present pore pressure will increase resulting in an overpressured zone.

Another aspect of this is formation alteration is what happens when drilling shales with water-base mud. This is often referred to as swelling shales. When studying this (with sonic logs) I was impressed when I realized that the thickness of the altered zone is commonly around 6 in (15cm). How can that be if the shale is impermeable? They are two factors to this:

The reduction of radial stress caused by the presence of the borehole and the capillary and osmotic pressures in the shale suck water from the borehole fluid. Often in petrophysics we tend to neglect the fundamental role of capillary pressures when talking about permeability and fluid saturation. This is again the case with shales.

Gary Muscio

Geologist

ALAIN BRIE, What pressures are you talking about? Is it both the confining pressure pre-drilling and the pressures of the fracking fluids? Re: altered zones, can you expand on this? Do you say that the drilling itself causes a change in the rock around 6 inch radius around the hole?

Alain Brie

Give occasional petrophysics training courses at Retired Petrrophysicist Expert

Gary, In a simplified way, a rock is subject to a total confining stress and, the pressure of the fluids in the pore space is the pore pressure. The difference between these two is the effective stress applied on the rock frame causeing compaction. In a normally pressured formation the pore pressure is close to hydrostatic with a gradient of 0.45 psi/ft. The total vertical stress gradient is approximately 1 psi/ft. During drilling the well pressure depends on the mud weight and is adjusted slightly above the pore pressure (overbalanced). Horizontal stresses are less than the vertical and usually vary with azimuth so we talk about minimum and maximum horizontal stresses. The pressure in the well is less than any of those formation stresses away from the well. The well disturbs the stress field and causes variations in the formation stresses in its vicinity; the radial stress reduces to the well pressure and conjointly, the tangential stress increases. This always occurs during drilling and is different from fracking.

When drilling through shales, especially when they contain smectite (swelling clay), the reduction of the radial stress due to drilling associated with the high capillary pressure and osmotic pressure in the shale causes the water from the mud to penetrate into and swell the clay. This softens the shale reducing its elastic moduli and sound velocity. This is the altered zone. Again nothing related to fracking. This alteration may result in chunks of shale falling into the wellbore causing borehole enlargement and wellbore stability problems. Modern sonic logs with a wide range of transmitter-receiver spacing show this altered zone. The dispersion of the dipole flexural wave also shows it. As I said the depth of alteration in shales is often as much as 6 in, which is surprising considering that shales have no permeability - unless you change the confining pressure. For hydraulic fracturing you increase the wellbore pressure by heavy pumping until the tangential stress in the direction of maximum horizontal stress becomes tensile, then the rock breaks open creating a fracture. You control the well pressure to avoid fracture propagation in nearby formations. This is easy in sandstones, but more difficult in shales.

Gary Muscio

Geologist

Alain, Great explanation. By azimuthal etc, are you saying in effect it is not a nice cylindrically (axis metric) symmetry problem? So now you have got us to the stage where we are ready to start fracking. So having drilled the hole and having swelled the rock into nice wet ductile clay, we start pumping high pressure fluids into the borehole. Even if we make cracks, how does the gas/liquid HC get extracted? How far around the cracks does the rock get affected (it was 6 inches when we drilled the borehole), so how do cracks help? Where does the HC go and how does it get there?

Alain Brie

Give occasional petrophysics training courses at Retired Petrrophysicist Expert

Gary, Effectively rock stresses and more generally elasticity is a 3D problem. Usually the vertical stress (the overburden) is largest, and the stress in the horizontal direction is different. We talk about the minimum and maximum horizontal stresses. In rock mechanics they are often called Sigma1 for the largest, Sigma2 for the intermediate and Sigma3 for the smallest. The differences in the horizontal direction are related to the tectonic context, mountain ranges, major faults, etc. Usually, altered shales are at intermediate depth, say 5000 to 10000 ft. Deeper than that smectite has converted to illite and generally shales are more compact and less prone to swelling and alteration. Mature source rocks are usually found at depth greater than 8000 ft and therefore tend to be rather compact. Whatever the situation, alteration occurs close to the wellbore and has little impact on later fracturing. To initiate fracturing you need to increase the well pressure until the tangential stress (in the direction of the maximum stress) becomes tensile. Then for that fracture to propagate in the formation the well pressure needs to be larger than the minimum horizontal stress. In normal reservoirs fracture wings extend to 1000ft or more. The fracturing fluid contains some calibrated grains of sand, but increasingly now ceramic grain called propant. The propant insures that the fracture does not close when the pressure is released allowing fluid flow. Fracturing shales is more complex, shales tend to be ductile, hence harder to break, and the horizontal stress in shales is often higher than in nearby formations making fracture containment more complicated. Because of that, fracturing shales is done in multiple sections and with limited penetration, say 100ft or less. Hydrocarbons come from organic material transforming to kerogen in anaerobic conditions and maturing to oil and gas with increased temperature. Most organic material deposits in deep sea basins with shales, which then become source rocks. When conditions are right the hydrocarbons migrate to a reservoir, but if they do not migrate, they remain in the shale, giving oil or gas shales. In this case we need to fracture the shale to create a conduit to produce these hydrocarbons as shales are effectively impermeable.

Tar sands

Tar sands (also referred to as oil sands) are a type of unconventional reservoir of petroleum deposit (Fig. 1.10). Oil sands are partially consolidated sandstone containing a combination of clay, sand, water, and bitumen (heavy black viscous oil). Oil produced from bitumen sands is often referred to as unconventional oil or crude bitumen, to distinguish it from liquid hydrocarbons produced from conventional oil wells. Crude bitumen is a thick, sticky form of crude oil, so heavy and viscous (thick) that it will not flow unless heated or diluted with lighter hydrocarbons such as light crude oil or natural gas condensate. At room temperature, it is much like cold molasses. Extremely large quantities of tar sands (oil sands) are found in Canada. Other large reserves are located in Kazakhstan and Russia.

FIG. 1.10

Tar sands specimen close-up photo. From Asphalt Ridge near Vernal, Utah. Image by Argonne National Laboratory.

Data from http://geology.com/articles/oil-sands/.

Tar sands extraction and processing

Tar sands can be mined and processed to extract the oil-rich bitumen to generate oil similar to oil produced from conventional oil wells, but extracting oil from tar sands is more complex than conventional oil recovery. Oil sands recovery processes include extraction and separation systems to separate the bitumen from the clay, sand, and water that make up the tar sands. Bitumen also requires additional upgrading before it can be refined. Because it is so viscous (thick), it also requires dilution with lighter hydrocarbons to make it transportable by pipelines.

Tar sands deposits near the surface can be recovered by open pit mining techniques. New methods introduced in the 1990s considerably improved the efficiency of tar sands mining, thus reducing the cost. These systems use large hydraulic and electrically powered shovels to dig up tar sands and load them into enormous trucks that can carry up to 320 tons of tar sands per load. After mining, the tar sands are transported to an extraction plant, where a hot-water process separates the bitumen from sand, water, and minerals. The separation takes place in separation cells. Hot water is added to the sand, and the resulting slurry is piped to the extraction plant where it is agitated. The combination of hot water and agitation releases bitumen from the oil sand, and causes tiny air bubbles to attach to the bitumen droplets, which float to the top of the separation vessel where the bitumen can be skimmed off. Further processing removes residual water and solids. The bitumen is then transported and eventually upgraded into synthetic crude oil.

Coal-bed methane

Another type of unconventional reservoir is coal-bed methane. Coal seams contain associated methane that can be produced in commercial quantities if it is dewatered in a manner that is economically and environmentally responsible. The problem with coal-bed methane is that the volumes and pressures tend to be low, and the amount of water produced can be an issue. The water is not saline but fresh or brackish. It is often fresh enough to be used in surface impoundments such as lakes or marshes. Coal-bed methane is produced in commercial quantities in Wyoming, Colorado, Oklahoma, Poland, China, and other regions that have significant coal reserves. One of the great advantages of coal-bed methane is that the coal itself is usually fairly uniform, and the repeatable units, including the "cleats" and natural fracture networks, repeat, as do the cyclothems in the coal beds themselves, which allows fairly accurate reservoir characterization.

It is important to understand the differences in the process for unconventional gas extraction, illustrated on the left of Fig. 1.11, and conventional gas extraction, illustrated on the right side of that figure. In conventional gas extraction, a well is drilled into a cavity underground that holds the natural gas. Essentially a straw is placed into the gas reservoir, and the gas then comes to the surface under pressure. In unconventional gas extraction, the natural gas is locked up inside tiny cavities in the shale rock. Think of the shale rock as a petrified sponge. The holes in the sponge contain the gas, and unless the sponge is "squeezed" the gas cannot come out.

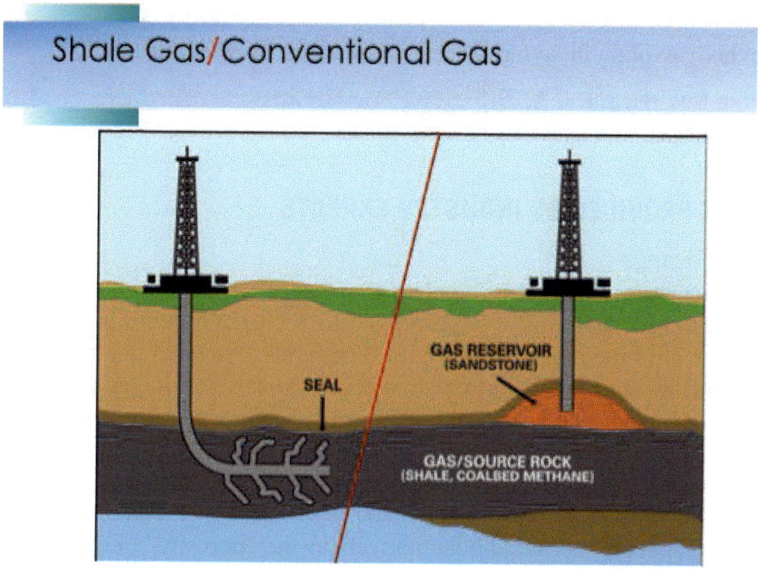

FIG. 1.11

Coalbed methane extraction.

Data from Jerome, A.P. The Health Consequences of Unconventional Gas Extraction ("Fracking").

In reality, the shale rock is pulverized underground, and then the gas comes out and rises to the surface.

Tight gas sands

Another type of unconventional reservoir is tight gas sands. There is often an overlap between shales and tight gas sands, with the difference often being a matter of permeability. Tight gas sands exhibit permeability in the microdarcy range, while shale gas permeability can be in the nanodarcy range. In each case, economic volumes cannot be recovered without horizontal wells and multistage hydraulic fracture stimulation. Low permeability sands are found in almost all sedimentary basins, and they can be adjacent to highly fractured formations where the gas migrated but the oil remained trapped. Thus, tight gas sands can function both as seals and as reservoir rocks, with the challenge that, if they do not demonstrate good fracability, it may be very difficult to recover commercially viable quantities. Tight gas sands can be stratigraphically discontinuous, with dramatic pressure variability. Understanding the depositional environment and modeling the reservoir in a way that incorporates the geological history, which includes sedimentation and then subsequent diagenesis, are essential to fully exploiting a tight gas sand reservoir. Tight gas sands occur in many North American basins, and in Venezuela, Russia, Indonesia, Saudi Arabia, Egypt, Argentina, Mexico, and China.

QUESTION 11

Can anybody explain in layman's terms what makes a reservoir tight?

Sebastiao Nogueira, R.G.A., B.S. Geo
Private Gemologist, New Jerusalem Precious Metals

ANSWERS PROVIDED BY INDUSTRY EXPERTS

Emmanuel Usanga
Technical Coordinator at Peak Petroleum Industries Nigeria Limited

Sebastiao, In layman terms, a reservoir is a rock buried in the subsurface such that the spaces within the rock serve as a trap for fluids. Provided that the fluid (gas or liquid) can flow out of that space, then that reservoir is considered productive since it contains moveable fluids. The fluids occupy pore-spaces or fractures in the rock. If the pore-spaces or fractures are too tiny or the fluids are so viscous that the fluids cannot move out of the rock, then that reservoir may be classified as "tight" relative to the type of fluid in it. The quality of the reservoir is then determined by the size of the "pore-spaces" measured quantitatively as the "porosity" of the reservoir. The original porosity of the reservoir may be destroyed or reduced by secondary factors such as "cementing" which is a chemical coating of the rock particles that originally made pore spaces for the accumulation of fluids. In a sand-stone reservoir, for instance, the "cementing" chemical substance can be calcium carbonate called calcite. When there is so much calcite acting as "cement" within the sandstone, then this excess of calcite diminishes the pore-spaces (porosity) thus making it difficult for fluid-flow. This situation is one scenario in which a reservoir may be classified as "tight".

A tight reservoir could become a productive reservoir if mechanical energy is applied to expand the pore-spaces, for instance by a process called "fracturing". The chemical alternative to turn a tight reservoir into productive is also possible by either decreasing the "viscosity" of the reservoir fluids or by chemical injection that dissolves the "cement". These artificial interventions are generally called "secondary" recovery techniques because they applied after some production has occurred. A tight reservoir for heavy oil may not be a tight reservoir for light oil. A producing reservoir may become tight if fine particles are moved into the pore-spaces as production continues. This is called "skin". Thus, tightness is, in some sense, a relative terminology. Fracturing technology ("fracking") when applied to tight formations such as oil-shales, carbonates, etc. is now making primary recovery of gas possible in basins around the world. Again, tightness becomes subjective in view of production technology available in the field or region.

Hossam El Fouly
Lead Geologist at TPS, Technical Petroleum Services, HEC Montreal, Egypt

Sebastiao, A low permeable reservoir is considered "a tight reservoir". The low permeability "K" may be due to the textural and fabric character of the rock when pores in between grains are very small or even worst when clays or silt is deposited within the sand or even later pore throats size are smaller and hence fluid flow is more constrained.

Damien Ndzanga
Completion Engineer, Schlumberger Oil and Gas Company
Sebastiao, Basically, a reservoir with low permeability and porosity will be considered a tight reservoir. Tight reservoirs are most common in Limestone areas like most of the inland environments, while most porous reservoirs mostly occur offshore where sands make up most of the reservoir rock.

Bruno Van Moerkerken
VU University of Amsterdam
Sebastiao, A tight reservoir is caused by either or both low porosity and/or low permeability. Therefore it is difficult and expensive to develop.

Sebastiao Nogueira, R.G.A., B.S. Geo
Private Gemologist, New Jerusalem Precious Metals
Bruno, Since you said that low porosity/permeability reservoirs are too difficult to develop I would like to know what are the challenges attached to its production?

Bruno Van Moerkerken
VU University of Amsterdam
Sebastiao, Basically, for tight reservoirs you need more wells, much more money and more geologists to do the job.

Damien Ndzanga
Completion Engineer, Schlumberger Oil and Gas Company
Sebastiao, In addition to what Bruno said, you might need to frackpac / gravelpac in tight reservoirs as well. That is very expensive as well. For production you may also need more wells and perhaps some gaslift pumps down hole.

Early Shale Play Production Projections and Subsequent Adjustments

Early projections about the life of the wells and recoverable reserves proved to be rather optimistic. Significant deviations from expectations emerged as shale plays were developed. The variances contributed to the oil industry collapse which started in 2014 and is projected to extend through much of 2017.

As a result, operators and researchers have focused efforts on increasing efficiency, better identifying locations of preferential hydrocarbon enrichment (sweet spots), implementing refracturing programs, and improving the use of fluids and proppants in the drilling, completion, and well-stimulation phases.

The two main issues are:

Steep decline rates: Even the most prolific wells decline precipitously, and after initially producing as many as 4000 bbls/day, can decline to virtually zero within 18 months.

Delayed completions: Because of the low commodities prices and the steep decline rates, it is important to encounter flush production while the oil and gas prices are high. As a result, operators have drilled but not completed horizontal wells. They shut in the well, and then return to conduct multistage hydraulic fracturing later. By delaying the completion and stimulation, they have an opportunity to plan the completion and use the latest technologies and techniques to optimize recovery. Additional issues include:

Water issues: Dealing with produced water as well as frac fluid by putting it down disposal wells can lead to induced seismicity, with economic, political, and environmental negative consequences. A number of new strategies have appeared to attempt to mitigate the problem:

- *Reusing the fluid*: Effective but only in limited quantities, due to the fact that the amount produced is many more times the amount that can be used.
- *Processing to the level of agricultural use*: Oil and gas producing companies in drought-stricken, water-hungry parts of California agricultural areas have been successful (Veolia is one of the companies providing services), but in general the costs are very high.
- *Processing produced water for use in roads and other industrial uses*: Reverse osmosis, crystallization, and evaporation have been used with different levels of success. The volumes are problematic, along with the costs.

Breakthroughs leading to better understanding of shales and tight gas sands have occurred in several different technical areas and fields, as described in the following sections. All are contributing to the optimization of reservoirs and pinpointing understimulated zones in order to refracture.

Geochemistry

TOC/kerogen-typing: Many plays that seemed promising at the advent of the "shale revolution" have proved to be uneconomic. Many times they are uneconomic because they simply do not contain enough hydrocarbons in a uniform manner across the basin. As a result, it is necessary to evaluate the play and find the TOC and to identify the type of kerogen in order to determine whether or not economic quantities have been generated. TOC can be calculated using indirect methods such as well-logs (the Passey method), but laboratory evaluations of cores are always best.

Gas fingerprinting (Utica/Marcellus): Migration pathways and the migration history of gas can often be traced by determining the composition of the gas and in particular the isotopes it contains. In addition to the direction of migration, fingerprinting can be used to determine the timing of generation and expulsion.

Geomechanics

Fracability: Poisson's ratio and Young's modulus can be calculated by indirect methods and can be used to determine a rock's "fracability."

Determining fracture networks (large and nanoscale): Understanding the stress regime and the geomechanical profile of the formation can help in characterizing fracture networks and predicting induced fracture heights and fracture network directions.

Expulsion upon generation processes: The generation of methane when the temperature and pressure reach the levels required also creates pressure due to the phase changes. As a result of pressure changes, the rock deforms and often fractures. These nanofractures can connect and function as conduits in the migration of natural gas.

Migration pathways: Geomechanical modeling is often used to determine the most likely fracture patterns, and to create a model of fracture architecture. The fractures, if there is sufficient conductivity, can stay open for long periods of time and constitute important migration patterns from the "kitchen" to reservoir rocks.

Induced fracture heights: Fracture propagation models are both two-dimensional and three-dimensional. In the case of shale plays, it has been relatively surprising to those studying fracture propagation to recognize the importance of the rock layers that overlay and underlie the zone of interest. Fracture propagation models are critical in predicting fracture interference.

Induced fracture behavior: Production declines quite rapidly in some cases due to the fact that induced fractures close or "heal" very quickly, despite the presence of proppants. Understanding the behavior of induced fractures has a great deal to do with fracture mechanics, and the compressive, anisotropic, and nonhomogeneous in situ stresses.

Borehole stability: Determining borehole stability involves pore pressure, rock properties (degree of anisotropy), and the stress states of the three principal stresses. Although the only way to accurately determine pore pressure is through direct measurement, there are models that bring together seismic data, and information from well-logs (including density and resistivity).

Thermal profiles

Heat flow for petroleum generation: The basin history and an analysis of the petroleum system can yield important insights into the likelihood of hydrocarbon generation, and the level of maturation, which could influence whether or not a reservoir has dry gas or oil. Chronostratigraphic markers that indicate timing of heat flow can be correlated across basins.

Potential for diagenesis due to high-temperature, supersaturated reservoir fluids: Understanding the historical heat flow, coupled with the tectonic history (which includes sea levels and crustal movements) can help predict where there will be diagenesis.

Reservoir architecture/determining ideal lithologies for reservoir development:

Depositional sequences: Sequence stratigraphy is used to map sequences between unconformities so that it is possible to predict thickness and facies changes, and to correlate them across the basin.

"Maximum rattiness" (optimized storage): In the case of shale plays, heterogeneity presents an opportunity because there is more storage in the mini-packets of relatively well sorted, grainy sand. The finely interbedded sand shale sequences are often intercalated with other formations, particularly in the cases of organic-rich black shales in fining-upward sequences.

Data mining/mathematical modeling on a large scale

Reservoir modeling: Reservoir models have been developed that utilize pattern recognition technologies, algorithms used in artificial intelligence and data mining. The workflows are often used in mature fields to identify patterns that correlate with pressure and production data. However, they are now being used in shale plays to identify sweet spots and the best places to refract.

Likely migration pathways: Numerical simulations can be used to model likely migration pathways that can point to probable areas of enrichment.

Evaluating results of drilling and completions: Production data analysis (decline curves, type curve matching, history matching, volumetric reserve calculation, and recovery factor estimation) can be tied to rock properties, fracture properties, with additional spatiotemporal data to determine which wells (and stages) are underperforming, and where there may be sweet spots (lenses and fracture zones with recoverable oil).

QUESTION 12

Recently I had the chance to study a reservoir which was highly affected by stylolitic processes (serrated and tooth-shaped as well, approx. on a 0.5-5 cm scale). It seems that this pressure dissolution related phenomena seriously reduced the porosity of the limestone. However, based on the cores, these quite often are filled with oil. How does this work? In spite of the reduced porosity, the edges of this stylolitic limestone do not act as a barrier? Should we deal with micro fractures or micro porous systems, which actually lead the hydrocarbons into these traps? What are your experiences with these types of reservoirs?

Istvan Nagy-Korodi, Ph.D.

Well site Geologist at MOL Hungarian Oil and Gas PLC, Romania

ANSWERS PROVIDED BY INDUSTRY EXPERTS

Tajeddin Benissa

Senior Geoscientist, PETRONAS, Malaysia

Istvan, It is really a complicated reservoir. Having a lot of stylolites indicates high stresses which usually results in breaks in the rock such as faults. However, to make things clear for you I suggest that you check the Paragenesis history of the reservoir or you can check the petroleum system history to know the timing of stylolite occurrence. Is it before oil accumulation in the reservoir or after? Then you can make an interpretation.

Istvan Nagy-Korodi, Ph.D.

Well site Geologist at MOL Hungarian Oil and Gas PLC, Romania

Tajeddin, We are just beginning to understand these reservoirs. However, based on the initial observations, I would say that these stylolites are from pre-oil accumulation period. But it is still questionable.

Sagnik Basu Roy

Geologist at GAIL (India) Limited

Istvan, Same observations were made while carrying out a petrographic study on a well in Middle East. The oil accumulation in these stylolites is due to the fact that not only has permeability developed in these small opening but also the porosity has increased. This is due to the dolomitization and ultimate generation of spars at the boundary of the stylolites. I presume that due to burial and due to faulting these foliations have developed. During their further burial saline phreatic water passed through these foliations causing a change of calcite to dolomite and thereby increasing the porosity. After dolomitization the oil accumulated. I suggest you carry out a petrographic study and use alizarine red to see the paragenesis.

Istvan Nagy-Korodi, Ph.D.

Well site Geologist at MOL Hungarian Oil and Gas PLC, Romania

Sagnik, The comments from the experts are really useful. In some places, I have seen a few white fenocrystals as well. The interesting part is that it seems that after the white fenocrystals (calcite and/or dolomite) are formed, the stylilotic process was still active, because it dissolved partly them.

Tajeddin Benissa

Senior Geoscientist, PETRONAS, Malaysia

Istavan, Is there severe cementing process within the rock? Or any sign for re-crystallization process? Sometimes the intense growth of large crystals within the texture of the rock during burial reduces the effect of physical stress such as compaction and then porosity can be preserved.

Istvan Nagy-Korodi, Ph.D.

Well site Geologist at MOL Hungarian Oil and Gas PLC, Romania

Tajeddin, Yes, in few places the re-crystallization is visible (I found 2x3 to 3x4 cm big dolomite crystals - pseudomorphs after calcite). However, I believe that in the presence of such big crystals the change of the former rock structure was inevitable. I am very much interested in the outcome of the thin section - petrographic studies.

Stephen Franks

Senoir geological consultant, RockFluid System, Inc

Istvan, I liked the comment made by Tajeddin. Could the oil have been in place prior to stylolitization? Is there oil in non-stylolitized porous portions nearby? For stylolitization to proceed there must be a way for dissolved constituents to move out of the stylolite- either by fluid flow or diffusion. But it doesn't require displacement of another phase for this to occur. If oil were in place previously, stylolitization might still occur by removal of ions along water films around the grains (assumes a water-wet reservoir). I can imagine oil moving into fractures or enlarged pores associated with stylolites, but I find it difficult to see how oil could easily migrate into a water-wet stylolite. I would think that the entry pressure for oil to displace water from a stylolite would be rather high, requiring a large column of oil.

Istvan Nagy-Korodi, Ph.D.

Well site Geologist at MOL Hungarian Oil and Gas PLC, Romania

Stephen, Based on the oil shows I would say that the oil was in place prior to stylolitization. And the latter mentioned process would just trap a part of the oil, or helped it to migrate. It is really difficult to say what exactly happened, as we are just at the beginning of characterizing a new basin with its reservoir section. What is interesting is that, in few places, there are 4-5 stylolites in 1 m, and they do not contain large oil accumulations (due to migration?). However, in case of larger (5 cm scale) stylolites they show relevant HC content.

Yves Geraud

Professor, Universite de Lorraine

Istvan, When did stylolitization occur? Are they parallel to the strata orientation or not? That is, were they induced by compaction or tectonic event (or events)? They could be opened under a new stress field (or change or stress axis) just prior to the oil migration. The tectonic calendar and the oil migration timing must be compared.

Istvan Nagy-Korodi, Ph.D.

Well site Geologist at MOL Hungarian Oil and Gas PLC, Romania

Yves, The stylolitization appears to have occurred prior the Mesozoic. The stylolites are more or less parallel to the lamination (they are compaction related). In some places they are serrated (zig-zag shaped) while in other layers they are tooth-shaped (like the teeth of a Halloween pumpkin).

QUESTION 13

What are some of the reservoir characterization challenges in the Eagle Ford?

Susan Nash

Director of Innovation & Emerging Science and Technology/AAPG, University of Oklahoma, USA

ANSWERS PROVIDED BY INDUSTRY EXPERTS

Hooman Laali

Geological Consultant at Blade Energy Partners, Dallas/Fort worth Area, USA

Susan, As with other unconventional reservoirs, the key parameters are the mineralogical make up for indications of brittleness, measures of organic matter, presence and maturity (TOC, Ro, etc.), and issues pertaining to principal stress conditions and rock mechanics. Prediction of response to hydraulic fracturing is very important for all unconventional reservoirs.

Marianne Rauch-Davies, Ph.D.

Geophysical Advisor at Devon Energy

Susan, Pertaining to information on brittleness, TOC (total carbon content) and information on natural fracture swarms is vital, like in all unconventional reservoirs. Conventional reflection seismic can be used by calculating elastic parameters from pre-stack data. Although the PSTM stack that I have seen over the Eagle Ford looks good, the pre-stack gathers are pretty noisy. That causes an issue with the angle stacks, which are also noisy, which in turn results in weak elastic attribute, like Vp/Vs and Poisson's ratio prediction. A better pre-conditioning of the gathers is essential.

Susan Nash

Director of Innovation & Emerging Science and Technology/AAPG, University of Oklahoma, USA

Marianne, This is a very interesting observation. I would imagine that there is quite a bit of variability over the extent of the Eagle Ford which further complicates things. Do you find your feel confident with micro-seismic surveys?

Marianne Rauch-Davies, Ph.D.

Geophysical Advisor at Devon Energy

Susan, You are correct. There is quite a bit of variability over the Eagle Ford. I also think that micro-seismic can be a good tool when you are already drilling but it cannot be used for pre-drill analysis. So you are stuck with good old seismic as the highest resolution dataset if you are at the well planning stage.

Thomas Klopf

Senior Geophysicist at KUFPEC

Susan, There was a 2011 thesis on the Oklahoma University website: http://mpge.ou.edu/research/documents/2011%20thesis/Namrita%20Sondhi_thesis.pdf

This thesis entitled "Petrophysical Characterization of the Eagle Ford Shale" had some good observations on the reservoir properties like on page 8:" Up-dip of the Stuart City Edwards reef margin, the Eagle Ford is normally pressured and primarily an oil reservoir (Fig 1.3). Down-dip of the reef margin is a modestly to strongly over-pressured and is primarily gas reservoir. The up-dip region has organic rich shale variability mostly due to reef margin controlled depocenters. Down-dip of the reef margin is characterized by lateral reservoir variability and primary permeability controlled by the location of distal turbidite deposition. Reactivation of faults underlying the Stuart City trend produced barriers to up-dip migration and areas of focused gas accumulation."

Susan Nash

Director of Innovation & Emerging Science and Technology/AAPG, University of Oklahoma, USA

Thank you, Thomas. These are great insights and the work is valuable. Now, the challenge is to determine how laterally variable (or extensive) the findings are. It would be fabulous if the reservoir was laterally homogeneous, but we know that's not the case (unfortunately).

QUESTION 14

In a single oil reservoir, two oil wells are present and crude oil properties are totally different for these two wells. What are the reasons behind that?

Biswajit Thander

Geophysicist/ Seismic Interpreter at Gujarat State Petroleum Corporation Ltd

ANSWERS PROVIDED BY INDUSTRY EXPERTS

Randy Pochel

Geologist/Chemist, Site Manager, Nation View

Biswajit, The first thing to determine is if they are indeed the same reservoir. From personal experience in an Illinois field, the first reservoir zone in this field was at the top of a cherty limestone with low porosity and heavy oil. Twenty feet lower was another reservoir with dolomite and lighter oil. Yet this field was listed singular. It was not until later publications came out that the two were differentiated. If you are on the fringes of the field where pinch outs and sporadic depositional beds occur then you could have the same time of emplacement of the oil. But different environments during emplacement and afterward could cause the differences you are seeing. If you are dealing with discordant beds and compartmentalized horizons then defiantly, you can have this situation occur. In the Permian of west Texas one of my duties was to assign discoveries to either known fields or a new discovery.

Rajesh Soni

Lead QI Geophysicist- Reservoir Development at Cairn India Ltd

Biswajit, In my experience, it is not very common in a single reservoir. Check for compartmentalization, and ensure that the samples taken are not contaminated.

Biswajit Thander

Geophysicist/ Seismic Interpreter at Gujarat State Petroleum Corporation Ltd

Rajesh, If these two wells are in different compartment then what are the possible reasons?

Rajesh Soni

Lead QI Geophysicist- Reservoir Development at Cairn India Ltd

Biswajit, It could be due to the presence of a sub-seismic fault barrier, or any permeability barrier like a tight facies or shale etc.

Biswajit Thander

Geophysicist/ Seismic Interpreter at Gujarat State Petroleum Corporation Ltd

Is it possible that source rocks for these two wells are different?

Erik Hoffmann

Petroleum system analyst at BP

Biswajit, Yes, it is possible that the source rocks are different - that, with two different compartments, you may also have two different fetch cells and migration pathways. So I will ask the next question: do you know of multiple source rocks in your basin? There are multiple ways to address this. A straightforward one would be to run GC and GC-MS, and maybe GC-MS-MS on your oil samples, then ask your friendly neighborhood geochemist to help you interpret the results (specifically, if you are looking at the biomarkers).Other possible causes of differing fluid properties would be biodegradation or gravity segregation. Again, a geochemist could help you determine the answer. If you don't have a geochemist on staff to help you, the laboratory that ran your samples might be willing to let you contract one for a brief project, or might be willing to help you interpret the results.

Thomas Klopf

Senior Geophysicist at KUFPEC

Biswajit, Please post more details on the depositional environment of the reservoir, distance between the two wells, whether there are faults or other barriers between the wells, structural position of the reservoirs to each other and how the oil samples differ.

Biswajit Thander

Geophysicist/ Seismic Interpreter at Gujarat State Petroleum Corporation Ltd

Thomas, yes, there is a fault in between two wells and the distance between two wells is 400 m.

Janine Sturdavant

Geologist/Partner at Shelby Resources, Greater Denver Area

Biswajit, The fault blocks may have different structural histories, one with later or earlier movement. Then the timing of oil migration into the fault blocks may be different, with possibly lighter oil migrating into the younger block, depending on conditions of the source rocks later in time. 400m is not very far, but if the faults are sealing, the later structural trap could have either re-migrated oil in it, or newly migrated oil. Isochron/isopachs should show timing differences.

Mohamed Saleh

Mud Logger

Biswajit, I can share with you from my experience. First, if this problem relates to the same reservoir in different wells it could be due to stratigraphic or structure reasons that it created different compartments even in the same reservoir. This could include a vertical barrier like shales, in which case you can check the log analysis with the petrophysicist. Otherwise there may be two different source rocks.

From the production point of view you should talk with the reservoir engineer to check the perforation intervals in the wells as regard these reservoirs to see if these wells are not producing as dual completions from another reservoir. The produced oil could be a mixture. Check the CBL and VDL logs for casing and cement. If there is a channel behind the casing or another reservoir below your reservoir of interest may be mixing oil into the production in the upper reservoirs. Finally, I think you should also get a geochemical analysis of physical and chemical properties of the oil samples.

Rafael Rodriguez

Petrophysicist Engineer-Sedimentologist, Schlumberger

Biswajit, Following the ideas previously mentioned, make sure that both wells are into the same hydraulic unit. To do this, you could apply the first principles of sequential stratigraphy. That is to say, identify key surfaces like Sequence Boundaries (SB) and maximum flooding surfaces (MFS). Even if the two reservoir units look to be in the same reservoir or rock unit (in lateral correlation), they may not have the same genetic origin, and that is important because this determines the evolution of each rock type in terms of diagenesis and its effect upon the stored fluids. The correlation of stratigraphic units must be accurate and calibrated with cores and seismic data. Evaluate the structural settings of the area and finally determine they are in the same geological unit. The petrophysical studies are helpful in an interpretation.

Nadeson Ganesan

Reservoir Advisor at Tatweer Petroleum

Biswajit, What differences in crude properties did you see between these two wells, in terms of API, viscosity, density etc? These differences when understood, can lead to applying the comments given by experts on narrowing things down to a couple of 'sources' for these differences. For instance, if different API and geochemical analysis shows no presence of alkanes, paraffins and possibly aromatics, you would narrow it down and say the crude went through rigorous biodegradation, then that leads to possibly an older oil, or migration. Cross-check with rock properties and you may see compartmentalization due to rocks pushed up or down in different times. All the above mentioned reasons from the experts are valid. The task is to acquire datanecessary to resolve this. List them down and cross them out as you can, then draw a hypothesis and bring other disciplines in to help you iron it out. There must be other clues present to be brought in. If compartmentalization is present within the 400m or more radius, then look at pressure relationship. Take pressure survey in these and surrounding wells too. I would tend to believe it is the case of compartmentalization rather than two different source rocks.

Biswajit Thander

Geophysicist/ Seismic Interpreter at Gujarat State Petroleum Corporation Ltd

Nadeson, Yes, API gravity, viscosity, density are different for the two wells. The reservoir is silty and multiple source rocks are also present in the depositional environment.

QUESTION 15

How important is thin section analysis to reservoir characterization?

Leonardo Bazzano

Sales Executive at Endeeper, Geological Knowledge Management Porto Alegre, Rio Grande do Sul, Brazil

ANSWERS PROVIDED BY INDUSTRY EXPERTS

Stephen Franks

RockFluid Systems, Inc

Leonardo, Thin section analysis is critical to reservoir characterization. Hydrocarbons are stored in the pores of rocks, so pore-scale examination and evaluation is extremely important. For example, what minerals and textures are controlling reservoir quality (porosity/permeability)? Can this be used to predict reservoir quality in other exploration prospects or in other locations within a field? What minerals in the reservoir affect log response (e.g. conductive minerals that may reduce resistivity and make reserves appear lower)? There are far too many things to list here, but understanding the rocks that make up a reservoir is absolutely critical. Thin section examination (along with SEM, XRD, etc.) is an important part of the evaluation. Tying this petrologic information to porosity/permeability data, log response, etc. is a very valuable exercise!

Rohit Arora

Geoscientist

Leonardo, Basic thin section analysis on oriented slices can give an idea of porosity, the type of cement, if bimodal quartz is present or not, sorting, weathering etc. High end techniques like cathodoluminescence can distinguish between dolomite and calcite.

Jyoti Shah

Senior Geoscientist at PETRONAS

Leonardo, Various workers have used thin section analysis results to delineate depositional environments. The grain size and rock composition variation laterally and vertically in a reservoir serve as important parameter for understanding diverse environments. Doing petrography with a microscope provides the most accurate results but is a tedious process. Now, there are various software program which come handy in determining the grain size and shape. Software 'Image J' is available free, online. This also helps determine the percent area porosity of a rock sample in thin section. Of course logs give us the data all along the well but thin section results can always be used to understand the finer details of rock texture and structure followed by depositional environment, etc. especially in a reservoir.

Matt Anaraki

Operation Geologist

Leonardo, Sometimes during drilling, you need to get more information about porosity and no wireline data is available. The most common applications of thin sections is Biostratigraphy and Danham classification in carbonate.

Dalila Mebbani

Geologist/Reservoir Engineer

SONATRACH PED (Production Engineering Development)

Leonardo, Adding to what was said, on thin section closed (by minerals) / open micro fractures can be examined in detail and oriented from thin sections as Rohit said. Also, images taken with an electronic microscope can show how mineral such as salt obstructs pores.

QUESTION 16

What is the best way to correlate deep water turbidite reservoir? My XRD data reveal barite & dolomite at a deeper depth, what could be the potential source of these minerals especially barite?

Ugwueze Charles

Lecturer, Department of Geology, University of Port Harcourt, Nigeria

ANSWERS PROVIDED BY INDUSTRY EXPERTS

Diane Jorgensen

Senior Geologist, Origin Energy

Ugwueze, I have observed barite as a late stage cement in turbidite deposits from Australia. Do you have thin sections or SEM to verify character? Correlating can be difficult. Integration of data is the key. Useful data include: palynology, seismic interpretation, sequence stratigraphy on accompanying slope fans, analysis of sediment source, etc.

Nuno Couto

Geologist/Consultant PETROBAS

Ugwueze, It may be a strange thing, though I have seen it occur in carbonates but, is there a chance that some deep faults could have carried some hydrothermal fluids from the deeper part of your basin to your reservoir? Is the dolomite you describe saddle dolomite? I have seen that saddle dolomite associated with barite associated with hydrothermalism. If that is not true in your case it will be necessary to look for a different explanation.

Ugwueze Charles

Lecturer, Department of Geology, University of Port Harcourt, Nigeria

I also discovered that it could be biogeneic, diagenetic, or related to hydrothermalism as you rightly observed.

Garry Kolafa

Owner, Exploration Systems International Ltd.

Ugwueze, The barite has a high specific gravity; therefore this gives the barite a chance to migrate deep during either a settling phase or diagenetically. The dolomite may have also been born via diagenesis, but it could have had come from a previous source that was formed closer to the surface (i.e. an atoll carbonate environment) and then transported to its present site.

Moussa Bari

CFM Geophysicist and Petrophysicst, SIGMA3

Ugwueze, I have seen barite in limestone nodules (cocretion septarian). Barite deposited in the fractures could be the result of hydrothermalism as you said or mineralization from low temperature solutions known to circulate in certain sedimentary basins.

Nuno Couto

Geologist/Consultant PETROBAS

UGWUEZE, Only the presence of barite is not enough to say that hydrothermalism had occurred. But if that is associated with saddle dolomite, then there is a strong chance that that is the answer.

Kiran Kant Prasad

Independent Oil and Gas Consultant/Petrophysicist

Ugwueze, I see your question in two parts mentioned above. For first part, if you assimilate all the available information, properly segregate, analyze and synthesize with well log data including a dip meter, it can be done. I am saying this based on my experience from 30 years ago. A setting which was seen as a tidal channel and associated regime earlier, turned out to be a submarine fan setting from the study and remains the accepted model.

CHAPTER

OVERVIEW OF QUALITATIVE SEISMIC INTERPRETATION

2

CHAPTER OUTLINE

Practical Solutions to Integrated Oil and Gas Reservoir Analysis. http://dx.doi.org/10.1016/B978-0-12-805464-2.00002-0

There are two major tasks in the interpretation and analysis of processed seismic data. First, one needs to develop a geologic framework, which includes identification and mapping the main faults and the main stratigraphic horizons. Once we have developed the geologic framework, the second task is to examine the area's hydrocarbon (HC) system components and potential:

- Where are potential reservoirs, traps, seals, mature source rocks, and HC migration pathways?
- Then, where would we drill to make a new discovery?

Also, for development and production, we need a geologic framework (if it does not already exist or does not have the level of detail that we need). Having that, we can consider where to drill development wells to answer questions on the spatial distribution of fluids and reservoir rock properties, or we can consider where to drill to increase hydrocarbon production from the field. Generally, the seismic interpretation process can be subdivided into three interrelated categories: structural, stratigraphic, and lithologic.

Structural interpretation: Structural seismic interpretation is directed toward the creation of structural maps of the subsurface from the observed 3D configuration of arrival times.

Sequence stratigraphic: Seismic sequence stratigraphic interpretation relates the pattern of reflections observed to a model of cyclic episodes of deposition. The aim is to develop a chronostratigraphic framework of cyclic, genetically related strata.

Lithology interpretation: Lithology interpretation is aimed at determining subsurface rock properties (e.g., porosity and lithology), reservoir fluid changes in time and space, fracture intensity, and so on from seismic data. Direct hydrocarbon indicators (DHIs or HCIs), such as bright spots or dim-outs, are elements employed in this lithologic interpretation process.

This process requires the best possible acquisition and processing having been performed on the seismic data, and it also requires knowledge of local geology (study area) from outcrop and preexisting wells. Also, a good picture of the geologic history of the area to be studied is important in making quality decisions during interpretation of the seismic data. It is also important for the seismic interpreters to have developed an understanding of the factors influencing regional tectonic sedimentation in the basin (area).

After the geology of the area to be studied is known, a data gathering and collation phase commences. This phase is very important to the success of the study. The quality of the data set determines the quality of the seismic interpretation produced after the study is completed (Fig. 2.1).

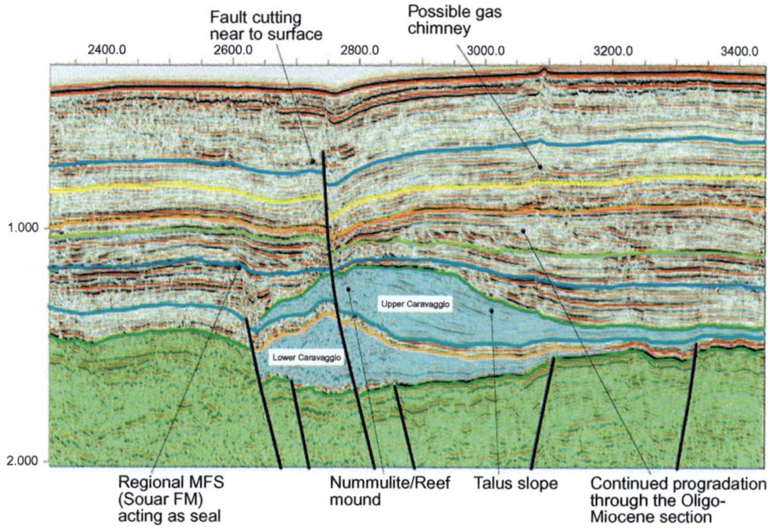

FIG. 2.1

Interpreted seismic section.

Courtesy: Heritage Oil, http://www.heritageoilltd.com.

DATA SET NEEDED FOR SEISMIC INTERPRETATION

Generally, a seismic interpretation study requires the following data sets:

- 2D, 3D, or 4D seismic data cube (or even a merged data set).
- Well logs with a full array of the lithology logs, resistivity logs, porosity logs, density logs and composite logs.
- Core data from the wells in or around the study area plus well-log reports if available.
- Petrophysical reports and petrophysical parameters, if available, are also important.
- Check-shot data set (velocity data) is also required.
- Well deviation data if the wells are deviated are also required. A drill stem test (DST) report will be needed if available.

Usually, the type of study to be undertaken determines the amount and type of data to be collated for the study, because the data set determines the output.

SEISMIC SECTION

A seismic section is a display of seismic data along a line, such as a 2D seismic profile or profile extracted from a volume of 3D seismic data. A seismic section consists of numerous traces with location given along the x-axis and a two-way reflection time or depth along the y-axis. A seismic section is also called a seismic line (Fig. 2.2).

FIG. 2.2

Seismic section.

Data from Virtual Seismic Atlas, http://www.seismicatlas.org.

A 3D seismic volume can be sliced in any vertical direction to create 2D lines, or sliced in horizontal planes to create time slices, which represent constant time. The volume can also be sliced along reflection boundaries to create "horizon slices" (Fig. 2.3).

QUESTION 16

I want to carry out seismic interpretation project. As an interpreter, why should I ask for anisotropic migration data to carry out my project? (Fig. 2.4).

Masoud Aali

Seismic Interpreter

ANSWERS PROVIDED BY INDUSTRY EXPERTS

Peter Wang, P.G.

Geophysical Technical Advisor at PARADIGM

Masoud, Because your seismic depths will be way too deep, and the shape of the structure you are mapping will change also. If you have astigmatism, you need eyeglasses that correct for it, or else you won't see properly. Anisotropy is "earth astigmatism", essentially. Isotropic depth migration is outdated. Try not to use it.

Jerry Coggins

Principal Geophysicist at Coggins Geosciences, Inc.

Masoud, It is a bit simplistic to say that every migration should be an anisotropic depth migration. How important the effects are to your interpretation problem is of fundamental importance. Some amazing quantitative results have been obtained without anisotropic migration. Often the structure doesn't change very much, and the

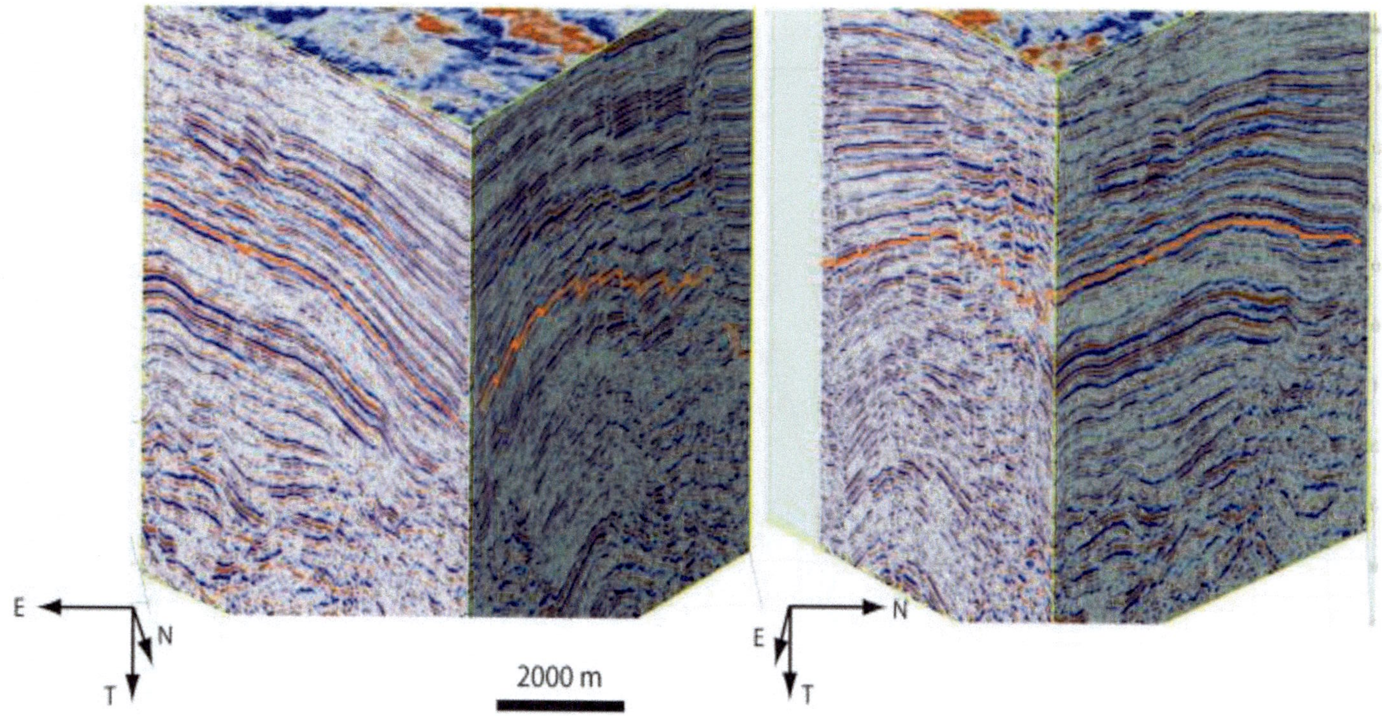

FIG. 2.3

Seismic interpretation data.

Courtesy: Energy & Mineral Resources Group (EMR), http://www.emr.rwth-aachen.de/aw/EMR/Energy_Mineral_Resources/Zielgruppen/gia/links/projects_research/~wab/seismic_interpretation_and_geological_mo/?lang=en.

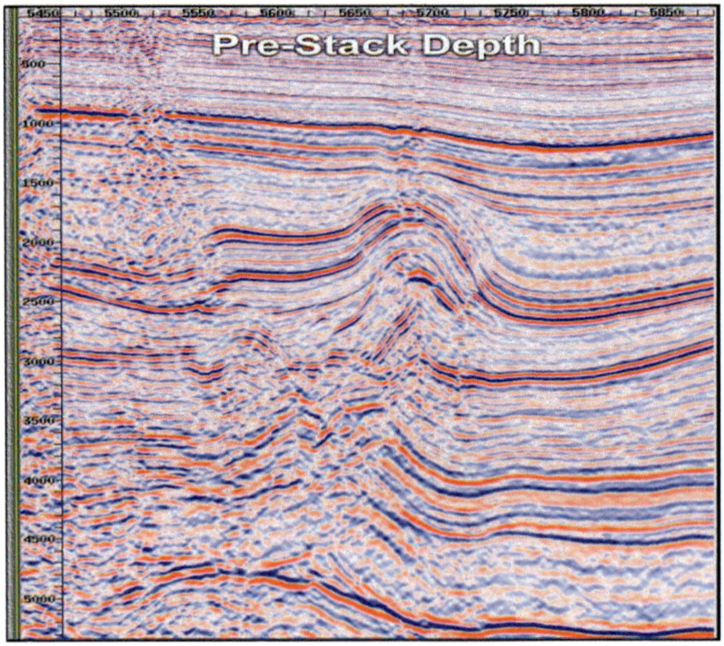

FIG. 2.4

Kirchhoff migration seismic data.

Courtesy: Paradigm, http://www.pdgm.com/getmedia/6f131640-7280-4e86-89a1-79e788bdaac8/Kirchhoff-migration_lg.jpg.aspx?width=1024&height=650&ext=.jpg.

industry has successfully been converting seismic depth to "true" depth for quite a while. Having said all that, anisotropic depth migration is an important new part of the imaging toolkit. It is not the solution for every problem. Always consider your objectives, cost, and schedule. I have interpreted the same fields on time migrated and anisotropically migrated data, and the differences were not significant. The difficulty is always how to determine how much tech is needed. There's no one answer.

QUESTION 17

What is the best way to do a 3D seismic multi-component interpretation?

Andres Espeso

Reservoir Geophysicist, Seismic Interpreter

ANSWERS PROVIDED BY INDUSTRY EXPERTS

Steve Adcock

Geophysical Consultant at dGB Earth Sciences

Andres, The challenges are mainly with the processing. It is very difficult to obtain a consistent and robust shear volume from the multi-component acquisition. A number of companies are acquiring 3-Component or 4-Component now as a standard, but only processing the compressional P-wave component. So just because it was acquired multi-component does not necessarily mean you will have a shear volume. Most of the inversion techniques that attempt to utilize the shear, don't use actual shear from seismic, but modeled shear from logs. Even so, they can produce good results. How much better if we actually get the shear from the seismic, who knows? If you have the shear data, calibration can be tricky. **Murray Roth's** company, Transform, recently acquired by Drilling Info, has a number of good interpretation tools for P-S interpretation workflows. In general:

- Load and calibrate (time tie the events/horizons between the P and S volumes). Tying synthetics and ray trace models from the wells will assist greatly. Inversion and velocity modeling are useful. This is mostly a stretch and squeeze process. The next step is a real problem if the volumes are not aligned in time or depth in this fashion.
- Divide volumes (i.e. create a Poisson's Ratio volume), and other rock property volumes if desired. If step one was too difficult, try to accomplish this over a sequence or between two horizons.
- Diverge into whatever workflow is appropriate to your goal.

Lynn Engel

Senior Geophysicist and President, Lynn Engel Consulting Ltd

Andres, it also depends on what your objective is. For reservoir characterization, using multi-component data in a Quantitative Interpretation workflow works very well. We have even observed improvements that are very high resolution at the end of the workflow, which was a nice surprise since the PS PSTM conventional stack looks low frequency when compared to the p-wave PSTM stacks.

QUESTION 18

I am curious to know how many seismic interpreters routinely use the pre-stack gathers themselves in building their interpretation or do you leave that to a QI specialist? If you do use them I would appreciate if you would share your views on the benefits and if you don't, then why not?

Andrew Forbes McIntosh

Senior Exploration Geophysicist Specialist

ANSWERS PROVIDED BY INDUSTRY EXPERTS

Robert McGrory

Senior Geophysicist at Crescent Point Energy

Andrew, I find doing my own QI work using pre-stack data improves my interpretations. What is important is not so much using the pre-stack data directly as much as it is important to do your own modeling to understand what attributes are important to quantifying the quality of reservoir rock. Another reason I like to work directly with the gathers is to be able to create near, mid and far angle stacks.

Specifically, I find the near stack generally ties my wells more reliably. Provided of course I have decent S/N (signal to noise ratio). Also having the gathers (NMO corrected) allows me to determine whether or not an amplitude anomaly is created from a legitimate AVO or due to an artifact related to residual NMO. I have seen cases where over corrected and under corrected gathers (poor velocity) analysis have created Fluid Factor anomalies or Product Stack anomalies. Recently, another project I worked saw a hideous multiple create a booming Class II anomaly. The multiples where barely visible on the gathers but by shear coincidence the true energy from the multiple line up with the true energy of the primary at mid to far offset and peak energy from a different multiple event lined up with the true from the same event on the near offsets. Coincident destructive interference on the near offset and constructive interference on the mid to far offset created a Class II anomaly that was very robust looking. After running some simple radon de-multiple I was able to get a more reliable set of gathers. I would not have notice if I was only looking at near mid and far stacks. It is so important to look at the gathers.

Hesham Shalaby

Senior Geophysicist at Dragon Oil

Andrew, I used to use pre-stack gathers for my interpretation in one of the operators I worked for. Pre-stack gathers aid strong understanding of the reservoir for the interpreter so one can easily assess the potential of a mapped prospect.

Cesar Delgado

Geophysicist at PEMEX

Andrew, I only use pre-stack gathers for calibrating logs and horizon, and work on PSDM for new location or development field.

QUESTION 19

What are the parameters needed to invert a seismic volume?

Messaoud Boukhenifra

Géophysicien chez Sonatrach_Agip

ANSWERS PROVIDED BY INDUSTRY EXPERTS

Hesham Shalaby

Senior Geophysicist at Dragon Oil

Messaoud, If you mean elastic inversion, you need the pre-stack seismic gathers, compressional and shear sonic and density logs to build the low frequency model and to extract the wavelet.

Messaoud Boukhenifra

Géophysicien chez Sonatrach_Agip

Thank you **Mr Hesham**. So, after that we get the AI volume for more information about the reservoir and objectives. How about the multiples effect? Is there any impact on the volume inverted?

Hesham Shalaby

Senior Geophysicist at Dragon Oil

Messaoud, the multiples and noises will also show up as they are in the input data.

Messaoud Boukhenifra

Géophysicien chez Sonatrach_Agip

Hesham, the reprocessing is recommended for multiples suppression before the inversion is to be done.

Anonymous

Senior Exploration Geoscientist

Messaoud, If the data is noisy or contains some multiples, that will have a big effect on the inverted volumes, because the inversion algorithm will take the 3D seismic volume, and build the impedance model that is most nearly responsible for the creation of that 3D seismic volume. And if the multiples persist among the primary reflections, they will be taken as Primaries, and they will contribute for the inversion process, and their presence will contribute for the creation of some unwanted impedance signatures, that will affect by the way: the Horizon slices and for sure the reservoir characterization will be totally out of reality (missing reserves or wrong well positions). If your seismic data is noisy or contains multiples; a reprocessing is highly recommended.

Son Le

Senior Geophysicist/Lead seismic Interpreter-Asia Pacific region, Schlumberger

Messaoud, if the data is noisy and multiple contaminated you have to do reprocessing first (Messaoud et al are right). Then the logs have to be "conditioned" meaning all the errors have to fix before doing inversion, because "any errors in the well log data are propagated throughout the integrated processing sequence and remain within the final reservoir model". Depending on what type of inversion you want to, you may use pre or post stack data for it.

QUESTION 20

Vertical and Horizontal velocity gradients, what are they? And how do they affect velocity. Which one has more impact on seismic data?

Raouf Nasr

Geophysicist

ANSWERS PROVIDED BY INDUSTRY EXPERTS

Guy Maslen

Chief Operations Officer at Globe Claritas

RAOUF - Many of our seismic processing algorithms contain approximations and simplifications, usually to make them run in a reasonable time frame. Some of the more common assumptions include things like ignoring refraction effects that cause a ray-path to bend. Ray paths are refracted when the velocity changes - either vertically, or horizontally. The gradient is a measure of this velocity change. While we usually think refractions as taking place at an interface where the velocity changes (for example the velocity of light in glass is slower than the velocity of light in air), you can also get refraction where there is a gradual change in the velocity within a material. This is essentially what happens when you see a "mirage" over a hot, plane surface - the thermal gradient changes the density of the air, and hence the velocity, causing the optical ray-path to be bent and creating a false image.

We have the same effects in rocks; vertical velocity gradients commonly arise with compaction, however, changes in depositional facies can also create gradients as can diagenetic changes. More complex situations occur when rocks with a vertical velocity gradient are uplifted or rotated through tectonic activity. We can also have horizontal velocity gradients within a layer as a result of these processes, as well as things like faulting or salt intrusions. The NMO equation and things like statics all ignore refraction effects, as do most time migrations. These are only well behaved with the velocities change relatively slowly horizontally, and/or the horizon layers are more-or-less planar. There are some time migrations that have asymmetric operators and so cope with velocity gradients better. Refraction can also be important when considering AVO (or AVA) effects, as again refraction may or may not be taken into account when calculating angles. Depth migration takes refraction into account, however, only if the vertical and horizontal velocity changes are built into the model. See also:

http://www.spe.org/glossary/wiki/doku.php/terms:velocity_gradient_seismic.

This is also a nice reference showing the issues caused by refraction events:

http://www.iongeo.com/content/documents/Resource Center/Technical Papers/EAGE_Chalk_tutorial_Jones_130601.pdf.

QUESTION 21

Which velocity reflects the lithology of the subsurface, NMO, STACK or RMS velocity?

Rakesh Yadav

Geoscience students

ANSWERS PROVIDED BY INDUSTRY EXPERTS

Grant Hanson

Consultant- Seismic Data Processing Software Testing at DISYS

Rakesh, none of the above velocity reflects the lithology of the subsurface. In the original seismic model of the earth of yesteryear, where the earth was flat layers of homogeneous material, the Normal Move-out velocity, Stack Velocity and Root Mean Square Velocity, were all the same thing and the actual velocity of the individual layers we called the interval velocity and it could be computed from the Dix Equation. As our science and processing algorithms have gotten more sophisticated, finding the real velocity of the lithology is today the major thrust of all the velocity model building methods of Seismic Imaging. So we hope it is the depth migration velocity model that actually reflects the lithology of the subsurface.

Adell Hakim

QC Geophysicist, Magnum Consultant for Exploration

Rakesh, the answer is exactly as Mr. Grant mentioned here "None of the above". The t/x reflection curve says exactly this solid fact. It shows an "apparent" velocity along its different x distances. Far away at the theoretical infinity distance the true velocity could be found. Bear in mind that all the handled velocities are apparent one until you obtain the interval velocity from a VSP operation or else, this will be the true velocity (Fig. 2.5).

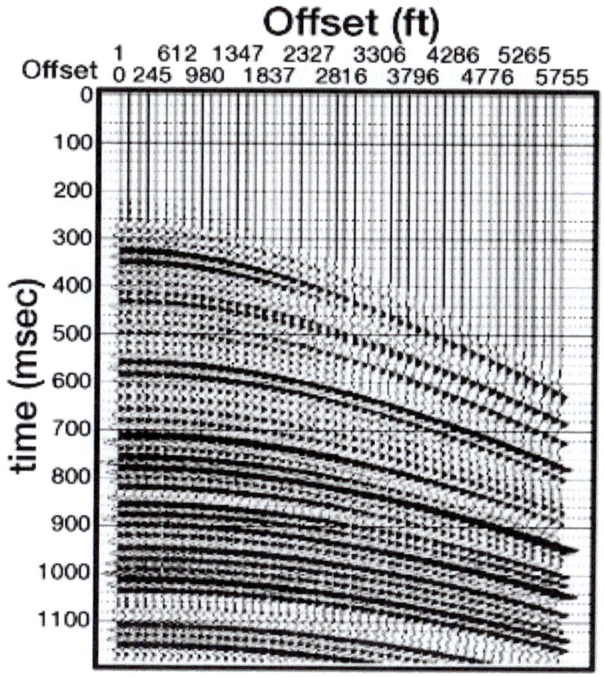

FIG. 2.5

Travel-time curve of the reflections for different offset between source and the receiver.

Steve Adcock

Senior Geophysicist at Resource Evaluations

RAKESH, I think Grant's "None of the above" is the correct response, even for modern seismic data. My reasoning: Seismic data is a function of velocity and density. In practice, we fiddle with the velocity during processing until the sharpest image is achieved. If we want to calibrate the seismic to lithology, we need to use velocity and density data from well and rock studies, as neither of those data (NMO, STACK or RMS velocity) directly provides the best imaging velocities. Even depth migration velocities need to be calibrated to tie lithology, so don't trust the seismic velocities indiscriminately. Always test the calibration to well/rock data. And don't forget to consider offset and shear velocities.

Quang BN

Technical Consultant at BITEXCO O&G Com.Ltd

RAKESH, general answer is "All and None"! If you are targeting at sand/shale scale then all of them (NMO, Stack and RMS Velocities) would give you the interval velocities of the lithological units via Dix equation. None of them reflects the lithology of the subsurface since you want to match these velocities to those of sonic logs. This is because theoretically the above are simply a mathematical averaging without any physical meaning of the wave velocities through the lithologies. The same is true with up-scaling of sonic logs. So solve this or wait till a new equation is found.

José Adauto de Souza

Senior Geophysicist, Consultant at Exploratory Vision

RAKESH, None of these velocities (NMO, STACK or RMS velocity) reflect the lithology of the subsurface. There is no chance in seismic wave propagation to travel across homogeneous layers along earth subsurface. Stack velocity is a number dependent on rock velocities, travel path, noise superposition, inaccurate NMO equation, computational algorithm, interpreter experience, etc. Seismic velocities from reflection surveys have minimum error at shallow depths and the error increase with increasing depth. More structured the geology is (faults, domes, anticlines, synclines), more difference in rock to stack velocities. At medium depths (+- 3000m) 5% difference is usual.

QUESTION 22

I am trying to understand the different iterations of the velocity model used for depth conversion usually undergoes in a typical seismic interpretation project.

1. Firstly, velocity model made from stacking velocities and possibly well check-shots. Vertical stretch volume based used as input to PSTM.
2. New velocity model needed after PSTM, made by similar analysis as NMO after PSTM.
3. Horizons interpreted with more well information available. Moving to vertical stretch horizon based and stacking like information less important. Regression of V=V0+kz in zones from checkshots and well tops.

What have I misunderstood and what steps are missing in a typical seismic interpretation project with relatively simple structure?

Andreas Paulsen

Senior Software Engineer

ANSWERS PROVIDED BY INDUSTRY EXPERTS

Mike Stone

Geoscience Technician at LUKOIL International Upstream West Inc.

Andreas, it depends on what data you have. If you have a ton of check-shots and just a grid from 2D, you just make average velocity maps and run with them. But let run through the theoretical for a second. Why do we use V0+Kz? The answer is that it works really well when all you are dealing with is a standard compaction curve. But if you have something like a carbonate which has a higher value of Z or an unconformity which will probably have a lower value. This is when you need a layer cake model. Basically you end up with the base of your upper layer as V1+Kz with a new K value for the new layer. This method works pretty well, it works with little data, and it actually sounds a lot like layer based tomography. So let's talk about the seismic Velocities. You mentioned NMO velocities, Why not just use those? Well a couple of problems. First think of how they are picked, if you have two clouds at the same level which one do you pick? The faster velocity, right? Why is that, well the slower velocity is probably a multiple. So Seismic velocities are probably fast. Also how did you get from RMS to Interval velocities? Did you do full blown ray-tracing? Or did you do what 95% of use do and rely on the DIX approximation? Also NMO Velocities are noisy, so most people who use them smooth them out. So why not use migration velocities? They are already smooth. I am still on the fence about this one. Stacking velocities are pretty much direct measurements. And migration velocities are still fast for some reason (I like to use anisotropy as a scapegoat but it's not enough alone). An Ideal velocity model would probably be a hybrid of seismic velocities slowed down to match a lower frequency layer cake model built from your check-shots. But you also have to ask why you are doing depth conversion? If you are generating a prognosis tops maybe you do all of this. It you are just seeing if you still have structure in depth, just use the migration velocities and be done with it. The difference between a comprehensive depth conversion and a quick and dirty are both going to result in very similar structure maps.

Mohand Cherif Ammour

Geophysicist at BP-SH-Statoil Join Venture

Andreas, a simple layer cake method that I used quite often consist of:

- Select horizons that correspond to major changes in AI (acoustic impedance). Those horizons will determine the layers of your model.
- Calculate pseudo interval well velocities at well locations for each layer.
- Extrapolate pseudo velocities across the area of interest using cross-plots or Geostatistics. Here you need to find a mapable parameter that correlates with your pseudo velocities. It can be filtered seismic Dix velocities, TWT (two-way-time) maps, Isochrones, etc. If you have decent well coverage you can just Krige your pseudo well velocities.
- Generate Velocity cubes by populating the model with the pseudo velocity maps generated previously.

Edgar Leonardo Galvan Aguilar

Geophysicist at Schlumberger Tech. Co

Andreas, velocity modeling and depth conversion is very different at the interpretation stage, comparing with seismic data processing. Every dataset is different and requires a different approach, this depend on the development stage of your basin;

- In exploration stage with few well data with 2D and 3D interpretation and seismic velocities usually doing depth conversion for prospect generation or play analysis or basin modeling with very coarse detail.

- In Reservoir characterization, doing depth conversion for delimiting wells and reservoir delimitation uses a different approach and different resolution on your velocity model,
- And finally on production stage with lots of data in your field requires depth conversion with very high detail to create geomechanical models in order to control the drilling of the field.

But all this velocity modeling has certain stages in common, like the use of well tops to calibrate your seismic interpretation (please do not forget to do synthetic seismograms prior to interpreting or velocity modeling). As we know the well data is limited, that's where the seismic velocities could be very helpful in order to increase the detail in our model, but remember the well data has more detailed information on velocities than 3D seismic velocities, that's why stacking velocities require to be calibrated with check-shot data, and there is a point where the correlation between them is very poor so using stacking velocities could cause more problems than benefits.

ADVANTAGES OF SEISMIC DATA

The advantages of seismic data are listed here:

- Seismic interpretation is inherently 3D, even if the data are 2D.
- Even if we only have a 2D seismic grid, we map a fault on several lines and interpolate a 3D fault plane.
- We are able to image traps that are the scale we would be interested in.
- We are able to image stratigraphy and to identify structural markers that tell us the sense and magnitude of fault offsets.
- Our structural framework provides a 3D context for understanding other data such as surface geology, well data and potential field data.

DISADVANTAGES OF SEISMIC DATA

Listed here are the disadvantages/weaknesses of seismic data:

- Seismic data has limited resolution: it can't resolve "small" features, such as a horst block 1 km by 1 km (especially in offshore, but in some land areas such as North Dakota smaller structures can be mapped).
- Steep dips (>about 35 degrees) can be very difficult to image.
- Acquisition can be difficult, e.g., in areas of variable topography, variable surface geology, or "hard" water bottom.
- Vertical axis is typically (migrated) time, not depth.
- Velocity variations distort geometries (especially in the near surface or near the water bottom interface).
- Display scales are commonly not V:H = 1:1, which results in distortions of geometries.

STRUCTURAL ANALYSIS

Structural analysis is the consideration of all of the significant processes that formed a basin and deformed its sedimentary fill. These processes range from basin-scale processes (e.g., plate tectonics) to centimeter-scale processes (e.g., fracturing). Typically, what we focus on is: How did the basin form? This includes mapping the major fault systems, looking at sediment deformation for potential traps, and identifying present-day traps and factors that may control their fill level (trap size). In some cases, we may also need to work out when the trap formed and its size history.

What type of information can we get from seismic data in terms of structural geology? We can identify and map faults, folds, uplifts, and other structural elements. We can also interpret structural settings and structural styles, divergent, convergent, etc. We also need to insure 3D geometric consistency in an interpretation so that it is structurally valid. We can determine timing relationships, such as the timing of trap formation (especially important if HC migration occurred about the same time as the trap formed). Finally, we can decide whether our picture is admissible—that is, have we violated any laws of nature?

Note that there is a synergistic relationship between structural analysis and stratigraphic analysis. You cannot get all of the structural information without working the stratigraphy. Similarly, you cannot get all of the stratigraphic information without working the structure.

FAULT INTERPRETATION

What are some of the clues to look for in seismic data to recognize faults? They are listed here (Fig. 2.6):

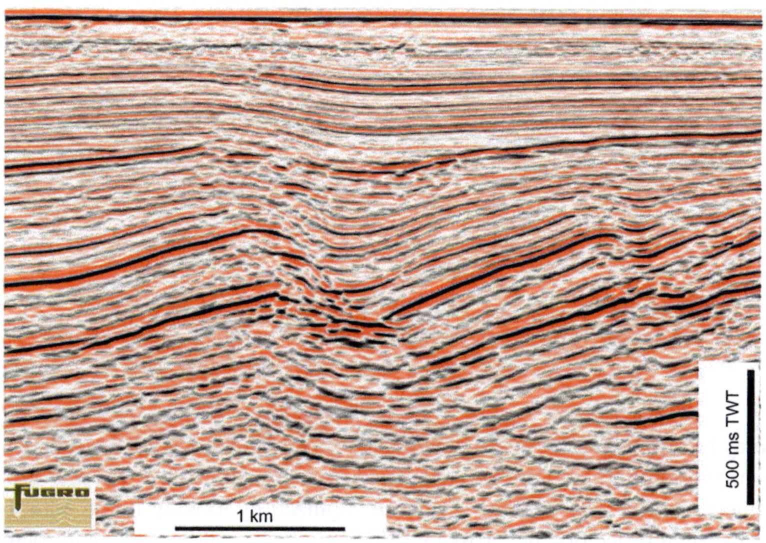

FIG. 2.6

Normal fault.

Data from Virtual Seismic Atlas, http://www.seismicatlas.org.

- Termination of reflections
- Offset in stratigraphic markers
- Abrupt changes in dip
- Abrupt changes in seismic patterns—e.g., a strong, continuous reflection turns into a low amplitude region.
- Fault plane reflections—only when fault dips less than about 30 degrees
- Associated folding or sag (some of this above the red fault)
- Coherency

Faults are recognized most often on a seismic section by the terminations of the strata reflections at the fault. These reflection terminations are not abrupt.

When the displaced horizon can be identified on both sides of the fault, the fault throw can be measured by noting the difference of travel time between the two horizons. Note that the best measure of the throw of the fault is given by the two strata reflections. And the best indication of the fault location comes from the fault plane reflections. When reflectors terminate abruptly, the corresponding reflections tail away smoothly into diffraction curves and, as a result, the actual fault location may be difficult to pinpoint on an unmigrated seismic section.

Seismic migration collapses the diffraction curves and the faults become more apparent on the migrated seismic section (left of Fig. 2.7).

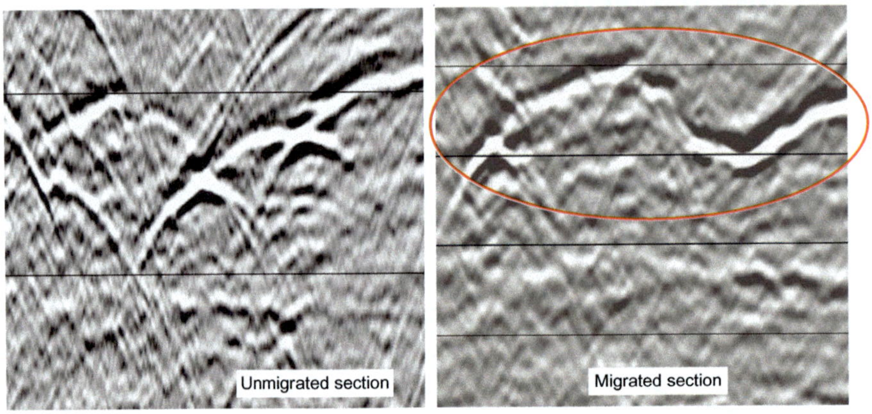

FIG. 2.7

Unmigrated seismic section with diffraction curves and the same data after migration.

Courtesy: Shell Exploration and Production Company International.

It may be difficult to identify a fault plane reflection on a seismic section due to ray-path problems around the fault and also because many faults have no clear breaks and no diffractions (Fig. 2.8). These faults are recognized by looking for an abrupt change in dip of several reflectors that cross the section, which could mean a fault plane. Note that seismic migration techniques also laterally reposition all inclined reflectors to their correct spatial location, whether they are stratigraphic interfaces or fault planes.

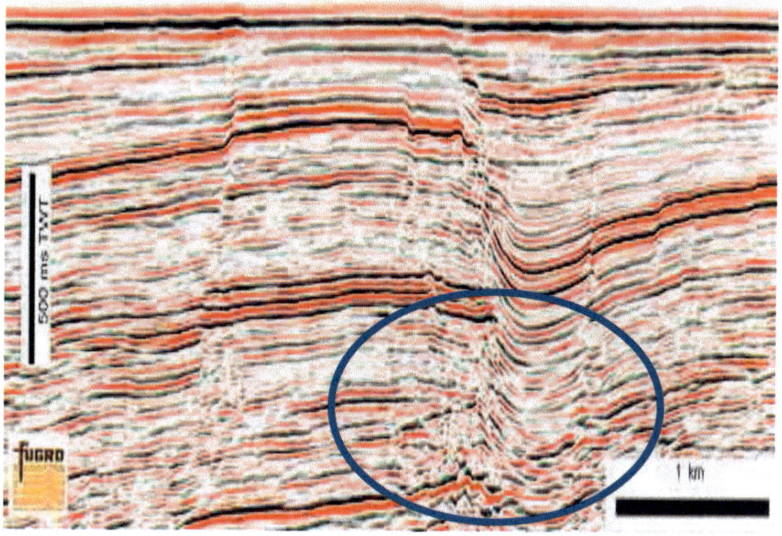

FIG. 2.8

Normal fault-related syncline, Offshore Tunisian.

Data from Virtual Seismic Atlas

FAULT PICKING

A fault block on a seismic section is defined as a subsurface area bounded on either side by faults. A fault block could be an up thrown (foot wall) or down thrown (hanging wall) side of a fault. Faults are picked on the seismic section to help delineate the geological structural trend in the study area. Fault mapping is carried out with the nature of the geology of the basin, that is, the structural style of the area, firmly in mind. Structural styles often provide a broad context for understanding the pattern of faulting that may be expected in a region. Fault patterns' basic utility lies in identifying certain basic patterns of deformation that are repeated in geologic provinces (Fig. 2.9).

Guide to Fault Picking

The seismic interpreter must have a good knowledge of the regional tectonic framework of the study area and apply the structural styles in interpreting faulting within the basin.

- The seismic interpreter must use the unmigrated and migrated section when interpreting faults. The migrated section is used to clarify and locate the fault where the section is perpendicular to the fault. On 2D seismic data, the unmigrated section is used to check the ties on the intersections.
- The seismic interpreter must focus attention on the seismic lines perpendicular to the major faults. Typically, these are the lines parallel to dip.

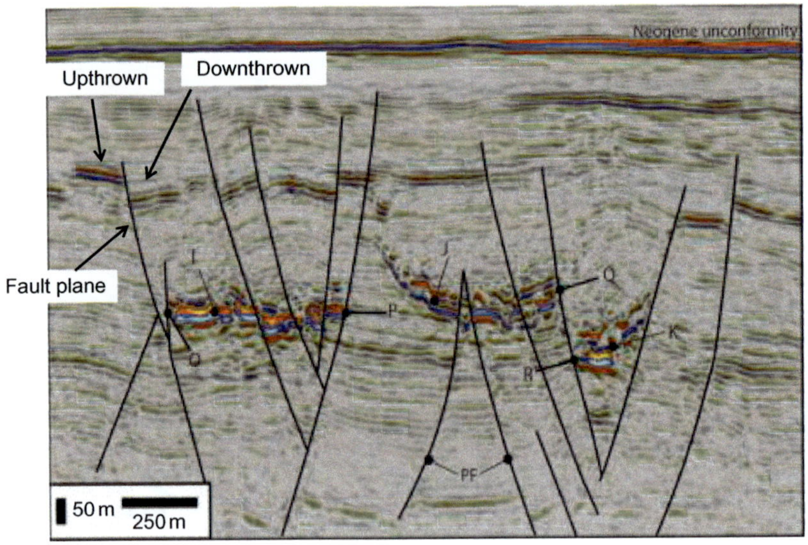

FIG. 2.9

Seismic line shows picked faults.

Courtesy: ResearchGate, https://www.researchgate.net/figure/258801807_fig5_Fig-5-Seismic-characterization-of-sandstone-intrusions-arrested-by-faults-in.

- The direction of the fault is obtained by correlating the fault from line to line. The interpreters ensure that the faults correlated are of the same type (coherency), stress mechanism and compatible throw.
- For 2D seismic data, the interpreter must tie fault picks at the line intersection on the unmigrated sections.
- The interpreters anchor the interpretation at the places where there is no dip (Fig. 2.10).

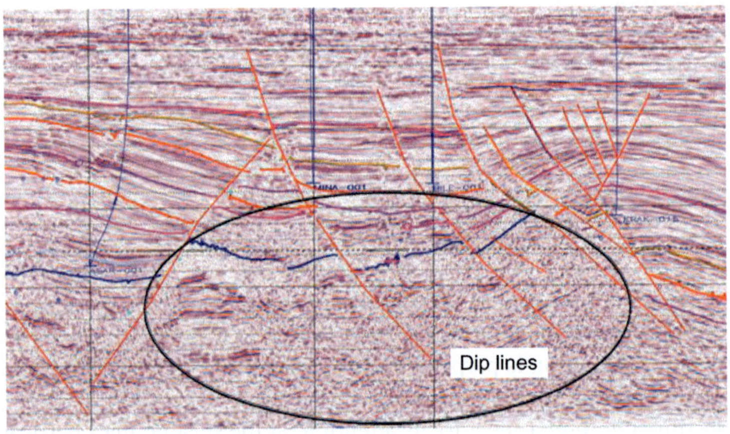

FIG. 2.10

Dip line correlation on the seismic line.

QUESTION 23

What is the different between listric and detachment faults in terms of geometry and associated structure? (Fig. 2.11).

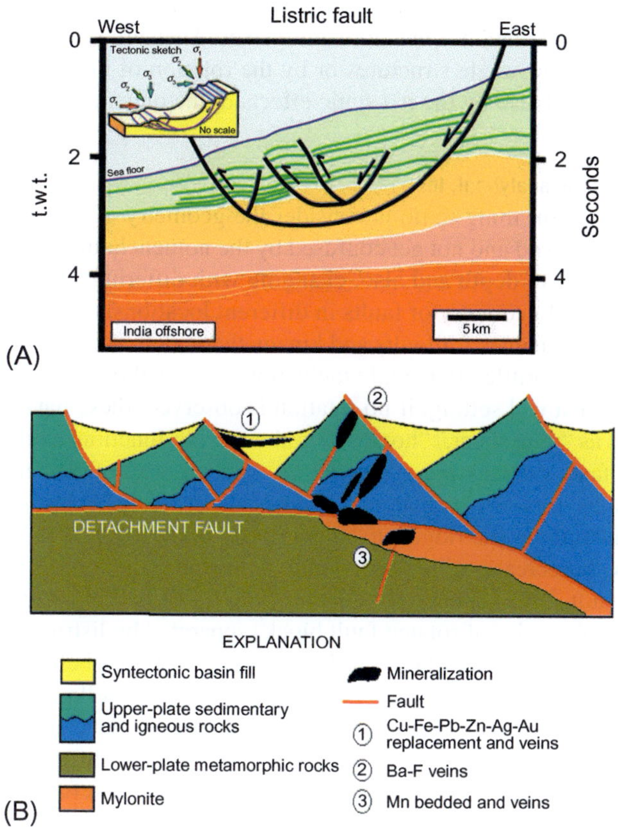

FIG. 2.11

(A) Listric fault and (B) detachment fault.

Data from Preliminary Descriptive Deposit Model for Detachment-Fault-Related Mineralization (Keith and Others, 1983).

ANSWERS PROVIDED BY INDUSTRY EXPERTS

Abd Elhefeith Mohammed Nour

Senior Geophysicist at GNPOC, Sudan

Listric faults can be defined as curved normal faults in which the fault surface is concave upwards; its dip decreases with depth. These faults also occur in extension zones where there is a main detachment fracture following a curved path rather than a planar path. Hanging wall blocks may either rotate or slide along the fault plane (e.g slumps), or they may pull away from the main fault, slipping instead only along the low dipping part of the fault. Roll-over anticlines will often form between bedding planes and the main fault plane as a result of the flexing between the two.

Detachment faulting is associated with large-scale extensional tectonics. Detachment faults often have very large displacements (tens of km) and juxtapose un-metamorphosed hanging walls against medium to high-grade metamorphic foot-walls that are called metamorphic core complexes. They are thought to have formed as either initially low-angle structures or by the rotation of initially high-angle normal faults modified also by the iso-static effects of tectonic denudation.

Steven Jusczuk

PLM Analyst at EMSL analytical, Inc

I think the important thing to do is consider the geometry of the fault surfaces and the processes involved and not get confused by the nomenclature. Geometry of faults usually changes with depth and also, changing with differing rock properties. Over the years, observed geometry of faults in different locations, and at different depths, has led to different names for faults and kinematic models. At shallow depths, where rocks behave most brittle, high angle faults (either normal or reverse) are most common. In an extensional setting, if no rotation is observed, these have been described over the years as "pure shear," "horst and graben, triple-junction" fault blocks. "Horst and graben" fault systems are symmetrical. Another way of modeling extensional systems is an asymmetrical rotational "simple shear," ridge-transform fault model. Here, perpendicular to the ridge, an "upper" plate is separated by a "listric" normal-offset fault with decrease of dip with depth. At its base, the fault is a sub-horizontal detachment fault. The "upper" plate has a broad, roll-over anticline. The "lower" plate consists of a series of rotated fault blocks separated by listric normal faults that root in the same detachment horizon. In a convergent thrust belt setting, low-angle "thrust" faults, or higher-angle "reverse" faults (sometimes collectively referred to as "ramps") root in horizontal large-offset decollements, or "flats." Decollements form in relatively thin, weak formations and tend to ramp up to different weak horizons where lithologies vary. In wrench settings, be extremely careful when interpreting kinematics. All in all, always observe the structure in as large an area as possible, and then deduce the kinematics.

QUESTION 24

I am interpreting a concession area, in which I am getting fault bounded closures. The trend of one fault and consequently the axis of closures along it are NNW-SSE. The second fault trends NNE-SSW/NE-SW and the axes of closures against this fault trend in that direction. Wells were drilled in these two trends of closures and it was observed that the one trending NNW-SSE had oil in a dolomite section. However, the results were not the same in the NNE-SSW trending closures. The petrography shows no variation in diagenesis and lithology is same. It is now assumed that the accumulation occurred after the structures trending NNW-SSE were formed. The subsequent deformation occurred normal to NNE-SSW direction on the pre existing structure causing oil to escape and these structures are water bearing as confirmed by duel packer MDT and petrophysics. Also in the TWT map one can see the NNW-SSE fault terminating against the NNE-SSW fault. Further the structural closures against NNW-SSE faults are steeper and shallower and almost elliptical while the newer closures are gently sloping and at a greater depth/time. So my question against this back drop is whether this hypothesis is correct and how to validate it? Are there any known fields where a similar phenomenon is seen? If yes kindly give some reference and details.

ANSWERS PROVIDED BY INDUSTRY EXPERTS

E.B. Setyo

We had a situation where from basin modeling the migration took place around 15mya whilst tectonic movement occurred 5mya and that changed and tilted the whole area. Our approach was by generating isochronic map from a datum near the 15mya marker to indicate where the old structure was at the time of migration, and then compare with present day structure. This technique allows us to identify early accumulations and remigration route when an old structure was destroyed by subsequent tectonics. Even at old structures that still exist today, we can find oil shows below current OWC suggesting that it was old accumulation that tilted and some of the oil spilled away.

Thomas Klopf

Senior Geophysicist at KUFPEC

It would help to know what the structural framework was for each episode.Are these compressional, strike-slip thrust faults or extensional, basement block faults? As you know, faults can be re-activated over a period of time. The maximum upper limit (MUL) of a fault on seismic can give an idea of the age of the faulting. The latest faulting would generally have the shallowest MUL. Isopachs can also give an idea of the age of the faulting as noted in the last reply by Mr. Setyo. The basin analysis of the migration pathway and timing relative to the structural deformation could explain why some structures are not filled. Juxtaposition fault seal could also be an issue depending on whether these are up-thrown or downthrown fault closures. There is not sufficient information to validate your hypothesis. You don't indicate how many structures of both trends have been drilled.

Mohamed Elnemr

Senior Geophysicist at PETRONAS

You need to define the migration time using BasinMod or PetroMod.

Also define the time of structures using Lithotech or 3DMove. You will need some extra details to feed these software with in order to confirm your idea.

FAULT INTERPRETATION IN CARBONATE RESERVOIR

QUESTION 25

Which techniques are best for fault mapping in carbonate? (Fig. 2.12)

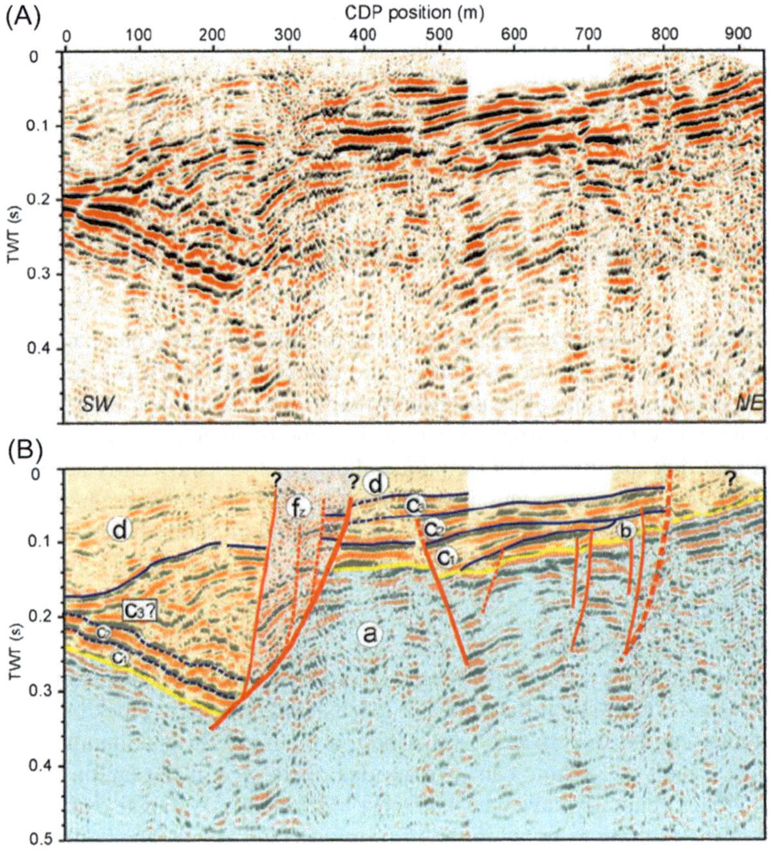

FIG. 2.12

(A) Seismic reflection section and (B) interpretation. Symbols: a, mesozoic carbonates; b, c, c1, c2, c3, and d, quaternary continental sequences; fz, fault zone; yellow line, top of the bedrock; blue line, main unconformities; red line, faults; and dashed line.

Data from ResearchGate.

Vinicius Ramos Pinto

Sr. Geoscientist at Emerson Process Management

ANSWERS PROVIDED BY INDUSTRY EXPERTS

Azer Mustaqeem

Petro-Explorers Inc.

Vinicius, Azimuthal AVO is one of the techniques used successfully for fault mapping although the work is a bit laborious. For post-stack data, coherence, volume curvature, fault enhancement filters and neural network applications can provide reliable answers.

Hector Andrade

Geophysical Consultant at Baker Hughes

Vinicius, I have found the technique described in this paper to be very helpful: Bulhoes Oliveira, 2005 SBGf: "Principio da SismoCamada Elementar e sua aplicação a Técnica Volume de Amplitudes (tecVA)". Right here in the Gulf of Mexico we handle a lot fractured carbonate datasets. Applying 3D color blending and transparencies, you should be able to identify regional trends of fractures (within the range of the seismic resolution).

Vinicius Ramos Pinto

Sr. Geoscientist at Emerson Process Management

Thank you, Hector. I know about the Tecva technique, I read about it long time ago. You are right, that was a good advice. I can try spectral decomposition and 3D blending. Let me ask you: which application do you use for 3D blending? And about attributes, have you tried to apply some others?

Hector Andrade

Geophysical Consultant at Baker Hughes

Vinicius, (RGB) you have a few options for 3D blending: Commercial: GeoTeric (Great software for seismic analysis and attributes visualization) and free for academic purposes: OpendTect which has spectral decomposition and very nice attributes options. And of course if you have access to Petrel or LandMark Suite (GeoProbe/Decision Space) you should be able to achieve some color transparencies that may help you out.

Azer Mustaqeem

Petro-Explorers Inc.

Hector, OpendTect software is also free for commercial purpose and has spectral decomposition and color blending (RGBO) where similarity can be used as opacity. The capabilities are built into the free version. Volume Curvature attributes is available with a commercial plug-in that can further enhance fault detection.

Amr Moslim

Lead Geophysicist at Pico International Petroleum

Vinicius, I have worked on fractured carbonates in Mexico. For Petrel flowchart, you may try structural smoothing "with narrow window", then Variance or Chaos "which is nice for detecting small fractures" and then apply anti-tracking, which needs more (understanding) for faults orientations, throws and apertures. You can also use the same workflow for curvatures attributes and then Neural Network to make your own beautiful blend or "Cocktail".

Jerry Coggins

Principal Geophysicist at Coggins Geosciences, Inc

Vinicius, I have found that vertical sections used in conjunction with coherence time-slices are very effective. Very small faulting is best found by horizon property extractions like coherence, amplitude, and dip/dip-azimuth. We are often interested in fracture detection, which is generally quite different from fault mapping. Lots of good suggestions for fracture attributes here, but we need to keep in mind that attributing any variations to fracturing is only implied and needs confirmation through local calibration to well control.

Vinicius Ramos Pinto

Sr. Geoscientist at Emerson Process Management

Some time ago I reached some reliable results using spectral decomposition and RGB color-blend for carbonate illumination. Nowadays, I like to do it using Open Works suite. I agree with Jerry Coggins about the use of vertical sections and time slices. For me, the attributes based on the similarity of the seismic trace seem to be better in the map view like coherency and structure cube, than in section view. Following Hector's suggestion, I have tried running a TECVA cube and then submitting to structural cube in order to highlight the discontinuities. I am still trying to analyze the effects on the data, but the visual result looks really good.

Finally, I like Amr Moslim's idea about the use of neural networks. I have some interest in working with neural networks but at this moment I believe it's going to be a little difficult due to my platform. What kind of approach are you used to?

Amr Moslim

Lead Geophysicist at Pico International Petroleum

Vinicius, I use Petrel and I have some experience with Kingdom and OpendTect for neural networking. I think this technique gives great results in fault (trend) mapping. You can QC your results by time-slices and x-sections in 2D or 3D viewers. I now work on a fracture carbonates field with 2D seismic data so I can't use these techniques. But I have some other "magic" I can work-by depending on the well data more than the 2D seismic. I have gotten good results for fault and fracture mapping especially when you relate them to production rates, pressure data and WC levels in the wells.

SEISMIC HORIZON MAPPING

Seismic horizon mapping involves picking and tracking laterally consistent seismic reflectors (geological structures, stratigraphy, and reservoir architectures) as shown in Fig. 2.13. The aim is to create both a structural and stratigraphic model of the area and to detect hydrocarbon accumulations, delineating their extent and calculating their volumes. The best place to start horizon mapping is at the top of the section, where definition is usually best. One can then work down the section toward the zones where the signal-to-noise ratio is reduced and the reflector definition is less clear.

Given a modern data volume, seismic horizon mapping is carried out for cross-lines and in-lines. The interval spacing for the mapping depends on the detail required of the study and the time available to execute it. The whole field of interest is mapped on the

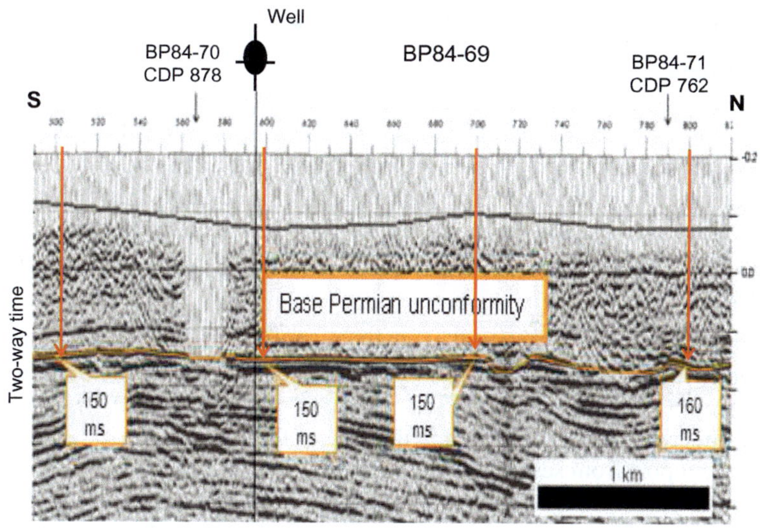

FIG. 2.13

The picked reflection time for the Base Permian horizon.

Courtesy: UK Onshore Geophysical Library, Martin Whiteley and Dorothy Satterfield.

seismic section. Depending on what you hope to achieve with the interpretation, amplitude extraction maps can be generated. A time depth conversion is carried out to convert the seismic reflection times to the depth domain. In a search for drillable prospects, as opposed to a general basin analysis, a time and depth structural map is produced and drillable prospects are identified on the maps and color coded as they relate to oil (red color), gas (green color), water (blue color) or condensate. The stock tank oil initially in place is then calculated, and reserve estimates are then carried out on the study.

Seismic Horizon Interpretation Steps

There are four basic steps in seismic horizon interpretation:

- Picking: To identify and follow the horizon (seismic reflector) to be mapped.
- Timing: To measure the reflection time from datum to the picked reflection (horizon).
- Posting: To transfer the measured reflection times (horizon) to the map.
- Contouring: To show the structure, relief and closure of the chosen horizon.

QUESTION 26

During seismic interpretation, which must be interpreted first, Faults or Horizons?

Sigit Wibowo

Exploration & Development Geophysicist, MontD' Oil Tungkal Ltd & MontD'Or Salawati Ltd

ANSWERS PROVIDED BY INDUSTRY EXPERTS

Graham Dryden

CEO & Exploration Manager, Danuma Oil Fields Ltd.

Sigit, the larger or more visible faults should be interpreted first. To interpret horizons you must first create some synthetic seismograms in order to identify the target reservoir, whether for exploration, characterization or field development, although interpretation is a joint process, you must simultaneously interpret both faults and horizons. A basic workflow for seismic interpretation would be:

1 Reading the reports for regional geology structural, sedimentological and petroleum system, this would give you an idea of what you might find in the seismic.
2 Read final report of wells drilled in the area, if there is any.
3 Read the report from seismic data acquisition, processing workflow, special analysis, dynamic range, frequency spectrum, resolution.
4 Construction of synthetic seismograms, identify the top of the reservoir or potential reservoirs. Extract wavelets, frequency spectrum and phase of the seismic, calculate the seismic resolution. Use GR logs to identify rock types use resistivity logs for fluid identification. Try to incorporate as much geological information to your synthetic seismograms as you can. Identify geological sequences with strong changes in the velocities to be included in the velocity model.

Once done all this research stage, you will have a better idea of how to begin your seismic interpretation study.

Graham Dryden

Senior Geophysicist – Prolific Oil Finder, SCOUT

Sigit, this is how I do it in SMT on a 3D survey after many years of rapid interpretation. We usually have only two days in a data room to understand the survey area which might be as big as 500 sq miles, so we have to work quickly. SMT is available in data rooms in most developing countries or in data rooms run by RPS. So you interpret your horizon first before you interpret your fault. These are the step/process on how to do your basic interpretation:

- Extract smoothed dip of maximum similarity (SDOMS) using the module RSI. This will show all faults and channels in time slice view very clearly.
- Create a slice cube using the SDOMS attribute result cube. Be patient! This makes it fast to scan up and down the faults.
- Extract a time slice and scan it through the cube looking at how the faults track laterally. This will give you a very accurate idea of how the faults split and interfere with each other.
- Scan through the cube in IL or XL view looking for the reflectors which will allow you to identify the best candidates for auto-picking.

- Auto-pick a horizon on one line, using the 3D pick option. It is slow but accurate, and will cross faults very nicely. It makes a shadow cube for ray tracing the first time you use 3D picking and this will take a long time. Be patient as it will make everything faster and more accurate from this point on. Open a window in plan view showing that picked line. Now do a horizon auto-pick on a box 10 lines either side of your picked line so that it uses the line that was picked as seeds and rapidly makes a swath 20 lines wide the length of your picked line. Note areas where the pick fails due to poor data.
- Now pick a line at right angles to your swath in an area of good data, and repeat as above.
- Move 50 to 100 lines and repeat. In this way you should be able to rapidly be able to pick a horizon in good areas and identify poor data areas where you will have to manually pick.
- Quite quickly you will have a detailed horizon picked every line and x-line. You can extract amplitude on this horizon to see channels and possible bright spots DHIs.
- Do the same thing for another good auto tracking horizon about 1 second deeper or shallower. Keep going for as many horizons as you need. The more the horizons the better. Now you have two horizons where the fault breaks will be clearly visible and can be tracked vertically from one horizon to the other. You should never try to pick faults unless you have the approx. top and bottom nailed in XY by your two horizons (at least) plus your cheat sheet which is the SDOMS cube. Otherwise you are just guessing. Let the data show you the way. Also, with tools available, the old way of picking every 20 lines then smoothing via a grid is not appropriate. The surfaces you build this way are faster and more accurate and will show faults and channels in exquisite detail. Very important for stratigraphy traps and fault leakage scenarios (Fig. 2.14).

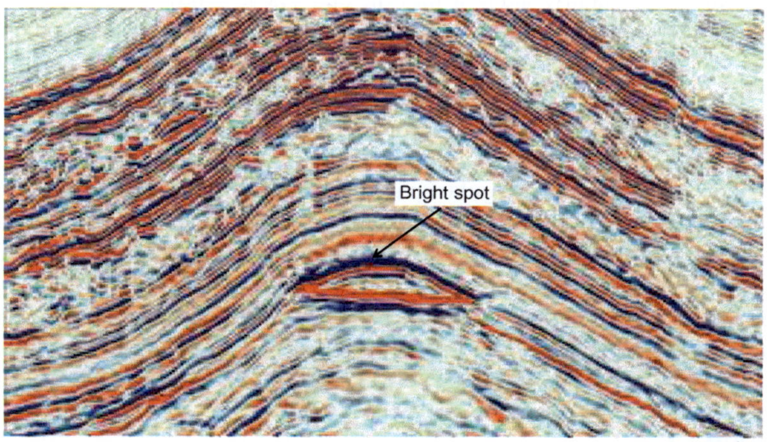

FIG. 2.14

"Bright spot" in seismic section.

Data from PGS Geophysical, copied from Seismic Atlas of Southeast Asian Basin: November 2008.

Now you are ready for fault picking which should be blindingly obvious to you. Since SDOMS will show you the strike of the faults, align your fault picking arb lines at right angles to the faults so you get the most accurate cut.

On fault picking, you might want to pick them every 10 or 20 lines. So pick the first one on a vertical line. Now don't jump 20 lines to the next pick! Instead scan one line at a time till you get to 20. This will help you know that you are on the same fault. If before 20 the fault vanishes, or dies, back up till you see the fault again and make that the last pick. Now go the other direction and repeat.

Two final tips;

(a) You can pick a fault on a horizontal time slice. This makes it easy to see where to put your vertical picks.

(b) You can export your surface as a grid to petrel. If you make the grid size the same as your bin size, you will get a high resolution, unsmoothed surface in petrel. Now you can add or tweak the faults in petrel using modeling while interpreting.

Ali Cad

Geologist

How can we interpret the seismic?

Graham Dryden

Senior Geophysicist – Prolific Oil Finder, SCOUT, Houston, Texas Area

Ali, first create a 3D cube which shows all faults and channels, that is, smoothed dip of maximum similarity. Now, while navigating in 3D you can see where the faults are cutting across the horizons and by scanning through the cube vertically, you can track the fault in 3D. But before mapping the faults, pick strong horizons which can be auto-picked across the whole cube. Extract amplitudes along these surfaces and study to see the fault breaks in 3D. Now you have an excellent idea of how the faults trend vertically and horizontally, and how they die out, and also, how the cross or join each other. Now interpret the faults. Next fill in less reliable horizons. Rinse and repeat.

GRIDDING

Gridding is done to convert irregular spaced data points to a regularly spaced grid by interpolation. It is done by determining values for grid elements on a map and using values chosen from nearby data points. Methods are deterministic and use linear and nonlinear interpolation methods, or may be statistical and use geostatistical approaches such as Kriging. Gridding is usually applied to one characteristic per map, such as structure, thickness, porosity, permeability or saturation.

QUESTION 27

Which gridding algorithm do you use for gridding a seismic 2D/3D horizon? Why do you use/prefer this algorithm? What are the pitfalls of the algorithm that you use and what would you like to add in it?

Farrukh Qayyum
Sr. Geoscientist at dGB Earth Sciences

ANSWER PROVIDED BY INDUSTRY EXPERT

Mike Cline
Consulting Geophysicist

Farrukh, a few of the commonly-used gridding algorithms are (not in any particular order): Inverse Distance, Krigging, Triangulation, Nearest Neighbor, and Minimum Curvature. Each has their own benefits and drawbacks, and should be used according to their features/attributes in a particular situation - popularity should not be a primary factor. The main objective is that it accurately defines a horizon in a geologically realistic manner. Data spacing and distribution are primary influencers of which method to use. For example, is the data from an auto-picked 3D horizon? Or from a widely, and irregularly spaced set of 2D seismic lines? Another consideration is whether the horizon is faulted, either into separate fault blocks, or smaller faults that die out, or terminate within the area to be gridded. Preventing the "smearing" across a fault requires a gridding method that can handle faults (with the appropriate parameters that are optimum for that case), or gridding the fault blocks individually (and then combining the separate blocks back into one horizon, after the gridding). There may also be stratigraphic features, such as channels, or reefs, which need to be considered. I would suggest testing your data, with different algorithms, and parameters, to determine which method produces the most geological results. I have also found that using visualization is much better than a map view, when checking the results. Often, gridding artifacts can be more obvious with visualization.

FAULT SHADOW ON SEISMIC SECTION

QUESTION 28

Fault shadow areas on seismic data are a velocity-related problem across the fault. How can we improve the data below the faults? Is it true that this issue can be solved by static corrections? Please can you give more information on this issue and the ways used to reduce its effects.

Raouf Nasr
Geophysicist

ANSWERS PROVIDED BY INDUSTRY EXPERTS

Guy Maslen
Chief Operations Officer at GLOBE Claritas

Raouf, fault-shadows are usually an imaging problem caused by refraction (ray-path bending) at the fault boundary. If it is minor, I suppose you might get some improvement from residual statics (as this would correct some of the non-hyperbolic nature of the reflection) although it is not something I have played about with much. Refraction issues usually require a pre-stack migration that has an algorithm that handles refraction; a pre-stack depth or hybrid depth-time migration would be the normal solution.

Anonymous

Senior Geophysicist

Raouf, every commercial data processing system has its own strengths and weaknesses when it comes to imaging complex geology. The best way to understand how well we can image the fault-shadow zones with a data processing package is to create full-wave field synthetics and process the synthetic data. This would create an absolute judgment with respect to imaged field data that you would make a judgment about proposed algorithms such as trim static (not worth to try in this case), tomography, or full wave field inversion (FWI) etc. 3D full wave field modeling algorithms are expensive, but 2D solutions are quite inexpensive and provide very valuable insights about your proposed algorithms. At the end of the day, you will fully understand the capability of your data processing package; velocity inversion, imaging or the combined effects of these two algorithms.

Fritz Damianus

Multicomponent Geophysicist at CGG

Raouf, fault shadow is a specific case of velocity depth model building. It is observed as problem with sub-fault image focusing and sagging due to localized low velocity-pore pressure related along/at some part of fault plane. I personally think trim static wouldn't help, of course depending on level of severity, as it won't restore image defocus which caused by velocity anomaly. It requires a special form of tomography to solve that, which in CGG usually called Fault-Constraint Tomography, there is paper publication describing this method.

Julian Brizzi

Geophysicist Consultant, Tullow Oil

Raouf, seismic imaging has their limits. If you have a velocity model you can trust, you may do ray tracing using the acquisition offset ranges and azimuths and at least try to understand if you was able to illuminate your shadow zones. You can also play a bit with your model to scan within your model error expectations if there is any extra chance due to inaccurate model used. Normally to get any non-conventional data (like refractions) from your seismic, very careful preprocessing should be carried out as we normally understand those as noise and they end up attenuated or totally removed.

QUESTION 29

If I have uncorrected structure on seismic, that means that we most likely applied false velocity on the velocity picking in processing. Or it is normal because this layer has faster velocity than the overburden. So how can I check the velocity picking if I have no processing software?

ANSWERS PROVIDED BY INDUSTRY EXPERTS

José Adauto de Souza

Senior Geophysicist, Consultant at Exploratory Vision

Frequently it is not possible to pick again the velocities. The input pre-stack data used to pick velocities is discarded as soon as possible in seismic processing procedure, due to HD high space needing. After finishing seismic processing, rarely the pre-stack data is preserved. Depth conversion is the Geophysicist Achilles heel. Seismic velocity is the value that better stacks a reflection; it is not the true rock velocity. At shallow depths it goes close to rock velocities. More depth, more imprecise the seismic velocities are, if compared to rock velocities. Depth errors around 5% are usual in medium depth petroleum targets (+- 3000m). Any depth conversion based on picked seismic velocities is inaccurate. Any PSDM is inaccurate in depth, because it is seismic velocity dependent. Wells are the key to improve depth conversion: effort should be done to match seismic depth conversion to well depths. More wells, more accurate are the conversion. You may obtain more information in paper linked in green box SEISMIC VELOCITY QUALITY CONTROL at http://www.exploratoryvision.com/#!velocity-suit/crbm.

Bob Avakian

Geophysicist, Instructor, Oklahoma State University Institute of Technology

Excellent points from Mr. Jose, however, if you are working a fair distance, physically or geologically, from well control you will need to work with what you have and communicate this uncertainty to your supervisor/client. Completely aside from stacking velocities, it is imperative that near surface geology and known structure be routinely incorporated into the interpretation and accounted for. Near surface and shallow velocity variations and structure have far more potential to mess up an interpretation than stacking velocities. This is more than just a statics problem. For 3D, I find this easier to check out on seismic profiles. In fact, I recommend interpreters generate several 2D profiles over critical areas of a prospect and plot the known geology across the top of the section, (by hand if necessary), to check this potential problem.

Istvan Gyorfi

Chief Geophysicist at Pacific LNG Operations Limited

Good points from experts above. Velocities picked in time in the PSTM process are erroneous depending on the geological complexity of the area in question. What I do usually is that I request from the processors that they supply the picked velocities in SEG Y format that can be plotted against the seismic data. Not always clear answers, but it is the minimum you can do. If you are lucky to have PSDM data, than cf. with the suggestions of Bob, use surface geology, known structures, dip domains etc. Faster velocities in near-surface domains usually are caused by structures (usually faulting) and in some cases can be recognized on surface maps (not always!).

Mangat Thapar
President, International Geophysical Company, Inc.
In addition to the above answers from experts, the error between interval velocities calculated from stacking velocities and rock interval velocity is strongly related to the magnitude of dip of your structure, i.e., larger error for steeper dips and smaller error for smaller dips. In the case of nearly flat layers, Dix's equation gives good estimates of interval velocities using stacking velocities. The best way to estimate the errors or correction is through ray trace modeling.

SEISMIC ATTRIBUTE ANALYSIS

In exploration and reservoir seismic surveys, the main objectives are, first, to correctly image the structure in time and depth and, second, to correctly characterize the amplitudes of the reflections. Assuming that the amplitudes are accurate, a host of additional features can be derived from them and used in interpretation. Collectively, these features are referred to as seismic attributes. Therefore, seismic attributes are quantities extracted or derived from seismic data that can be analyzed in order to enhance information that might be more subtle in a traditional seismic image. This should lead to a better geological or geophysical interpretation of the seismic data. Examples of seismic attributes are wavelet shape, frequency, amplitude, phase and complex trace information contained in transforms of the real trace.

Why Are Seismic Attributes Important?

Due to the increased use of seismic data in development and production of oil and gas, it is necessary that the geoscientist extract the most information available from the seismic response. Seismic attributes enable interpreters to extract more information from the seismic data.

Applications include:

- Hydrocarbon presence and identification
- Prospect identification and risking
- Reservoir characterization
- Well planning and field development

An attribute that is often used by seismic interpreters is the seismic amplitude. It is usually reported as the maximum (positive or negative) amplitude value at each sample along a horizon picked from a 3D volume. In many cases, the amplitude of reflection corresponds directly to the porosity or to the saturation of the underlying formation. Most seismic attributes are poststack, but those that use common midpoint (CMP) gathers, such as amplitude versus offset (AVO), must be analyzed prestack.

SEISMIC ATTRIBUTE REFLECTION STRENGTH

QUESTION 30

What is seismic attribute reflection strength used for and how we can extract information from seismic image?

Zareef

Exploration Geophysicist

ANSWERS PROVIDED BY INDUSTRY EXPERTS

Mohamed Shihata

Geoscience Business Development Specialist

Zareef, Average Energy integrates the envelope between para-phase events. This attribute is useful for mapping and visualizing lateral changes within structural and stratigraphic units as it highlights stratigraphic detail through energy fluctuations across traces. Values are in degrees. There are a lot of uses for average strength but the important one is DHI which sometimes can help in identifying lateral lithology changes. Average energy attributes can be used to identify gas chimneys and to predict gas migration passage where high average energy attribute values indicate probabilities of gas accumulation in shallow depth by comparing average energy attributes inside the gas chimney with a shallow anomaly (Fig. 2.15).

There are a lot of methods that can be used for extracting information from seismic image

Visualization Methods

1. Co-blending geometric attributes (similarity, curvature, dip) with physical attributes.
2. Transparency
3. RGB (red-green-below)

Segmentation Methods

1. Geobody extraction using an adaptive algorithm (auto tracking geobodies)
2. Fault auto tracking
3. Likelihood fault tracking (Dave Haley)
4. Semi auto tracking faults in 3D visualization.
5. Salt segmentation (growth algorithm)

Statistical Methods

1. Neural network
2. PDF (probability Density Function)
3. Cluster techniques unsupervised (K-mean, self-organization map)

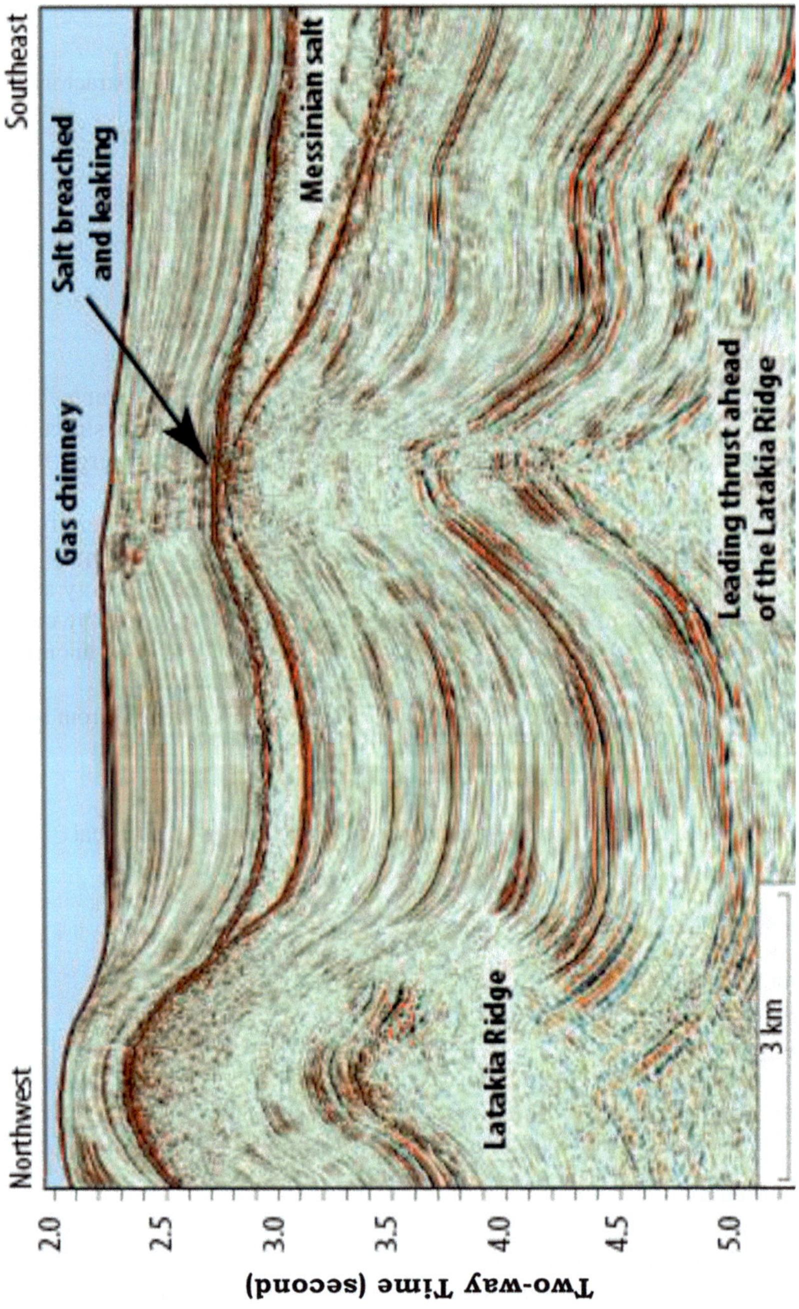

FIG. 2.15

Gas chimney present above the leading thrust of the Latakia Ridge.

From Bowman (2011); Data from: http://www.saltworkconsultants.com/blog/tag/salt_tectonics/.

Zareef

Exploration Geophysicist

Can you also tell me about instantaneous/response/cosine phase, instantaneous frequency (Low and High) and apparent polarity? How can we detect these in seismic images and why they are used?

David Paige

Seismic Inversion Developer

Zareef, First, one should be aware that the normal seismic section does not directly represent lithology. Instead it is a mixture of multiple primary reflections and coherent noise. To make any assumptions on frequency, amplitude and polarity, advanced inversion and integration are needed to identify attributes of the actual beds.

THE USE OF INTERVAL ATTRIBUTES ON AN INTERFACE PROPERTY

QUESTION 31

How do we justify the use of interval attributes like RMS and other averages on an interface property like simple seismic amplitude data? Are the extrema values a better representation or is RMS?

Rajeshwaran Dandapani

Director Geophysics at Telesto Energy

ANSWERS PROVIDED BY INDUSTRY EXPERTS

Shiladitya Sengupta

New Ventures & Asset Geophysicist – Cyprus Exploration Team at Total

Rajeshwaran, I guess you are trying to find if it is right to use RMS or extrema value attributes for quantitative reservoir characterization. Please correct me if I am wrong.

I gave your question some thought and with my limited experience I would answer it from two different perspectives:

(1) RMS computation is very sensitive to extrema amplitude values, so if these extrema values where noise, it would square these values as well. Therein lies one aspect of your answer. In such a case, where the data is noisy, one should be very careful using RMS amplitude as a seismic attribute for quantitative reservoir characterization. Perhaps in such a case, it is better to use extrema values extracted directly from the zero-phased amplitude dataset. The convolutional model and the reflection coefficient formula show that a reflector's zero phase amplitude can be directly related to the reservoir's impedance. A thin-bed tuning curve model shows that zero phase amplitude is also directly related to reservoir thickness. Additionally, gas substitution modeling shows that a reservoir's zero phase amplitude can be influenced by changes in pore fluids. Therefore a convincing theoretical conclusion is that changes in zero phase amplitude are a function of changes in reservoir impedance, thickness and pore fluid. This conclusion has been proven by many successful quantitative reservoir characterizations done with zero phase amplitude.

(2) From a different point of view – if you are trying to find out if RMS seismic amplitude is a suitable 'interval attribute' or not, I would say that indeed theoretically RMS amplitude computed in a user defined window is not as reliable for quantitative reservoir characterization as absolute (or relative) impedance attributes, since the amplitudes in the absolute impedance dataset describe the impedance of the rocks, whereas the RMS seismic amplitudes describe the impedance contrast between rocks. Stated differently, the impedance attribute is related to the geology while the RMS attribute is related to the 'derivative of the geology'. The absolute impedance data is theoretically the best option, but it has drawbacks related to its low frequency content (and, it can be very expensive in terms of both money and time delays). If the low frequencies are removed, the result is a relative impedance dataset, which is in practice perhaps the best (and least expensive) seismic attribute.

Rajeshwaran Dandapani

Director Geophysics at Telesto Energy

Shiladiya, thanks for your detailed analysis of my question. The RMS values in fact average out the spikiness of the peak or trough amplitude. But the question is whether that RMS value is representative of the impedance of that interval. In my view it is not. When we average over the reservoir interval on reflectivity data we are studying the amplitude variations caused by the wavelet which needs to be eliminated as we rightly do in inversion. The peak or trough amplitude value is in fact, representative of the impedance contrast across the top or bottom interface of the reservoir. If the overburden and under burden shale impedances are assumed to be constant then the extrema values can give some idea about the reservoir properties if we are assuming perfect zero phase data. Many in the oil industry even today believe that if the data is not perfect zero phase and it varies laterally, then use of RMS in place of peak or trough value is justified which in my view, has no basis.

Mohamed Shihata

Geoscience Business Development Specialist

Rajeshwaran, I completely agree. We can't just use one seismic attribute when we deal with our geological target. We must think about what is available of seismic attributes. Sometimes is difficult to make inversion or if we can make an inversion using the wrong calibration it can destroy our model. So building your result depending on one attribute is dangerous. RMS is one of the best interval attributes but why do we deiced on one attribute? Although it works well in many cases, avoids summation errors and enables you to see changes in amplitude in your interval, there are other problems with this interval attribute that comes from your small interval. And remember to determine your target first. If you haven't enough time to generate many attributes and calibrate them with your target then RMS will be fast and give good starting result about your channel trend or DHI. But, if you have enough time you might think of other attributes, like the envelope, and combine many attributes with each other by opacity or RGB.

Raju Arasu

Senior Geophysicist & Lead Triassic Paleozoic Prospect Evaluation Unit at Kuwait Oil Company

Interval attributes computed in seismic amplitude data are the seismic properties or the seismic response to the rock, fluid and structure/stratigraphy lying within the interval, and not the rock properties themselves. What we infer from the seismic properties is the next level of interpretation. For the question on RMS, let me take an example of a layer cake consisting of 2 sand layers separated by a thin shale, say MFS, all lying within an overall shale sequence. Seismic may not resolve these sands, but due to the presence of the shale between, it may produce a broader laterally varying peak amplitude (L-to-H impedance +ve convention). Here RMS attribute calculation is more appropriate rather than mini/max to get the areal extent, sand-shale ratio and the depo pattern of both of the sands (together). In case inversion resolves the sands and you are able to map top and base of each perfectly (!) then you can extract the layer properties individually. If neither is achievable after inversion, the interval property result may be very similar to an amplitude seismic attribute. Another place where you may use RMS is when you want to capture an anomalous discontinuous event lying within a sequence.

Rajeshwaran Dandapani

Director Geophysics at Telesto Energy

May I add small correction to the above? The normal seismic response is the response of the seismic wave to the contrasts in elastic properties between two rock media across an interface. This is rightly depicted by the convolutional model where the contrast in elastic properties across the interface gives rise to the reflection coefficient which is the extrema value, either peak or trough, that is (unfortunately) convolved with a band limited. If the bandwidth of the wavelet were to be infinite or at least having the low frequency components from zero to 8Hz like inverted data, then the normal seismic data would be equal to the inverted data which represents an interval property. We would then possibly be justified in using the interval attribute like RMS on it.

In my view the RMS attribute of a data, which responds to the interface property, is going to dilute it by contaminating it with the wavelet component which actually needs to be eliminated. In the special cases, as you have mentioned, where there is a thin layer of shale within a reservoir which is well below the resolution limits of the seismic bandwidth, nothing works properly even inversion fails due to interference from the closely spaced interfaces. In that case the complex waveshape itself, (which is the result of the interference pattern), can be studied in Stratimagic to get a qualitative assessment of that effect. But the wavelet shape holds the key rather than the RMS amplitude. I am not totally against the use of the RMS attribute on data responding to interface property. In some cases where the noise level is very high, the extrema values which represent the reflection coefficients are contaminated with noise which can averaged out if we consider a small window across the interface rather than the extrema value at the interface itself. These are my views only and they can be contested by anyone. I am open to discussion and further learning on this topic.

Raju Arasu

Senior Geophysicist & Lead Triassic Paleozoic Prospect Evaluation Unit at Kuwait Oil Company

Rajeshwaran, I agree with you. RMS is more useful as a pattern recognizer in a sparsely interpreted volume. As mentioned by Mohammed, it is quick way to get qualitative results and is useful in some special cases such as the one I mentioned earlier. When you go to specifics - close to a calibrated reservoir event- you may like to use for mini/max amplitudes.

QUESTION 32

If we work with post-processing data, what is the best seismic attribute for reservoir characterization?

Andres G Espeso B

Reservoir Geophysicist

ANSWERS PROVIDED BY INDUSTRY EXPERTS

Ronny Meza

Geophysicist and Geomodeler

Andres, I don't believe in a perfect attribute for seismic characterization. However, there exist attribute combinations that make it possible to characterize a reservoir. For example, if you need to evaluate fractures, you need to use geometrical attributes. But if you need to evaluate lateral facies changes, you may need inversion attributes (Impedance P and PS), linked with rock physics analysis. Now, if you need to detect channels and their geometries; you need to try with intervallic attributes and spectral decomposition, Hilbert transform, etc. So, you could characterize your reservoir across the integration of several information from different attributes. But, all this is possible, if you have a amplitude cube with good quality.

Miguel Nunez

Andres, Ronny is right. There is not one and only one attribute for reservoir characterization but if you develop a good work-flow using different attributes combinations and techniques you will find the best combination to characterize each reservoir and its sedimentary environment.

Ryan Martin

Sr. Seismic Interpreter

Andres, Another option would be to use a whole host of seismic volumes in a multivariate analytic approach, thus improving confidence that your facies separations are highlighting discretely different zones. This can also be supplemented by any other type of data, including well logs and production/engineering data.

Mohamed Shihata

Geoscience Business Development Specialist

Andres, It is possible to generate at least 30 different seismic attributes from a given seismic data set. This presentation addresses the question of which of those post-stack attributes is most appropriate to use for a quantitative seismic reservoir characterization. Our conclusion is that an absolute impedance inversion is the best attribute in theory, but, in practice, a relative impedance inversion is much more practical.

http://marine.veritasdgc.com/WebSite/VTechDocWeb.nsf/all/CACC2C5FB7B461C687256CA500422803/$File/0075.pdf.

Poststack Attributes

The stacked seismic data volume is commonly used for interpretation of geologic structure and seismic attributes. The most common attribute is simply amplitude, although its interpretation in thin-layered beds is not necessarily straightforward. Amplitude is often found to correlate strongly with porosity and/or liquid saturation (oil/water versus gas) because those reservoir properties have a strong effect on both velocity and density, and seismic reflections are generated at boundaries where the acoustic impedance (the product of velocity and density) changes. Some examples of poststack attributes are coherence and dip.

1. *Coherence* is a poststack attribute that measures the continuity between seismic traces in a specified window along a picked horizon. It can be used to map the lateral extent of a formation. It can also be used to see faults, channels or other discontinuous features.
2. *Dip* is a poststack attribute that computes, for each trace, the best fit plane (3D) or line (2D) between adjacent traces on a horizon and outputs the magnitude of dip (gradient) of said plane or line measured in degrees. This can be used to create a pseudo-paleogeologic map on a horizon slice.

Fig. 2.16 shows a DIP attribute calculated along an interpreted horizon (Image: courtesy, ExxonMobil). Low dips are light; high dips are dark. Two types of geologic features are revealed:

- Linear segments of high dip are associated with fault offsetting the horizon.
- Curvilinear segments of high dip are associated with channels and other stratigraphic features.

How do we separate structural and stratigraphic features?

- Structural features are consistent over many time slices.
- Stratigraphic features change rapidly from one time slice to the next (shallower or deeper).

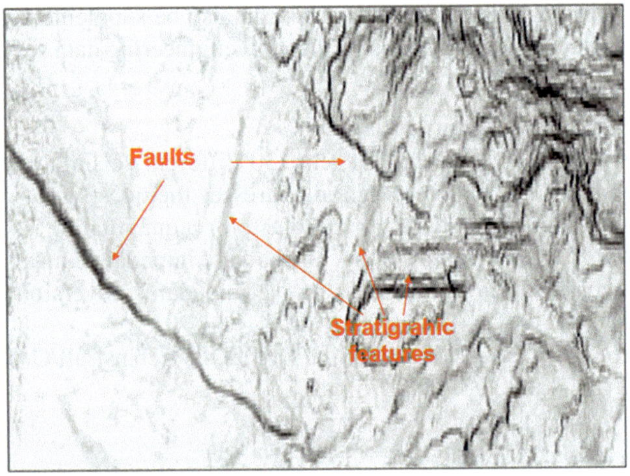

FIG. 2.16

A DIP attribute calculated along an interpreted horizon.

From APPG slide show. Courtesy: ExxonMobil.

FILTER APPLIED ON COHERENCE CUBE TO REDUCE NOISE

QUESTION 33

I am using a Landmark system for my seismic interpretation. My problem is there a great amount of noise in the generated coherence cube. Could you kindly recommend a filter that can be applied on the coherence cube to reduce the noise?

Hongdan Deng

Royal Holloway University of London

ANSWERS PROVIDED BY INDUSTRY EXPERTS

Gaynor Paton

Director of Geosciences at GeoTeric

Hongdan, The best way to quickly improve the quality of the coherence result (or any edge detection) is to run post stack noise attenuation on the reflectivity data. This should be structurally oriented and edge preserving to ensure the best subsequent attribute results. This blog post shows the effect different post stack noise attenuation filters can have on a semblance attribute, all created in GeoTeric: http://blog.geoteric.com/search?q=noise+attenuation.

Staffan Van Dyke

Senior Geophysicist

Hongdan, Post-stack filters are notoriously problematic. Ideally, if the data are truly that bad, perhaps you should consider a new pre-stack migration and deconvolution run (prestack processing, that is). However, if you do not have that ability, I have found a median filter with a 3-5 trace sampling bin works well enough to smear out the bad data.

May I recommend a technique that I developed to increase the resolution of a semblance volume? If you refer to the section called "Refined Semblance" at the end of an article I published in 2013, titled, "Seismic Meta-Attributes and the Illumination of the Internal Reservoir Architecture of a Deepwater Synthetic Channel Model," you will find an innovative solution to increasing the resolution without filtering the data.

Here is the link: http://www.slideshare.net/staffanvandyke/seismic-metaattributes-and-the-illumination-of-the-internal-reservoir-architecture.

Mohamed Shihata

Geoscience Business Development Specialist

Hangdon, First you need to create a good workflow for data preconditioning using filters with edge preserve (pca filters-dip steering), do you can generate SD (spectral decomposition) for best frequencies band and then applying coherence algorithm on the result. If there is still a lot of noise you can use a low pass filter or a Laplace filter as they may reduce the effect of noise.

Prestack Attributes

Most seismic attributes are poststack, but those that use CMP gathers such as AVO must be analyzed on prestack data. AVO attributes have as their basis the variation in amplitude of a seismic reflection with varying offset. These attributes include: AVO intercept, AVO gradient, intercept multiplied by gradient, far minus near, fluid factor, etc.

QUESTION 34

I am going to carry out seismic attributes studies on 3D seismic data. How should I start? Which software can I use?

ANSWER PROVIDED BY INDUSTRY EXPERT

Thomas Klopf

Consultant Geophysicist

The first step is to;

- Collect all the well data in the area which will have a bearing on the reservoir interpretation.
- A petrophysical analysis of the objective reservoir should be done and used in seismic modeling of the gathers. Some petrophysical analysis will have been done for any existing inversion of the 3D seismic data.
- The inversion wavelet would be used to generate the seismic gather model. If no inversion exists, the seismic modeler can estimate a wavelet. A comparison of the actual gathers and the modeled gathers at the wells will confirm if the 3D seismic data is showing a change in amplitude between wet reservoir and hydrocarbon filled reservoir.

- The first attribute plot should be an amplitude extraction of the reservoir event to see if there is conformance with higher amplitude and the field outline structure. In the exploration case, seeing what works on existing discoveries in the basin may indicate the prospectiveness of another, similar trap.
- If there is amplitude change, there should also be other attribute variations that may be plotted against reservoir porosity and other useful reservoir properties.

Note that all the seismic attributes like Vp/Vs and Poisson's ratio are a function of the slope and intercept of a reflection event plot of amplitude versus offset angle on a seismic gather. Finding the best attribute to show mapped reservoir extent or porosity on cross plots can be trial and error with color bars. Thickness resolution can be model with a wedge model to show below what reservoir thickness no attribute response is found.

QUESTION 35

What is the basic science behind Envelope Attribute?

Ammad Ali Qureshi

Exploration Geophysicist at Petroleum Exploration Ltd

ANSWERS PROVIDED BY INDUSTRY EXPERTS

Dr. Biswajit Thander

Geophysicist/Seismic Interpreter at Gujarat State Petroleum Corporation Ltd

Ammad, envelope amplitude attribute is basically the measure of reflection strength. That means the acoustic impedance difference between the overlying and underlying sedimentary layers. A complex seismic trace from which the attributes are derived is given as,

$$\mathrm{CT}(t) = \mathrm{T}(t) + i\mathrm{H}(t)$$

where:

CT(t) = complex trace
T(t) = seismic trace
H(t) = Hilbert's transform of T(t)
H(t) is a 90 degree phase shift of T(t)
The Signal Envelope (E) is calculated by using the formula:
$\mathrm{E}(t) = \mathrm{SQRT}\{\mathrm{T}(t)\wedge 2 + \mathrm{H}(t)\wedge 2\}$

The instantaneous energy of the signal is represented by envelope which is proportional to the magnitude of the reflection coefficient. This attribute represent mainly the acoustic impedance contrast i.e., reflectivity.

Ammad Ali Qureshi

Exploration Geophysicist at Petroleum Exploration Ltd

Biswajit, How can envelope attribute be helpful and useful for a seismic interpreter?

Dr. Biswajit Thander

Geophysicist/Seismic Interpreter at Gujarat State Petroleum Corporation Ltd

Ammad, envelope attribute is useful for identifying bright spot, thin bed tuning effect, sequence boundaries, unconformities, sills, channels, major changes in lithology etc.

Evan Bianco

Geoscientist at Agile Geoscience

Ammad, envelop attribute might give some sense of what the envelope of a seismic trace is, and how it can be useful. One way that I think of envelope, is that is the expression of reflectivity that includes all possible polarities of a wavelet, from -180 to 180. Zero phase data will have the peak of the seismic amplitude trace, aligned with the peak point on the envelope. All of other phases will show a time shift.

QUESTION 36

What are the advantages and disadvantages of the different amplitude value cubes for the same 3D data? We received the data from different processing centers and in various formats (16 bit/32 bit). Can this difference affect attribute study and others uses?

Zaw Win Aung

Assistant Geoscientist at MPRL E&P

ANSWERS PROVIDED BY INDUSTRY EXPERTS

Peter Rowbotham

Geophysical Advisor at Apache Corporation

Zaw Win, there are two views on this:

1. If we have put a lot of effort and expense to record data at a certain precision, then why throw it away just to save some disk space with a 32 to 16 bit conversion? The same could be said of sampling in x,y,t. Why do we often reduce our sampling in final processed products compared to the acquisition?
2. By staying in 32 bit we are carrying around huge quantities of data that have no impact on the decisions we are trying to make. As long as the conversion to 16 bit has been done competently (no shift applied - zero is still zero; no clipping of important amplitudes), then calibrating wells to 16 bit seismic and deriving attributes should work just as well as to 32 bit. The argument I make for 16 bit could also hold true for 8 bit but that would depend very much on the type of attributes you were considering. For example yes/no attributes would be less sensitive than continuous attributes. One word of caution though - the decisions/ interpretations we want to make now are not necessarily the same as what we want to make in the future. We should ensure we maintain future access to fully sampled, high precision data, even if we choose to work on lower precision, smaller volumes for efficiency.

Matt Hall

Geoscientist at Agile

Zaw Win, another thing I have seen in data deliveries is data with and without final 'cosmetic' filters (e.g., spectral shaping, fk, fxy). Even if the filtered volumes are better for interpretation, I think it is best to use the unfiltered volumes for quantitative work (and especially for anything to do with frequency). Smooth the result (attribute volumes, inversions, etc) if you have to but at least use raw input. On bit depth, I agree with Peter. Here's something I wrote once about the subject http://www.agilegeoscience.com/blog/2011/2/4/b-is-for-bit-depth.html.

Andrew Wilson

Geophysical Advisor

MATT, Imagine you've received a "final" volume which has significant steeply dipping coherent noise, perhaps because the processing contractor was requested by your in-house processing specialist to leave the dip limit open on the migration in the hope of imaging fault planes, the end result being a lot of "mig noise" which is clearly visible. Imagine that the contractor has also had enough pride in their work to attempt to reduce the noise in their "final" volume by applying a FK or similar "cosmetic" filter and this has been effective at removing the steeply dipping noise. You subtract one volume from the other and see that there is no geological signal in the difference.

Which volume would you use for QI work? Processes like well-based wavelet estimation are notoriously sensitive to noise, so would you just skip that and try to work with the noisy data using a zero phase statistical wavelet? Or move to use the seismic volume after mig noise removal?

Matt Hall

Geoscientist at Agile

Andrew, I would say that removing 'steeply dipping coherent noise' is, by definition, not a cosmetic adjustment. It's a judgment call of course. But I've seen lots of examples of final 'tweaks' by processors that make a volume easier to interpret, but what was removed is less equivocally 'noise'. I guess my point is that geology contains random noise too! Or, to put it another way, "no geological signal" is not always obvious.

ZERO CROSSING ATTRIBUTE

QUESTION 37

What is the theory behind, how zero crossing attributes works?

Jamal Ahmad

Deputy Manager Geophysics at Essel Group Middle East

ANSWERS PROVIDED BY INDUSTRY EXPERTS

Jeffrey Ross

Senior Geophysicist at Whiting Petroleum

Jamal, you need to be a bit more specific. Zero crossing is not an attribute in the classic sense. It is simply where the band-limited wavelet response crosses from positive-to-negative samples or the reverse. It is the presence of no amplitude for a given phase response. In a perfect world, with perfect minimum phase data, then the initial zero crossing would correspond to the actual T-zero of any impulse response from a reflection coefficient. This would only be true for non mixed-phase, minimum phase data. This is NOT true for mixed-phase or zero phase data. The presence of, and location of subsequent zero crossings after the first are dependent upon the dominant frequency of the wavelet and the number of side-lobes.

Tim Davies

Geophysical Consultant

Jamal, what were you specifically looking for from a zero crossing event? What is generally more interesting is where amplitude stops increasing or decreasing. The zero crossing is a somewhat arbitrary thing really. This is especially true where there is a significant amount of phase instability such as you might find in recovering a harmonic series. It might be more interesting to track zero crossings if there is no inherent wavelet imposed on the data.

Jamal Ahmad

Deputy Manager Geophysics at Essel Group Middle East

Jaffrey and Tim, thanks for your valuable comments. What I am looking for is the stratigraphic truncation and to delineate the extent of clinoform downlaps. If I understood correctly, zero crossing will see the curvature of wavelet, meaning positive or negative amplitude. The point where the interpreted horizon has neither negative nor positive amplitude, is treated as the zero crossing. It that so?

Actually I didn't see anything like this in Landmark suite of software but since it works more or less like some seismic attributes I wanted to learn more about the science behind it.

Jeffrey Ross

Senior Geophysicist at Whiting Petroleum

Jamal, if I understand what you are trying to do correctly, then I am not sure that zero crossings are the best way to get your desired results. A better way to do what you want would be to use an attribute like 'stratal convergence' as published by Chevron. Look it up in Google.

Brian Schulte

Geophysical Advisor at Respol Calgary, Canada

Jamal, When we pick horizons we generally want the data to be zero phase. Zero phase means the geological boundary is at the apex of the peak or trough in the seismic data and this is why we do well ties. We utilize zero phase because the principles of AVO are based upon zero phase data. When we do inversion we then integrate the data which will phase rotate the data 90 degrees and cause the inverted seismic wave to actually represent the rock properties so that in an inversion, the horizons are actually on the zero crossing. As I said the wave now represents the rock property we inverted for. As we all know by removing the wavelet we get higher resolution and we can see subtle events better.

In most cases we invert the near traces (0-15 degrees) which tend to reflect the reflection coefficient ((AI2-AI1)/(AI2+AI1)) since the reflection coefficient equation contains the AI or Acoustic Impedance terms. Acoustic Impedance is density times velocity for that rock layer.

Since we have removed the effect of the wavelet in the inversion we then have higher frequency data and we can then run volumetric curvature on the acoustic impedance to be able to bring out the flexures in the data. We assume where we have flexures we have fractures due to the bending of the rock. Check the literature for curvature especially that by Chopra and Marfurt, 2007.

Jerry Coggins

Principal Geophysicist at Coggins Geosciences, Inc

Brian, Inversion does not remove the Wavelet Effects, but reduces tuning and interference effects; but the data is still band-limited. Besides AVO inversion, there other typical seismic volumes integrated to be relative impedance. Zero crossing measurements on these yield top and base reservoir and reservoir thickness.

Brian Schulte

Geophysical Advisor at Respol Calgary, Canada

Jerry, I agree that we increase the resolution and reduce tuning and interference with inversion. But we do it by strongly attenuating the wavelet side lobes in the inversion. With the seismic inversion we do also broaden the frequency spectrum through adding in low frequency components from the well log data and seismic velocities. The high frequency is also derived from the well control. We are then able to see events in the inversion we could not see in the original seismic data.

In some cases we were unable to map a certain event, but after the seismic inversion we are able to do so. Maybe saying we increase the frequency content is not the right way to say it but we do increase the seismic resolution.

Jerry Coggins

Principal Geophysicist at Coggins Geosciences, Inc

Brian, Generally the well log is used to determine a wavelet and to build a low frequency model; but high frequencies are not derived from well control. This is for a JASON type inversion.

Brian Schulte

Geophysical Advisor at Respol Calgary, Canada

After thinking about it, I agree with you. We are not increasing the frequency content but we are tying the well data to the seismic data which is an important step. We are removing the side lobes of the wavelet which enhances the resolution and reduces tuning and interference.

We tend to equate improving seismic resolution with increased seismic frequency, but as mentioned we can do it by removing the wavelet side lobes in the seismic data. This improved resolution helps with attributes such as curvature and other attributes.

Thanks for the discussion, helps me to see where I have been lazy in my thinking.

Tim Davies

Geophysical Consultant

I think side lobe characterization is important. It's nice to beam form where the side lobes aren't. Harmonics are much simpler to recover in the absence of lobes. To me the only great zero crossing is that one which corresponds to the formation interface.

Jamal Ahmad

Deputy Manager Geophysics at Essel Group Middle East

It is hard to say that the data we are working on is hundred percent zero phase. But there are some attributes which help estimating zero phase volume. Assuming we have zero phase data, we can check its spectrum, compare it with the spectrum of well(s) and this way design an operator. We remove the wavelet response from our seismic volume and convolve that reflectivity volume (better to say like this) with designed operator. This way we get an inverted volume. And yes, we have to apply a phase shift before doing this (sometime we have the option of performing this step in end). And for that, we need to know the phase of data as the software assumes that data is zero phase.

The output result of this is a volume which looks more or less like the original except for a little broader reflections in lower part of section. But overall it gives the variation in trace format. How can we say the effect of side lobes has been removed? Because of the limitations of the geophones (now we have geophones in market which can measure frequency level up to 2Hz), the low velocity model is used optionally to add low frequencies in data. And this model, if used in inversion, can help detect fluids effect as well.

It is important to note that a low frequency model, is not low velocity.

ATTRIBUTE ANALYSIS IN CARBONATE RESERVOIR

QUESTION 38

Which attribute is best for delineating reef carbonate? We did post stack inversion, spectral decomposition and coherency for carbonate reservoir but none of the attributes helped us in understanding the lateral extent of the reservoir. We have some producing wells in the given area (Fig. 2.17).

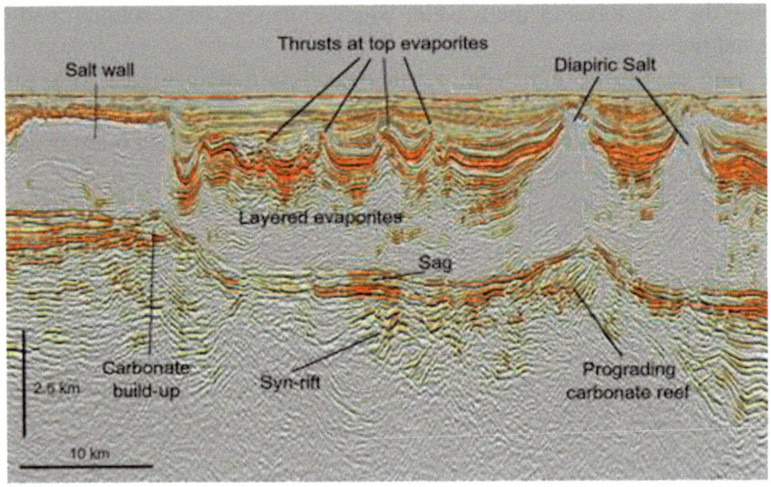

FIG. 2.17

RTM depth section highlighting pre-salt sequences, carbonate build-ups and post-salt features (CGGVeritas Cluster Extension Survey).

Courtesy: CGGVeritas.

Shweta Bankhwal
Sr. Geophysicist at ONGC Ltd

ANSWERS PROVIDED BY INDUSTRY EXPERTS

Jonny G. Pinto
Senior Geophysical Advisor

Shweta, Trace envelope/reflection strength should do an OK job on delineating the shape of the shoal/patch/reef. Spectral decomposition should do the trick looking at thickening and in this case when correlating it to edges of your carbonate bodies. I have seen areas where a low AI/High Porosity Carbonate rivals Shale having the same characteristics, therefore the same seismic response. As the Carbonate facies change (Oolite vs Vuggs) so does the AI response from low to High. In these situations use LMR attributes, they really help differentiate lithology. Of course keep in mind that it is all relative to the quality of your seismic.

Jamal Ahmad
Deputy Manager Geophysics at Essel Group Middle East

Shweta, I hope, you have already tried coherency and wave-field spectral decomposition. I think, "Duplex wave migration" will help (but originally it was designed for fracture zone identification).

Maria Beatriz deSantaAnna Camisuli
Seismic Interpreter – Independent Consultant

Shweta, You might be using the right attributes in the wrong interval! I would recommend you check on the stratigraphic interpretation to constrain the analysis window. You could play with flattened volumes and/or proportional slices to better visualize your attributes. It's also important to view the amplitudes histogram with the color palette to highlight the feature you want to characterize! Perhaps it's just a matter of fine tuning. Then if you have wells around, did you calibrate the seismic response of the reservoir?

Jonny G. Pinto

Senior Geophysical Advisor

Shweta, Maria is right. Every interpretation should start at the well site. Check your ties, synthetic responses and cross plots results. Get familiar with what type of response you are getting from the wells. Then proceed to using the seismic to get the same answer. If can't do that, you need to double check your workflow.

Alexander Mihai Popovici

Chief Geophysicist

Shweta, Have you tried Diffraction Imaging? Diffraction Imaging is the imaging of discontinuities in the earth. Diffractions are the seismic response of small elements (or diffractors) in the subsurface of the earth, such as fault edges, pinch outs, reef edges, small-scale faults, fracture zones, fluid fronts, caves, karsts, and channel edges, and in general all objects which are small compared to the wavelength of seismic waves.

Ryan Martin

Sr. Seismic Interpreter

Shweta, In my experience, in classical carbonate environments a good attribute to use is an algorithm that copes with lateral amplitude nuances for reef edge delineation. If you have a high-amplitude reef system but the edges are not represented by phase breaks, this algorithm performs beautifully. Another process to use is lateral stratigraphic analysis to highlight reef bed thinning, allowing you to accurately pinpoint the exact edges of these reefs. For reef identification itself, an amplitude envelope would be helpful, although I am concerned that spectral decomposition won't reveal much information. Most major interpretation packages will have the appropriate algorithms you need.

Joe Valenti

Geophysicist

Shweta, As Ryan said, an edge attribute that responds to amplitude breaks will often beautifully delineate carbonate reefs. GeoTeric has an edge attribute called "Tensor" which highlights amplitude breaks by calculating the gradient structure tensor throughout the entire seismic volume.

Spectral Decomposition and RGB color blending, also available within GeoTeric, cannot only delineate the boundaries of the reef but can also allow you to quickly and easily visualize heterogeneities within the reef due to changes in either lithology, thickness or pore fill.

Of course, the seismic response of carbonates can be highly variable and strongly depends on bounding rock types. Understanding the regional stratigraphy will allow you to better anticipate the seismic response from the reef system.

Jonny G. Pinto

Senior Geophysical Advisor

Joe, So one could tell the difference for a high porosity scenario, between a shale and a carbonate body, just by using color blending and spectral decomposition. Boy I'd like to see that work. In my book the only way to predict this is by properly modeling each scenario vs. lithology/thickness and fluid.

Jorge Reveron

Senior Geophysicist at Repsol

Shweta, I agree with Jonny. You must do well modeling with log data at seismic scale to understand elastic properties and response to see, for example, if acoustic impedance is capable of discriminating between carbonates and surrounding sediments.

What kind of lithology do you have underlying and above the carbonates? Maybe if you can't see a good response with acoustic impedance or coherency it's because sediments above carbonates are as hard as the carbonates, so the impedance contrast is weak. You should find the answers with well modeling.

Shweta Bankhwal

Sr. Geophysicist at ONGC Ltd

Thank you all for your good suggestions. In most of the wells we are getting high acoustic impedance within the reservoir but in few wells it is low. So I am not able to understand the reservoir property. Even in the data slice generated from the horizon (top of reservoir) and window (50 ms, 100 ms etc) we are not able to get any impedance pattern which can help us in delineating the lateral extent of the reservoir. We are not able to make any inferences from the inversion results. Though the RMS amplitude attribute helped us in marking the facies boundary, we couldn't detect the exact shape of the body. We have carbonate and shale sequences within the reservoir which is mentioned in the core data from the well. Interval velocity is also very high which varies from 3800 m/s to 5000m/s. So around 20-26 msec seismic window (which is a peak itself) covers the 60m (approx) thick reservoir.

Jonny G. Pinto

Senior Geophysical Advisor

Shweta, check your vertical resolution to see if the data is under-tuning which makes it a different ball game. Low AI is good, since it implies High Porosity. Hope your reservoir is high porosity, otherwise it would be difficult to identify carbonate/shale fill with fluid. If High AI goes along with High Resistivity and low Porosity, then your reservoir is tight. You need to integrate seismic interpretation with Geology. Attributes are nice to use, but you need to know what you are looking for based on your log/petrophysical data. The new generation of seismic interpreters tends to look at only seismic which is not good. One needs to pay attention to well data.

Mohamed Ouabel

Senior Geoscientist at Petronas

Shweta, Your main goal as I see it is to figure out the lateral extension of your reef and then find some internal patterns that correlate with your wells hydrocarbon results. So, we can subdivide your question in two main parts. Your first question is concerning the geometry or the shape and the extension of your reef. RMS amplitude helps you as you said to get the general anomaly, and you want to extend this image to the limit of what your seismic data can give you. That's really good. According to that, why don't you go for something simple but very helpful, such as creating a composite amplitude map of the top and base (usually reverse polarity exist between these two events), by picking the two horizons and just add the amplitudes (absolutes values) according to your seismic phase (is zero phase then you pick the min and the mix amplitudes)? Normally, if your seismic data is not so noisy and the relative true amplitude processing was done properly, you will get an amplitude map that points you to the tuning effect. Due to the thinning of your reef edge, you have just to remove the tuning effect by multiplying by a scalar and you will get the extension of your reef apparent thickness beyond the ¼ of the wave length seismic resolution. Or you can do the same exercise by using your AI inversion which should help more because you are dealing with the AI interval values of your reef, rather than an amplitude measurement from two reflecting boundaries. This technique will be more robust. (Just have a look at the paper Simple Net Pay Estimation from Seismic: A Modeling Study by Rob Simm first break volume 27, September 2009). But this technique requires that you run a model based on your wells' information to get the scalar. By the way, you can use the absolute amplitude map or the AI map extracted within the top and base of your event to compute porosity or any other attributes based on the behavior of your AI and the your well log results.

Raju Arasu

Senior Geophysicist & Lead Triassic Paleozoic Prospect Evaluation Unit at Kuwait Oil Company

Shweta, If the amplitude spectrum is flat and stable filter the data for 25-50 Hz. Since the body is surrounded by tight carbonates you are likely to get dim-spot. AI reduction is due to high porosity, HC and shale within. So you don't see conventional results in spectral decomposition. It will be interesting to see the pattern of the extreme negative amplitude in frequency slices. Also try phase output at different frequencies. Create PHIE logs and try Emerge of HRS. Display the seismic in wiggle form in full dynamic range (no clip of high amps and no-miss of the smallest amps) and look for anomaly such as waveform change, temporal frequency, additional zero-crossings just above or within the reservoir, etc. Accordingly run facies class or window attributes to delineate the extent. Try to find out the type of the reef. Being 60 m thick it has developed over a short geologic window. It could be a patch reef that grew on seafloor reliefs, in which case the size may be small, but will have company. You may see a colony of patch reefs in the area. Keep your eyes open.

QUESTION 39

Which of the seismic attributes is best in identifying carbonate build-ups (reefs)?

Mohammed Gumati

Lecturer at Benghazi University

ANSWERS PROVIDED BY INDUSTRY EXPERTS

Hesham Refayee

Geoscientist

Mohammed, When you have tons of attributes streamed from different mathematic algorithms, it is difficult to choose the right one in a carbonate environment. However, there are known attributes associated with specific geologic features. For instance, Karst features, reef edges and fractures are visible with similarity (coherency) attributes. Also, multi attribute Neural net work and waveform segmentation (UVQ) are good for picking up build-ups, shoals, channels, lobes,….. Please refer to this link for more: https://opendtect.org/opendtect-attributes-matrix/.

David Paige

Seismic Inversion Developer

Mohammed, The first problem is making sure the attribute applies to the bed itself, and not just to one interface.

Jonny G. Pinto

Senior Geophysical Advisor

Mohammed, Carbonates usually have a high AI contrast (when embedded by shale/sands) and are very distinct from their surroundings, so I would try velocity and coherence attributes to determine their geometry. I have used these two in the past with good results.

Augusto Garcia

Geophysicist

Mohammed, Curvature attributes are great for understanding sedimentology and sometimes geometry. Try using GeoProbe from LandMark.

Seismic Attribute Analysis in Seismic Stratigraphy

QUESTION 40

What are the uses of seismic attributes in seismic stratigraphy?

Akrem Soltani

Petroleum Engineer

ANSWERS PROVIDED BY INDUSTRY EXPERTS

Steve Adcock

Senior Geophysicist

Akrem, Attributes for seismic stratigraphy, vs other forms of seismic analysis, tend to be used to highlight seismic geomorphology. This may be done in single or cascaded attribute workflows, which are specific to the seismic characteristics of the geomorphology being visualized. In general, attributes which help highlight shape, such as various dip and curvature attributes, are particularly useful. Once the geomorphology is identified, usually with a combination of attribute selection and opacity rendering, other attributes are used to highlight gradations within and adjacent to the shape. For example, highlighting fluvial systems with anastomosed (interconnected – ed.) fans, channel bars, and cross-cutting channels may incorporate many different seismic attributes. Dealing with all the many attributes available for seismic interpretation can be confusing. Classification of all these attributes and their methods of use is challenging and still evolving. For reference book on this, I will suggest you read Chopra and Marfurt's 2007 book: Seismic Attributes for Prospect Identification and Reservoir Characterization. As a student you can buy it from either society for $70. Here's a link to the SEG bookstore version: http://library.seg.org/doi/book/10.1190/1.9781560801900.

Azer Mustaqeem

Petro-Explorers Inc.

Akrem, Seismic Geomorphology - Application to Hydrocarbon Exploration and Production by Davies et al. is a very good read: Geological Society Special Publication 277.

Mohamed Shihata

Geoscience Business Development Specialist

Akrem, Extract physical properties from your wells (net sand, average porosity, etc.)

Extract attributes from the seismic data at corresponding TWT (i.e., horizon slice)

Look for correlations between seismic attributes and physical property at well locations

Use correlations, if found, to produce a map of the physical property.

SEISMIC ATTRIBUTE FOR UNCONVENTIONAL RESERVOIR

QUESTION 41

I am trying to generate attributes on 3D seismic and then integrate results with Microseismic data, for unconventional reservoirs in central Alberta. I am using Transform software (Fig. 2.18).

Which attributes work better with tight sand reservoirs and which software generates better horizon based attributes? Transform seems to not be working well with them. I have access to Petrel, Seisware and Kingdom?

Aamir Rafiq

Consulting Geophysicist at 3DGeo Exploration

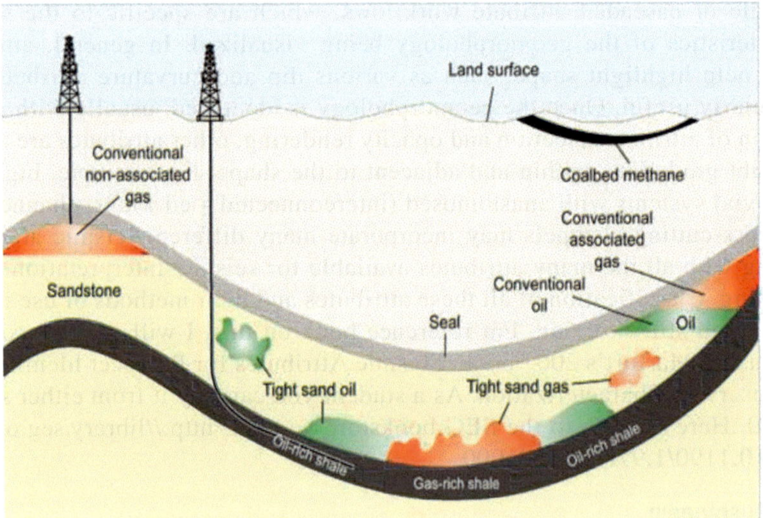

FIG. 2.18

Tight sand reservoirs (oil and gas).

Data modified from Energy Information Administration and United States Geological Survey.

ANSWERS PROVIDED BY INDUSTRY EXPERTS

Mohamed Shihata

Geoscience Business Development Specialist

Aamir, Think of the physical attributes that can help follow changes in lithology and don't forget curvature attributes which work very well with fractures. Don't depend just on horizon attributes because there may be some errors in interpretation. A second piece of advice I think kingdom will help you extract many attributes along the horizons. It is very fast but must you generate volume attributes first and you may still want to combine attributes with RGB as it may work well in opendetect and FFA Geotric. Also remember to combine geometric attributes, especially curvatures with physical attributes especially instantaneous attributes. Don't forget frequency decomposition and also combine low-med- high frequency information. The best software for this is FFA Geotric

Marianne Rauch-Davies, Ph.D.

VP Of Applied Technology for Geomage

Aamir, Have you looked into diffraction imaging? This is a new technology that extracts the diffractive part of the wave field and represses the reflections from pre-existing seismic data. Smooth surfaces generate spectacular events but small scale discontinuities like fracture corridors in tight rocks produce diffraction waves. These events can sometimes be seen on seismic stack data but are usually removed as "noise" or lost during the migration process. You can then use the migrated diffraction cube to map areas that are fractured and integrate them with your micro seismic events. The interpretation can be done in any software like SMT or Landmark for example. You can find articles on the SEG and EAGE websites explaining these techniques and showing examples how to interpret them.

Jonny G. Pinto

Senior Geophysical Advisor

Aamir, When integrating attributes with micro seismic, you should use Geology-Seismic and Engineering attributes. All three families together with multivariate statistics should be your best weapon.

SPECTRAL DECOMPOSITION

Frequency attributes used as DHIs involve separating and classifying seismic events within each trace based on their frequency content. The application of these attributes is called spectral decomposition. Spectral decomposition provides a means to use seismic data and the discrete Fourier transform to image and map temporal bed thickness and geologic discontinuities in 3D seismic surveys.

The concept behind spectral decomposition is that a reflection from a thin bed has a characteristic expression in the frequency domain that indicates temporal bed thickness. Two methods are supported to output spectral amplitudes. These are the frequency slice and frequency spectrum methods.

✓ In the frequency slice method, panels of *amplitude* spectra at specific frequencies are output.
✓ In the frequency spectrum method, a *frequency* spectrum at specific horizon or time is computed.

An additional tool called spectral decomposition scaling should be used to apply a spectral compensation factor to each frequency slice. The goal is to balance the wavelet spectrum by normalizing each frequency slice according to its average amplitude.

QUESTION 42

How do I distinguish between confined and unconfined channel systems in seismic sections and attribute characters?

Zaw Win Aung

Assistant Geoscientist at MPRL E&P

ANSWERS PROVIDED BY INDUSTRY EXPERTS

Jerry Coggins

Principal Geophysicist at Coggins Geosciences, Inc

Zaw Win, mostly this brings to mind turbidite depositional systems. The first step is to do some research. Most basins will have published papers defining the approximate locations of the shelf edge, slope, and basin floor. The slope is the best place to find leveed and unleveed confined channels. Generally, these geometries are very distinct on seismic traverses perpendicular to the channel. Semblance time slices flattened on nearby continuous shale will typically show the internal geometries well; things like spillover or channel migration show up well too. Based on the seismic geometries, you should be able to see any broader sand body expression on both types of data. Remember that you can have small slope fans that are unconfined, and that basin floor fans may be eroded by younger confined bypass channel systems. Remember to search for examples of these images to use as analogs.

Esuene Samson

Sr. Geoscientist

Zaw Win, I completely agree with Jerry. However, it shouldn't be easy to distinguish between confined and unconfined channel systems and their fill, using the typical seismic facies analyses methods. So, you now need to apply a robust sequence stratigraphic methodology to help you distinguish between these deposits. Even then, your degree of certainty would depend on the resolution of your data.

Alfredo Sánchez González

Ingeniero Geologo Especialista & Consulting

Zaw Win, have you tried to work in a time window with your horizons or horizons well interpreted? In this window you can generate pseudo horizons and the amount of these you will decide according to the seismic resolution. At this window you can apply multiple attributes as instantaneous amplitude, sweetness; in my case I apply spectral decomposition and semblance too with good response. This allow you to see sedimentary features depending on the analysis area (floodplain, delta, pro delta), otherwise is the combination of attributes in a color cube, where each color you assign an attribute, the result are high quality maps with better interpretation and understanding.

Zaw Win Aung

Assistant Geoscientist at MPRL E&P

Alfred, I had tried with my horizons on time window and also done with multi-attributes and that gave me a lot of idea as your point of view. Furthermore, I have studied with geo-body (X plot) to better understand the lithological variation of channel. Thank you so much Alfredo.

QUESTION 43

I have been working on PSDM data for a field and have applied spectral decomposition method for this purpose. But have not been able to see very good results although the area is supposed to contain a lot of channels. One of the reasons for this is that higher frequency data is not available for the region and also the area contains lot of faults. What else can I do to enhance the geomorphology from the seismic data and how else can spectral decomposition be used for seismic interpretation?

Tripti Kumari

Processing Geophysicist at CGG services India PVT Ltd

ANSWERS PROVIDED BY INDUSTRY EXPERTS

Yosef Khairil Amin

Sr. Geophysicist at PTTEP International, Bangkok

Tripti, have you tried to do some frequency filtering on your seismic volume? Split it into, say, 3 volumes. Of course you have to first check your seismic frequency bandwidth. First frequency volume contains the available lowest frequency. The 2nd contains the medium range, and the third contains the highest frequency in your data. Once ready, create the envelope frequency for each volume. Load all these three volumes simultaneously in 3D window. Play around with the color bar. In my experience it will reveal the subtle stratigraphic and structural features in your data.

Anand Prakash

Snr geophysicist at Kuwait Oil Company

Tripti, you can use spectral decomposition to enhance the visibility of faults. Simply analyze the frequency slices of the amplitude volume generated by application of spectral decomposition. You will observe that some lineaments appear only in some of the frequency slices and not in all. Retain the signal in these range in your data by applying appropriate filter and do fault interpretation on this filtered volume, you will be surprised by the clarity.

QUESTION 44

Can I use one volume of spectral decomposition for picking seismic reflectors?

Cesar Alcazar

Geologo at Cepcolsa

ANSWERS PROVIDED BY INDUSTRY EXPERTS

Farrukh Qayyum

Sr. Geoscientist at dGB Earth Sciences

Cesar, In theory, No. For certain cases, e.g. to map tuning regions, you may want to use the thin bed reflectivity or equivalent methods with a wider range of frequencies. However, if you have a fixed frequency (iso-frequency.) volume, then you cannot identify a reflector. This does not rule out the mapping of local seismic facies on the iso-frequency volume (e.g. bright spots, tuning, unconformities, pinch-outs etc).

Paulus Allo

Geophysicist at Statoil Indonesia

Cesar, off course you can. Please note that you may not see all the geological features in one iso-frequency volume. However, for the practical purposes of picking seismic reflector, sometimes it is easier to do it in iso-frequency volume rather than in normal seismic (again, it is not always the case but I have experienced many cases that it does help).

Farooq Riaz

Geophysicist at United Energy Pakistan Limited

Cesar, colored Inversion or Spectral bluing is better option rather than using spectral decomposition for seismic picking. Normally we use spectral decomposition volume for attribute study to see the different geological facies, that is, channel, compartmentalization etc. on particular frequencies.

QUESTION 45

How can we define these attributes: dominant frequency, spectral response and spectral decomposition? How can we interpret them in a different way?

Shweta Bankhwal

Sr. Geophysicist at ONGC Ltd

ANSWERS PROVIDED BY INDUSTRY EXPERTS

Mahmood Khan

Geophysicist

Shweta, by plotting the dominant frequency for a particular time window we can identify the particular frequency has got good amplitudes and that can be worked on in confidence.

Spectral response gives us all the frequencies available in the data with their amplitude values so, we can choose among the frequencies according to resolution required for our particular work. Then we can use the chosen frequencies to look closely at their behavior through spectral decomposition.

Mohamed Shihata

Geoscience Business Development Specialist

Spectral decomposition and its application in mapping stratigraphy and hydrocarbons.

https://www.dropbox.com/s/1oikiz772ysxnb2/spectral%20deco.pdf.

Mohammadreza Vanaki

Head of Geophysics at KPE

Shweta, by using popular processing software such as Promax or Vista, you can define a dominant frequency. Spectral decomposition can be done using different methods by using VVA software. Also you can play with frequency in Volume Attributes in Petrel. Spectral decomposition is a good hydrocarbon (HC) indicator especially in sand reservoirs.

Mohamed Shihata

Geoscience Business Development Specialist

The dominant frequency is usually determined by measuring the time between successive peaks or troughs and taking the reciprocal. Dominant frequency depends on wavelet period time so it indicates changes in change of frequency related to bed geometry and lithology itself. So my first advice to you is to calibrate your dominant frequency with well information.

Spectral decomposition has many applications, one of which is DHI. You can also use spectral decomposition to indicate thin beds or to enhance your data.

Mohamed Salah Sedek

Geophysicist

Mohamed Shihata, What you previously mentioned is not necessarily the dominant frequency; the reciprocal of the time between two successive trough/peaks is the frequency (not the dominant). The dominant frequency is the frequency that represents the strong amplitudes on your seismic section/cube. You can get it by plotting the amplitude spectrum.

Mohamed Shihata

Geoscience Business Development Specialist

Salah, Thanks for illustration. Your comment is true but I am confused as to how to display dominant frequency as seismic attributes along section, and how to display dominant frequency at time slice and along a cube. If we review complex traces analysis we find instantaneous frequency, computed as the time derivative of instantaneous phase, represents the mean frequency centroid of the power spectrum. If instantaneous frequency, computed as the time derivative of instantaneous phase, then the centroid of the second moment of the power spectrum, or the RMS frequency, is obtained as the vector sum (sum of squares) of average frequency and the variance. Barnes calls this the dominant frequency, which is similar to the instantaneous frequency (wavenumber), except this corresponds to the RMS frequency/wavenumber of the amplitude spectrum (or centroid of the power spectrum) of the seismic wavelet.

I usually make an amplitude spectrum analysis to determine dominant frequency also to get my best window for calculated attributes if you decide uses RSA attributes, but when we deal with dominant frequency as seismic attributes you try to find good and simple relation between dominant frequency and your geological target. If we deal with seismic attributes as processing steps we can't predict the response of seismic attributes to our target we try to calculate and compare attributes response with each other.

Mohamed Shihata

Geoscience Business Development Specialist

Salah, I found this definition from Hilterman, 1985.

The dominant period (τc) of the response corresponds to the time from one peak to the next or from one trough to the next. The reciprocal of this dominant period is a measure of the dominant frequency (fc) of the signal or wavelet spectrum. The reciprocal of the half-width of the response-envelope (τb) provides an estimate of the bandwidth (fb) of the signal spectrum.

https://www.dropbox.com/s/yg3u5ja1xlnhino/dominant%20frequency.jpg.

Mohamed Salah Sedek

Geophysicist

Mohamed Shihata, Hilterman was showing how to calculate it (which is the same method as the regular frequencies) but he highlighted the dominant frequencies based on amplitudes (the strong ones). Do you see my point?

Mohamed Shihata

Geoscience Business Development Specialist

Salah, Yes Hiltermn says that the dominant frequency is related to the dominant period (τc) of the response which corresponds to the time from one peak to the next or from one trough to the next. The reciprocal of this dominant period is a measure of the dominant frequency (fc) of the signal or wavelet spectrum as I said in my first comment.

And I also mentioned the reciprocal of the half-width of the response-envelop (τb) provides an estimate of the bandwidth (fb) of the signal spectrum as you said in your comment.

And maybe we are seeing the same thing but with different eyes, like semblance in velocity analysis processing and semblance in seismic attributes. Perhaps we are looking at dominant frequency in processing steps and dominant frequency in seismic attributes analysis so we are looking at the same idea but in different places and for different applications.

https://www.dropbox.com/s/aibof2qe4vsotrm/dom.jpg.

Seismic attributes are easy calculated but hard to interpret, and if you chose the wrong parameters they can indicate a wrong response.

Rajeshwaran Dandapani

Director Geophysics at Telesto Energy

Both the Mohameds Shihata and Salah are correct. One is describing it in time domain and the other in frequency domain respectively. Both are using the term dominant which refers to the maximum amplitude which is the key factor. If you select a window of interest and measure the average time period of the brightest reflectors within that window and take the reciprocal of it then it will be exactly equal to the dominant frequency which you obtain by doing a Fourier transform of that window. The bandwidth will be the frequency range where the amplitude drops down to 6dB level from the dominant frequency level which is equal to the expression in the time domain which states the reciprocal of the time period is where the amplitude level drops to half of the dominant period. In Petrel the dominant frequency is calculated as the hypotenuse of the instantaneous frequency and instantaneous bandwidth i.e. square root of IF square + IB square. All three quantities are very useful in predicting reservoir parameters including fluid fill.

The thickness variation of the reservoir whose other parameters like fluid fill, net-to-gross and porosity are more or less constant can be predicted by studying the first spectral response over the reservoir interval.

It is important to note that, spectral frequency is nothing but the frequency at which any portion of the reservoir gets tuned and exhibits the brightest amplitude. When you study the amplitude slices of various frequencies which are the outputs of spectral decomposition we see that various portions of the reservoirs exhibit high amplitude patches depending on their tuning thicknesses. If you have many wells then you can cross-plot these first spectral frequencies observed at the well locations with the actual thicknesses measured in the wells and they will be inversely proportional. A regression line can be fitted to the measured thickness versus first spectral frequency cross-plot which can be used to predict the thickness of reservoir away from well control. This method of course works well when the reservoir is pretty uniform in terms of other parameters like NTG, Porosity and Fluid saturations and the major variant is the thickness. If all the parameters are varying considerably then the quantity called “net pay” which is the product of all the above mentioned reservoir parameters can be cross-plotted against first spectral frequency in place of thickness. Then the regression analysis can be performed and the resulting equation can be used to predict the distribution of net pay away from well control.

Starting Point of Spectral Decomposition

The starting point of spectral decomposition is to decompose each 1D trace from the time domain into its corresponding 2D representation in the time-frequency domain by means of any method of time-frequency decomposition, such as:

- Short-time Fourier transform
- Continuous wavelet transform
- Wigner-Ville distribution
- Matching pursuit

Once each trace has been transformed into the time-frequency domain, a band-pass filter can be applied to view the amplitudes of seismic data at any frequency or range of frequencies. Technically, each individual frequency or band of frequencies could be considered an attribute. The seismic data is usually filtered at various frequency ranges in order to show certain geological patterns that may not be obvious in the other frequency bands. There is an inverse relationship between the thickness of a rock layer and the corresponding peak frequency of its seismic reflection. That is, thinner rock layers are much more apparent at higher frequencies and thicker rock layers are much more apparent at lower frequencies. This can be used to qualitatively identify thinning or thickening of a rock unit in different directions. Spectral decomposition has also been widely used as a DHI. Several examples of spectral decomposition are shown in Figs. 2.19 and 2.20.

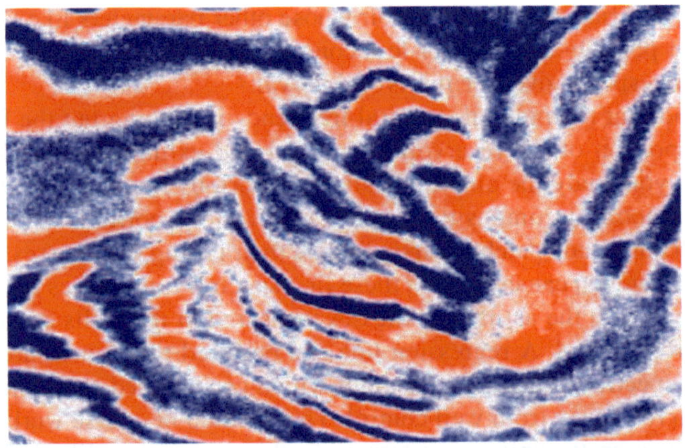

FIG. 2.19

Time-slice.

Courtesy: F. Schroeder.

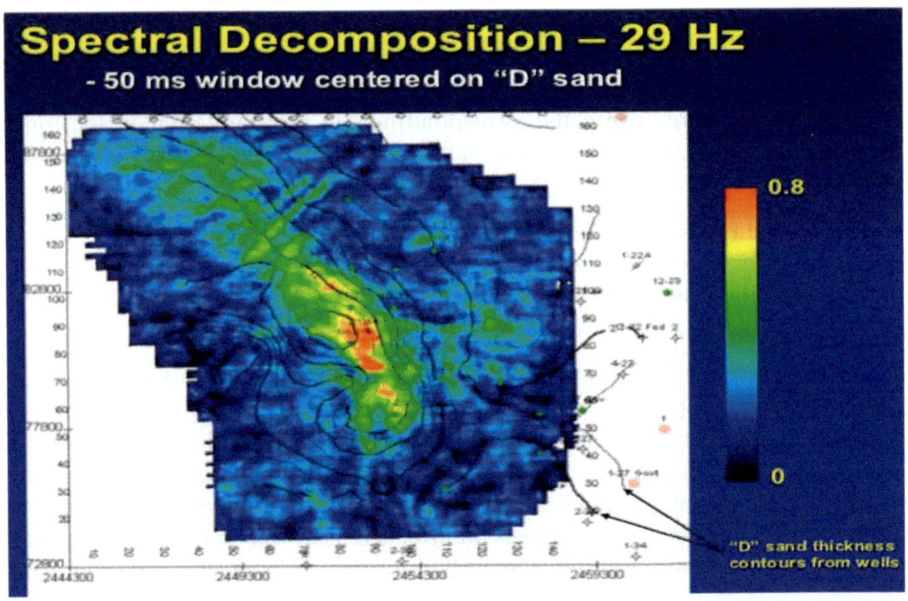

FIG. 2.20

Spectral decomposition.

Courtesy of True Oil Company, LLC, Casper, Wyoming.

SPECTRAL WHITENING/BLUEING

QUESTION 46

Is frequency enhancement through spectral whitening for boosting the continuity of the reflectors or it is part of signal. Is it useful or not? If not, then how and why?

Ammad Ali Qureshi
Exploration Geophysicist at Petroleum Exploration Ltd

ANSWERS PROVIDED BY INDUSTRY EXPERTS

Belal Khalil
Geophysicist at BP

Ammad, whitening might be a good procedure to test in the case that you are looking for more detail in the data. If your seismic lacks high frequencies or is too low on them and you think that having that part of the data enhanced will improve your visualization of the thinner, finer, inter-reflector features, Spectral Whitening may be worth trying. The reason why some might not prefer to use Spectral Whitening is that it causes a boost to "all" the data's frequency bandwidth (hence the name white, as in white light containing frequencies of all other colors). If your data is noisy, noise will be boosted just as well as the other parts of the data. Whitening is essentially a de-convolution (as any filtering process is) that involves applying a de-convolution operator with a very small gap (usually one sample interval in length) to the data. This effectively maximizes the effect of the operator on the data's wavelet and compresses it to a great extent, and therefore boosting high frequency (thinner, finer, inter-reflector) events. Bluing on the other hand boosts the higher end of the data's spectrum (hence the name, in analogy to the color blue). This may give a similar result to whitening; however, the difference between a Blue and a White dataset will be clearer when investigated, when looking at the data's amplitude spectra. Like our colleagues suggested, some applications use well data as a source to calculate the filter to be used for Bluing in an attempt to enrich the seismic data with frequencies present in Acoustic Impedance logs.

Awal Mandong
Geoscientist Project Leader at CGG

Ammad, to add to Belal's comments above, if you see seismic color map RED-WHITE-BLUE you will understand why it is called spectral whitening or bluing. Whitening means making the amplitude spectrum "having more white color" or increasing the mid-section of the amplitude spectrum while suppressing the rear (red and blue) which most likely low and high frequencies noise. Bluing means making the amplitude spectrum "having more blue color" or increasing the high section of the amplitude spectrum or enhancing/boosting the high frequency content. Bluing boost high frequency to acceptable level since it is constrained by well. Sometime high frequency/thin layer make it difficult to see continuity. You can try to interpret the data on low frequency first then redo the detail on high frequency version. Or filter down the seismic data first, if the data is already in high frequency.

QUESTION 47

My question is about frequency enhancing through spectral whitening. Is it useful or not? If it is useful, how can it be used and if not, then why?

Ammadi Ali Qureshi

Geophysicist at Oil & Gas Development Company Ltd. Pakistan

ANSWERS PROVIDED BY INDUSTRY EXPERTS

Azer Mustaqeem

Petro-Explorers Inc.

Ammadi, Frequency enhancement algorithms are common and some of them are very effective. The main issue is based on the type of data being enhanced; given a good algorithm with an experienced processor who knows the limits of the data and the algorithm can be very effective. An in-experience processor with the same algorithm can create very un-real results. Anisotropic effects and low S/N ratio at higher frequencies can cause distortion in amplitude and phase in the frequency enhanced seismic.

Raju Arasu

Senior Geophysicist & Lead Triassic Paleozoic Prospect Evaluation Unit at Kuwait Oil Company

Ammadi, This issue needs to be thoroughly discussed processing geophysicists. I feel the frequency enhancement gives a false impressions which in turn mislead the interpretation to a large extent. Multiples become dominant in the deeper section where the original frequency content is much lower. We lose a lot of information, which is amplitude-based in particular, which may not be recovered in the post stack processing.

Mike Currie

Geophysicist/Partner at Deep Gulf Energy

Ammadi, As Mr. Mustaqeem states, some algorithms do a fine job. I have been using one for quite some time now with excellent results. Literally hundreds of well ties have convinced me that I am achieving an average of 50% improvement in resolution with no artifacts and with minimal noise. An example is shown on the 'Case Studies' section of my website. This is from the Gulf of Mexico. There's little doubt in my mind that this is the most cost effective enhancement that I offer my clients. It should be obvious that this is very much a "garbage in, garbage out" situation. Nothing is going to compensate for bad input data.

Raju Arasu

Senior Geophysicist & Lead Triassic Paleozoic Prospect Evaluation Unit at Kuwait Oil Company

Mike, a couple of questions: what's your depth of interest and how good is the FE data beneath the salts in GOM? What I am dealing with is land data which has its own woes compared to marine. My interest lies below 2.5s, about 15k' & deeper. A formation of 1k' thick at 2s acts as terrific high freq. filter, allowing only <20Hz beneath it. Simple and inter-bed multiples take over and I feel that FE boosts this noise and obscures genuine reflectors.

Mike Currie

Geophysicist/Partner at Deep Gulf Energy

Raju, Without being too specific of course, depth is about 18,000'. This is not a subsalt image, but I have used the process somewhat successfully on a couple of subsalt prospects. Much of my experience with this software is onshore in the USA specifically in Utah, North Dakota, and South Texas. Resolution enhancement has always been at least 50%, if not more. Clients have been delighted at the improvement. As an example, the process allowed us to image the main, thin Middle Bakken reservoir for the first time from some spec data that I would characterize as 'just fair'. However, I fully understand your comment regarding the multiples, hence my comment about data quality. Having said that, if you are experiencing enhancement of the multiples at the expense of the primaries, you might try a different algorithm. That should not happen unless the multiples are quite low frequency to start with and you are seeing 'more' of them following the processing.

Rajeshwaran Dandapani

Director Geophysics at Telesto Energy

Spectral balancing is a good method for studying the response of the seismic wave at frequencies away from the dominant frequency. It should be carefully handled so it enhances only the signal component. Going by my experience as a seismic data processor it has given good results when I restricted the enhancement up to frequencies 20 dB down from the dominant frequency level. One good way of avoiding the multiples from getting enhanced is to restrict the enhancement only to frequencies higher than dominant frequency since the multiple energy remains mostly below the dominant frequency.

Mike Currie

Geophysicist/Partner at Deep Gulf Energy

"One good way of avoiding the multiples from getting enhanced is to restrict the enhancement only to frequencies higher than dominant frequency since the multiple energy remains mostly below the dominant frequency"- An excellent point. There is no good reason to treat all frequency bands equally. I will give this a try next time I have a project with data showing multiples. I am sure it will be a bit of a 'balancing act' in the lower frequencies - as always. Since I typically use these data as input for inversion, I cannot afford to lose much of the primary low frequency trend. Thanks for the insight!

Rajeshwaran Dandapani

Director Geophysics at Telesto Energy

Mike, you are right about inversion. You may have to use a dataset where this kind of spectral bluing is not applied. But for improving the resolution and multiple suppression (spectral bluing i.e) enhancing only the frequencies above dominant frequency, works quite well.

Murthy M Vemkata

Seismic Imaging Manager at PGS

There are recently developed processing tools that can be used to split the full spectrum of a single trace into different bands, treat the problematic band with an appropriate tool to reduce unwanted noise and then recombine the cleaned up band with the rest restoring the full bandwidth again. Secondly, I am not quite sure about the assumption that the multiple energy is always below the dominant frequency. That may be the case for land datasets where multiples are mostly inter-bed peg legs. But in deep water offshore data, the bounce is fully within the water column and there is neither loss of energy nor frequency as compared to the primary signal. The first order multiples in offshore data are always of high energy and high frequency.

Raju Arasu

Senior Geophysicist & Lead Triassic Paleozoic Prospect Evaluation Unit at Kuwait Oil Company

Murthy, I have a question. The formation I referred above is capable of entrapping energy and producing peg legs between its highly reflective top and base anhydrites. Having high amplitudes compared to primaries they were picked in the semblance plots for NMO correction. These high frequency multiples are likely to have been further enhanced while running FE, dominating over the low freq-low energy deeper data. Does it make sense?

Murthy M Vemkata

Seismic Imaging Manager at PGS

1. Typically high velocity lithological units like salt, anhydrites or basalts are the dominating monolithic units where they occur. The chances of energy bouncing peg legging within them is a long shot. Typically, energy gets dissipated in these types of rocks and there is considerable amplitude and frequency loss when the energy passes through, rather than the creation of multiples. I worked with anhydrites on top of salt in Gulf of Mexico and rarely observed this kind of peg-leg multiple above the anhydrites.
2. The other scenario is where one has a very shallow water column overlying a very hard water bottom. This can be deadly since the energy trapped within shallow column appears as a continuous train of short period multiples contaminating the entire record with the frequency decreasing as you move down. If the beds do not have a pronounced dip and are parallel, one of the cycles in mid record times could be misinterpreted as a peg-leg.
3. If it is a genuine peg-leg from between the two layers of anhydrite (difficult to imagine though) there is a pretty good chance FE is going to enhance the multiples unless the problematic band of frequencies is isolated and treated separately.
4. Looking at the multiple model generated from any of the multiple attenuation programs or using a ray tracing model to validate your geology with the primary model can help identifying the source of multiples.

QUESTION 48

I am working on a seismic volume with band-pass frequency 10-15:40-45. In order to see channel like features, I thought to increase the frequency content to 60Hz but output was blurred. How can spectral bluing add the content of frequency in a volume which it was not inherent in it? Spectral decomposition helped in similar circumstances but it is also ambiguous as I am not sure how the software divides the volume into frequency slices and creates a merged picture.

Jamal Ahmad

Deputy Manager Geophysics at Essel Group Middle East

ANSWERS PROVIDED BY INDUSTRY EXPERTS

Ramses Meza

Senior Geophysicist at BHP Billiton

Jamal, Spectral bluing will shape the seismic spectra to resemble that of the reflectivity series as seen on wells, which within the seismic bandwidth happens to be blue. Given that it is a shaping filter, the bluing operator will not do any bandwidth broadening so your seismic, after bluing will have the same bandwidth but with higher frequencies boosted compared to the lows within the original bandwidth. This alone can help you since events would be sharper and new reflectors can appear, but as always you will need to verify the validity of this enhanced response by performing new well ties.

Jamal Ahmad

Deputy Manager Geophysics at Essel Group Middle East

Thanks Ramses for explaining this. And sorry, there were couple of typos in my subject line, hope these didn't bother you in interpreting, what actually I meant.

If I may rephrase your comment; spectral bluing is a shaping filter which only boosts up the higher frequencies in data but does not add new frequencies. The software uses exponent values to address this; 0.1 means boosting lower frequencies and 0.4 means boosting higher ones. So if one has to apply spectral bluing, one should strictly play with frequencies which are present in data?

There is a related term, spectral whitening. It meant to broaden bandwidth. Will that bandwidth be amplitude related or would it help in broadening frequency content?

Jamal Ahmad

Deputy Manager Geophysics at Essel Group Middle East

Ramses, just a quick question. In order to verify the enhanced response of blued seismic, is only the synthetic seismogram used, or there are any other ways to check validity of newly appeared reflections?

Ramses Meza

Senior Geophysicist at BHP Billiton

Jamal: There is no other way to check the validity of your new events other than a well tie and/or a VSP. Regarding whitening, I strongly believe that is a valid processing resource since during processing you want to compress the "wavelet" on the whole volume by deconvolution, therefore whitening is implicit. However that would mean that all frequencies within the bandwidth contribute equally to the signal and that is a wrong assumption when extrapolating whitening assumptions into reservoir characterization where we know the RC spectra must be blue-ish and resemble interference patterns at the target. In summary, my opinion which might be wrong, is that whitening must be a "*don't*" for detailed quantitative seismic work.

Raul Carbonari

Seismic Interpreter

Jamal, Do you have the seismic cube without filter gain? Ask in the processing center. In it you can apply spectral whitening deconvolution post migration. In this way you can see more easily the amplitude in the high frequencies. As always you must work inside the real frequency range.

Jamal Ahmad

Deputy Manager Geophysics at Essel Group Middle East

Raul, I agree this is more towards processing but like spectral bluing, whitening can also be applied in real time interpretation to broaden spectrum. But let me check with processors as well.

Friso Brouwer

Senior Geoscientist and Technical Manager

Jamal, neither spectral bluing, nor spectral whitening will broaden your bandwidth; both methods reshape the frequency spectrum. The methods use different assumptions: spectral whitening assumes among other things, that the earth reflectivity spectrum is white (flat). Spectral bluing assumes among other things that the earth reflectivity spectrum is colored and biased to high frequencies (hence the term bluing).

For sedimentary patterns that are self-similar, the bluing assumption is better, (see literature references below). The correct bluing coefficient is normally derived from well logs, so it is a calibrated process. And as Ramses mentions, the process can be QC-ed post-bluing, by checking new synthetic to seismic tie(s) and correlation coefficient(s).

There should be plenty of literature on spectral whitening and I can recommend the books on seismic processing by Yilmaz, though that contains much more information than just spectral whitening. For a focused article I recommend a search in the SEG library http://library.seg.org/.

For the practical application of blueing, you can view this webinar: http://www.opendtect.org/tutorials/Webinar_SSB_SCI/.

If you want to go back to the basics read the following articles that lay out the foundation for the bluing method.

Walden and Hosken, 1985, an investigation of the spectral properties of primary reflection coefficients, Geophysical Prospecting 33, 400–435.

and

Velzeboer, 1981, The theoretical seismic reflection response of sedimentary sequences, Geophysics, Vol46, No 6, P8430-853.

Jamal Ahmad

Deputy Manager Geophysics at Essel Group Middle East

Friso, Which software would you recommend to perform this exercise? I used DSD and ARKCLS for spectral bluing and colored inversion. DSD uses mathematical reshaping of the seismic wavelet through exponent value (see my initial comment for detail) while ARKCLS uses well log impedance to generate reflectivity spectrum and the operator is designed to match seismic spectrum with the well log spectrum. I found both interesting.

Friso Brouwer

Senior Geoscientist and Technical Manager

Jamal, I had a quick look at the support webpage were the DSD spectral bluing is explained.

From what I read, the beta in the DSD power law R(f)=c*fβ is the same beta as in ARKCLS linear regression - in ARKCLS this beta is derived by linear regression of log R(f), against log f. In DSD that beta is set by hand, or can be determined outside the application using a well log. Conversely, I know that in ARKCLS, in the case there are no well (logs), beta can be set by the user.

While I am a bit biased (ARKCLS is one of our partner companies), I would still probably prefer the ARKCLS application because the well calibration of beta coefficient is integrated in the application - and with wells logs being available that would be the preferred workflow for me.

Bernal Manzanilla

Geophysicist

Have you explored Phase multipliers? A colleague of mine in college used them with pretty satisfying results in terms of resolution -what a tricky concept it is. I agree with the above comments in the sense that, what you aim at doing would violate the principle of the conservation of energy and matter. The frequencies you don't record will not be there and the simple operation of adding frequencies for a spectral "bluing" to occur, will lead to wrong conclusions since you lack phase information. The mean content of amplitudes will shift to higher frequencies but remember you miss the second value -phase- on the pair of the Fourier domain properties.

I would recommend a couple of papers which I hope can be of help. Please see: http://www.onepetro.org/mslib/servlet/onepetropreview?id=SEG-2009-1092.

Stark, Tracy. Frequency Enhancement via an Integer Multiplier or Just Another GeoWizardry Attribute?

Chavez-Perez, et al. Does Frequency Enhancement Imply Vertical Resolution Improvement?

Ramses Meza

Geophysicist

I agree with Bernal's comment about the phase. Actually phase is highly underrated and that is a mistake. Spectral shaping techniques (i.e. colored inversion, bluing, etc.)

assume that seismic data is zero phase, at least within the bulk of the bandwidth and, most of the time that's not true so care needs to be taken in that regard.

Friso: I support your comments about DSD and ArkCLS.

Jamal Ahmad

Deputy Manager Geophysics at Essel Group Middle East

Bernal, I would like to add here that some of the interpreters prefer to apply spectral bluing prior to inversion. If we apply inversion on the original seismic volume (with original frequency spectra), we might miss some valuable reflections which are tuned and not visible on seismic. One the other hand, seismic inversion assumes that the seismic is zero phase, as Ramses mentioned. Although it is hard to find zero phasing, one has to apply phase shift prior to performing inversion. So I believe both spectral bluing and seismic inversion do not involve any phase changes during the processing. Inversion involves a phase shift of −90 degree before applying the operator (estimated from the impedance log). In fact this is what I am not sure about, why do we apply this shift?

Bernal Manzanilla

Geophysicist

Jamal, My other two cents are:

The assumption that added frequencies will be constant phase or aka same phase as the ones existing is not what I refer to because this is an incorrect assumption. You simply cannot create frequencies where they were not recorded. The phase conditioning I talk about is not zero phasing (it has nothing to do with it). What I talk about is that what you aim at is enriching frequency content and most of the time you would have phase wrapping included. That is the phase conditioning I talk about. The problem of adding frequencies constitutes itself an inversion process in the sense that you would have to derive new frequencies from assumptions about the reflectivity series.

The other topic would be applying the so called "spike" deconvolution which in fact tends to flatten spectrum - and not to "blue" it - but to increase bandwidth as Ramses comments above. Of course this boosts all frequencies regardless of being noise or anything else. I would go for instantaneous attributes if what you want is to get rid of the hassle of figuring out the phase.

QUESTION 49

What is the importance of blending seismic volumes on the basis of red, green and blue color? How does this method work?

Hassan Abbas

Geophysical Data Analyst at Shell

ANSWERS PROVIDED BY INDUSTRY EXPERTS

Ryan Martin

Sr. Seismic Interpreter

Hassan, In terms of viewing multiple potentially helpful seismic volumes, it allows you to widen the color spectrum in a very dynamic and often visually-impressive way, at which point you're able to make various conclusions regarding the color changes indicating changing lithology, pore-fill, etc. A great result of this, as long as you're using appropriate volumes to blend with appropriate dynamic ranges, is that while it's an RGB blend, the user can see brilliant colors like purples, oranges, yellows, and so on. This subtle yet relevant color differentiation can make a big difference when you try to separate out producing zones from non-producing zones. There are also new industry techniques to mine through a plethora of seismic volumes and use analytics to discern which are the most relevant in terms of a particular reservoir or target. The blending technique specifically deals with a 3D colorbar, where high saturation of the first channel selected gives a result of bright red, and so on for the other 2 color channels. High response from all 3 volumes yields a white response (red + green + blue), and low response will yield black. A good starting point in terms of blending would be to experiment with some standard seismic attributes, like coherency, curvature, amplitude envelope, or any number of frequency magnitude volumes. In terms of data reconnaissance, it is a very time-efficient and easy way to get some visually fantastic results to add to your overall interpretation.

Charlotte Rowe

Seismologist

Ryan, Unless you know your intended audience, you might want to stay away from red/green. Plenty of people (men, mostly) are red/green colorblind.

Ryan Martin

Sr. Seismic Interpreter

Charlotte, While that's true, hopefully people exercise discretion when interpreting volumes with such color brilliance and contrast if they have a known deuteranopia-related deficiency. We're in a pretty visually-dependent sector of the industry; and hopefully seismic data analysts know what they're getting themselves into.

Charlotte Rowe

Seismologist

It wasn't that long ago that seismic graphics were all presented in black and white. Many of the older (senior.....hierarchically up there management) types may not have considered their color blindness when they got into this field. And in fact, only in some of the industrial gee-whiz graphical woo-woo imagery is color needed. For fundamental research and analysis, it's all equations and derivations, if not quite back to the slide rule. I myself have encountered situations where the top management was offended by unnecessarily "loud" graphics and images that were fancy and attractive but did not actually convey any additional information - which moreover were reliant upon red/green and these guys were colorblind. A colleague of mine who also answered to these individuals developed an "I hate green" color-scale just to use for presenting our results. Don't dismiss it out of hand; it's a consideration.

Federico Riart

Senior Geophysicist

Hassan, this is a wonderful way to mix different attributes, but you need to have software that can do it. I recommend OpendTect, because is cheap and easy to rent, and allows mixing of the three volumes. The most common use is to blend different bandwidths obtained from frequency deconvolution algorithms. It will be amazing when you go through your data and check the stratigraphy and sedimentary bodies that appear. And you can isolate them using different techniques.

Adriana Valerio

Sr. Geoscientist

Hassan, It is true that one of the most common uses is to combine 3 different frequencies volumes (extracted using spectral decomposition) but you can also combine any other 3 different seismic attributes i.e two edge/discontinuity attribute and rms/envelope. With GeoTeric you can easily do RGB blends and additionally create color blends with another attribute to accurately identify stratigraphic features and extract them using a more robust extracting technique named Adaptive Geobodies instead of using traditional threshold technique for geobody extraction. This technology uses statistics extracting the geobody even if it has different seismic responses along it allowing a better delineation of the stratigraphic features even for your geological model and/or volumetric calculations.

Hugo Garcia

Seismic Geophysicist at Chevron

Hassan, In GeoTeric Red Green and Blue are not the only options for color blending. For blending magnitude volumes from the frequency decomposition tool the RGB color scheme is the best one. But to enhance, for example, structural features the CMY (Cyan, Magenta and Yellow) is the best one just because of the way the human eye perceives colors. If we want to blend attributes that use a rotary color map (for example phase, azimuth, etc.) better suited color scheme is the HSV (Hue, Saturation and Value). So, depending on your objective and the nature of your analysis you have the option of choosing the color scheme that is more appropriate, the one that reveals more relevant information. One other valid option is the opacity blends. Opacity Blends is a layered blend of volumes based on opacity. The opacity of one layer affects the visibility of subsequent layers. For example, if the first layer is completely opaque then none of subsequent layers will contribute to the Opacity Blend. The Opacity Blend is based on real color values so the opacity and color map of each of the layers affects the output. In GeoTeric you can blend up to 3 attribute volumes or, one color blend and up to 2 attribute volumes.

Raju Arasu

Senior Geophysicist & Lead Triassic Paleozoic Prospect Evaluation Unit at Kuwait Oil Company

Hassan, Combinations of color-blend and opacity blend are used in optical stacking to visualize very subtle features. This can be done in VoxelGeo, Geoprobe or any 3D visualization tool, but it requires a lot of patience and time to get a fairly satisfactory result.

Pavel Jilinski

Senior Geoscientist

Hassan, For RGB and CMY it may be interesting to plot "similar but not equal volumes" - discrete frequencies, band pass, structural attributes (coherence, dip etc), and others. This will highlight small differences between them helping to get a better understanding of the heterogeneity.

ATTRIBUTE FOR FAULT DETECTION IN THIN BEDDED STRUCTURE

QUESTION 50

We always talk about vertical seismic resolution, but this concerns the bedding thickness. What about the fault offsets? How can I be sure whether it is possible to observe a fault or not?

ANSWERS PROVIDED BY INDUSTRY EXPERTS

S. Reddy Ravula

Manage Seismic Data Processing and Director at Alpha Geo India Ltd

Spatial sampling (resolution) is important to see "Faults" in the seismic data. I have done "Model" studies for a given geologic model setting, with "Faults" ranging in displacement of 25 to 100 feet at a depth of 8000 feet, and generated synthetic seismic sections. Geologic "Faults" generate diffractions. Diffractions have high frequency. To see the diffractions on unmigrated seismic stack data, preserve high frequency content while processing and keep spatial sampling to 6.25 to 12.5 m (Trace to Trace sampling). You could do few simple test displays of your existing seismic data. For example:

1. Display stack data without any filter and migrate the data and then display.
2. From the above steps, Display stack and migrated data, all traces and every other trace without filter. If there are "Faults", the diffraction pattern should be noticeable on the stack data.

Ross Huntley

President, Geodesign Ltd, Western University, Calgary, Canada

This is similar to vertical resolution except that you have to account for the dip limits imposed by both the earth and your design. Some areas allow 'turning rays' therefore 90 degrees dip imaging but most are limited by high velocity layers above the zone of interest. The survey also has to have migration apertures appropriate to the dips and sampling. I do not have the equations handy but most books on design have them (Vermeer or Galbraith et al.).

QUESTION 51

Do you have any suggestion about how to detect faults or fracture systems in thin bedded horizons? Variance and Chaos didn't help (noisy results without any major continuity of detected fractures). Is there any magical parameter setting I should apply to receive better results from these attributes?

Masoud Aali

Seismic Interpreter

ANSWERS PROVIDED BY INDUSTRY EXPERTS

Jonny G. Pinto

Senior Geophysical Advisor

Masoud, The problem with running either curvature or coherence to detect lineaments or faulting is the resolution of the tool. Most of the time the resolution is too high and the output shows too many details that tend to mask what you are looking for. Of all the software I have worked with, Transform seems to do the trick in controlling the amount of resolution in the output volume. The software will let you experiment with the default values, so depending on your data set you may need to do some trial and error testing. It really makes a substantial difference.

Renjun Wen

Geologist

Masoud, In addition to the difference due to attribute algorithms, two steps are important in controlling the quality of generated attributes for fault detection. The first step is data conditioning, e.g., filtering out noise while preserving desired features. The second step is the proper selection of window size and dip steering parameters. These parameters control the scale of the faults to be highlighted. All this can be done easily inside Attribute Studio in an interactive manner (see what you get immediately, even for a very large volume).

Here is a paper about interactive attribute analysis (http://www.geomodeling.ca/interactive-seismic-attribute-analysis-for-reservoir-characterization/).

Raju Arasu

Senior Geophysicist & Lead Triassic Paleozoic Prospect Evaluation Unit at Kuwait Oil Company

Masoud, Increase the filter size 7,7,7. or, Rerun Variance/chaos, followed by ant tracking again on the output volume of the first round. Analyze the existing output to determine the azimuthal orientation of noise, if it is not random.

HOW TO DETERMINE RESERVOIR THICKNESS USING SEISMIC ATTRIBUTE

QUESTION 52

How do I detect the variation in thickness in a reservoir using seismic attributes? Is the using of high & low frequency isolation enough to do so?

Mohamed Salah Sedek

Geophysicist

ANSWERS PROVIDED BY INDUSTRY EXPERTS

Renjun Wen

Fonder, Geomodeling Technology Corp.

Sedek, Spectral decomposition attributes are very useful mapping the thickness variation, because zones with different thickness distributions will display different spectral signature.

Mohamed Salah Sedek

Geophysicist

Renjun, The most important parameters are: amplitude and frequency. Spectral decomposition gives the values of these parameters but the question is: How to distinguish between the high and low thickness using this technique?

RM Iman Argakoesoemah

Senior Petroleum Exploration Geologist

I have three basic questions:

(1) Based on the methods discussed previously, which one is the most "precise". That is which method comes closest to the actual thickness defined from logs to calibrate and why?

(2) Is there any guidance or plot of statistics that we could follow to calibrate the "expected" thickness derived from seismic attribute? Say the thicknesses varies from 20' (tuning?) to 100' pay zone reservoir; how much should we add "plus and minus" to each thickness given the above limit?

(3) Any published papers about the two points above?

Rajeshwaran Dandapani

Director Geophysics at Telesto Energy

RM Iman, Seismic wave amplitude response to reservoir thickness is not simply solved by just cross-plotting amplitude versus thickness. Seismic amplitude is the combined response of reservoir thickness, porosity net-to-gross, type of fluid mixture and fluid saturation. Hence whenever we try to establish a relationship between the amplitude and reservoir thickness we need to make sure that all the other parameters involved are not changing much. The best thing would be to correlate the net pay, which is the combination of all the reservoir parameters mentioned above, with amplitude which will give better prediction rather than just varying the thickness alone.

Mohamed Shihata

Geoscience Business Development Specialist

Understanding Thin Beds Using 3D Seismic Analysis Workflows.

Henning, Alison* http://www.ffa-geosciences.com/download/GCSSEPM_2011_ThinBeds.pdf.

Ivan Marroquin, Ph.D

Senior Technical Geophysicist Advisor

I found this thesis on the use in spectral decomposition for mapping stratigraphy and hydrocarbons. The author also discusses some frequency based attributes. The link to the thesis is http://www.agl.uh.edu/pdf/theses/jianlei.pdf.

SEISMIC ATTRIBUTE AS DHI

A seismic attribute that can indicate the presence or absence of hydrocarbon is known as a DHI.

QUESTION 53

What are the methods or processing technology to enhance flat spots? (Fig. 2.21).

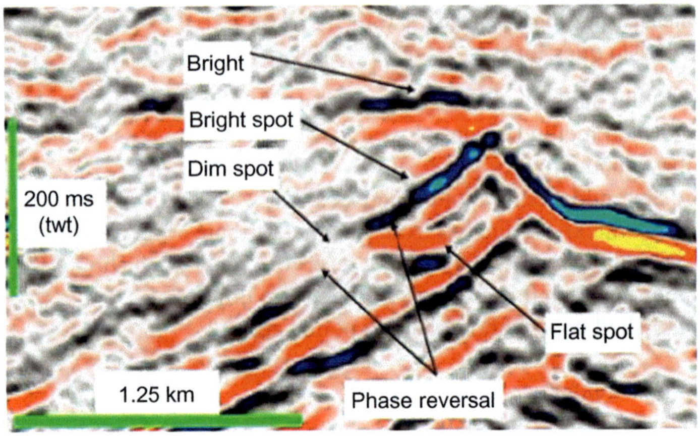

FIG. 2.21

Flat spot on the seismic section.

Source: Løseth, H., Gading, M., Wensaas, L., 2009. Mar. Pet. Geol. 26 (7), 1304–1319.

ANSWERS PROVIDED BY INDUSTRY EXPERTS

Fazrin Oktafian

Geophysicist at PT Elnusa, Tbk

If you just want to see the flat spot effect clearly, you have to increase the gain of amplitude before stack (gather), then stack the data.

Jeffrey Ross

Senior Geophysicist at Whiting Petroleum

The very best way to enhance 'flat spots' is to become a more observant interpreter. The amplitude distribution of a true flat spot and its relation to the host rock is a diagnostic feature that may not only tell you important things about your reservoir and its internal architecture but whether the flat spot is related to hydrocarbons at all.

When you blindly 'enhance' particular signals without properly understanding them you are just asking to drill dry holes on false anomalies (read multiples or any one of a number of flat coherent noise effects). Good interpretations take time and lots of observation of subtle signal variations. There are no shortcuts as shortcuts increase risk.

Lee Lawyer

Sr Geoscientist

Jeffery is correct. Clearly the impedance contrast at the oil/water or gas/water contact is of paramount importance and whether the oil-water contact is transitional. In other words the geological model is critical to the interpretation. A classic example is a structure near the large oil field off Nova Scotia or Newfoundland (name gone from memory). A beautiful flat spot was drilled on a closed structure with a resultant dry hole. Best explanation was a paleo oil-water contact with leaky reservoir. Other flat spots to be aware of are located at the Opal A to Opat CT boundary.

Steve Adcock

Geophysical Consultant at dGB Earth Sciences

You can enhance "flat spots" any number of ways with processing, but take a close look at the contributing gathers before move-out and remember: flat spots don't have to be flat except to other adjacent stratigraphic horizons, peg-leg multiples occur and can appear "flat", and as Lee and Jeffrey pointed out, interpretation in a geologic vacuum leads to dry holes.

D.N. Avasthi

President, SPS Consultants

A very good flat spot enhancement tool can be found in the Plug-in SFE (Seismic Feature enhancement) in OpendTect 4.4 software of dGB Earth Sciences of U.K. The latest version of OpendTect 4.4 can be downloaded at http://www.opendtect.org.

QUESTION 54

I am working on spectral decomposition and I discovered from my spectral analysis that the dominant frequency of the brine zone is lower than the dominant frequency of the either oil or gas zone. What might be possible reasons for this phenomenon?

Ayo Adebayo

Geoscientist at Essar

ANSWERS PROVIDED BY INDUSTRY EXPERTS

Jeffrey Ross

Senior Geophysicist at Whiting Petroleum

Ayo, I would be very wary of ascribing the effect you describe directly to the pore fluid content as that makes very little physical sense. I would suggest that a far more likely cause would be related to the distribution of reservoir properties that control your spectral response and also have a strong control on water saturation. I would suggest you talk to your geologist and find out if there is a strong relationship between the relative permeabilities to hydrocarbons and water and how that relates either to depositional facies or diagenetic history differences between hydrocarbon and water legs of your reservoir.

Jerry Coggins

Principal Geophysicist at Coggins Geosciences, Inc

Ayo, One of the problems with spectral decomposition is that it is difficult to translate the results into quantitative information. It is a tool for trying to detect fine scale features in your data by taking advantage of differential tuning behavior. You may try doing conventional spectral analysis, and make sure to check as above for the same effects.

Rajeshwaran Dandapani

Director Geophysics at Telesto Energy

Ayo, This can happen if the thickness of the hydrocarbon column from the top of reservoir to the contact is less than the thickness of the water column from the contact to the base of the reservoir. Moreover if the hydrocarbon column is sitting just on top of your water column as in the case of fourway dip closures, the underlying water column shows a lower dominant frequency because of attenuation of higher frequencies by the overlying hydrocarbon column.

Jeffrey Ross

Senior Geophysicist at Whiting Petroleum

Dandapani, I don't understand your response. Attenuation of high frequencies in thick gas columns has been often documented, but that would result in the spectral frequency response of the water being higher, not lower as described by the original post.

Rajeshwaran Dandapani

Director Geophysics at Telesto Energy

Jeffry, the seismic wave after passing through the hydrocarbon column loses high frequencies and the dominant frequency drops. There are many instances where these low frequency shadows are picked up by spectral decomposition around 40 to 50 ms below the reservoir zone and not at the reservoir level.

Jeffrey Ross

Senior Geophysicist at Whiting Petroleum

Dandapani, The gas frequency attenuation effect would only apply here if (1) the effect was also detected within the gas column itself and (2) the gas column was thick enough to generate the effect. I am not aware of any verified cases of this effect being observed from oil. It is also noteworthy that the original post said "in a particular reservoir" not below it. This would suggest that the author is seeing his frequency response to the sides of the HC accumulation as well, which would not be affected by gas shadowing. If, however, the reservoir is much thicker that the gas accumulation then you would be correct only directly below the gas. This point is unclear in the original question.

Rajeshwaran Dandapani

Director Geophysics at Telesto Energy

Jeffrey, This is due to the overlying gas column which is absorbing the high frequencies. You can verify this if there are any brine sands without the overlying gas column. They should show higher dominant frequency if they are at more or less the same depth and stratigraphic level.

Mohammed Farfour

Reservoir Geophysicist

Ayo, I agree with Mr. Dandapani. Indeed, you have to know that there are some key major factors that control frequency responses of formations: lithology (sand, shale, etc), thickness (thin, thick) and fluid content (oil, gas, water). Thus, to come up with a reliable interpretation, one should examine and consider all factors. In case thickness and lithology do not change, one can relate frequency response to fluid type and saturation. The frequency behavior you have observed is commonly seen when reservoirs are thin. The frequency response would be a result of both tuning and hydrocarbon presence. In this situation, hydrocarbons will make reservoir preferentially respond to higher frequencies and show anomalous spectral responses at these frequencies. I suggest you to read the following papers which will give you more details:

Burnett, M. D., Castagna, J. P., Méndez-Hernández, E., Rodríguez, G. Z., García, L. F., Martínez-Vázquez, J. T., Avilés, M. T., and Raúl Vila-Villaseñor, R., 2003, Application of spectral decomposition to gas basins in Mexico: The Leading Edge 22: 1130–1141.

Eugene Lichman, Gennady Goloshubin. Unified Approach to Gas and Fluid Detection on Instantaneous Seismic Wavelets, SEG 2003.

Michael Doty

Principal at Doty Geoscience

Ayo, If you have sufficient well control you can attempt to establish whether or not there is a statistically meaningful relationship between the frequency effects you are seeing and the pore fluids. As some have commented, the effects may be due to other geological factors. Live oils with high dissolved gas content can create attenuation affects similar to gas.

Ramy Fahmy

Geophysicist at RASHPETCO-BG Group

Thank you all. There is a different meaning between Spectral Decomposition (SD) and the dominant frequency. As we know the amplitude for hydrocarbon filled rock is offen brighter than water relative to surrounding. So if you apply spectral decomposition (SD), you will find the hydrocarbon filled zone is brighter than water filled zone and you can use SD to detect water leg for the reservoir more accurately than using absolute average magnitude. This is also a better a thickness indicator. If you need to use SD as frequency dominant for Hydrocarbon indicator, you need to present it as cross section not as map because you will see only three color blending regardless of the bright or mix of the color and so it is not widely used.

On the other side, considering the Dominant Frequency, you can use Instantaneous Frequency attribute and you should find the frequency lower for hydrocarbons as compared to water. But it may be not so simply identify the hydrocarbon zone, so need to magnify the attribute value so you can use sweetness attribute which it equal to instantaneous amplitude divided by square root of instantaneous frequency. At the hydrocarbon zone there is large amplitude effect due to impedance difference for hydrocarbon-sand and at the same time hydrocarbon causes frequency attention. So, one gets a large amplitude value for sand and larger value for hydrocarbon-sand. Take care if using sweetness. Really, applying Spectral Decomposition in conjunction with Sweetness is very interesting and tested for regional screening in turbidity slop-channel system.

QUESTION 55

Is that always the case that low frequency tells us the presence of hydrocarbon and the high frequency indicate high content of clay in the rock? Did anyone come across something different?

Joseph Gnapragasan

Seismic Interpreter

ANSWERS PROVIDED BY INDUSTRY EXPERTS

Manuel Zepeda

Geophysicist

Joseph, It depends on what kind of seismic are you working on: gathers, partials, PSTM, filtered, gain, etc. In my experience, first use the partial stack to get the full stack. Low frequencies have a strong potential for prediction but you need to calibrate and tie with your wells. Also, for more detail check spectral decomposition (PD, CWT, EPD, etc) but consider the geological factor as well. There are a lot of things you need to consider, seismic quality, shadow frequency, processing, resolution, and so on. More specialized rock physics and cross-plotting methods can help to relate lithology and separate fluids with logs and seismic attributes. Elastic properties are good tool to do this. For gas response my experience I prefer specdecomp and SI. Sometimes basic attributes based on the full stack data camouflage the real response.

Jian Zhou

Geophysicist

Joseph, Besides data quality reasons, low frequency could also be the result of a reflection coefficient ramp and high frequency signal may be caused by thin layers. Hydrocarbon effect could increase or decrease frequency depending on layer thickness, see: Chen et al. (2008, Geophysics, doi: 10.1190/1.2978337). An example of high frequency in a carbonate reservoir is shown by: Li et al. (2011, Geophysics, doi: 10.1190/1.3554383). Modeling with a tied well may be a better way to solve the specific problem.

Jonny G. Pinto

Senior Geophysical Advisor

Joseph, Good answers. The basic question you should ask yourself is: if I see it on the logs (cross-plots), then there is a good chance I should see it in the seismic (assuming average OK seismic). So as suggested by most experts; do your cross-plot first.

Mohamed Shihata

Geoscience Business Development Specialist

Joseph, There are numerous laboratory and field examples where low-frequency components of reflected seismic waves show surprising imaging capabilities. Ironically, such components are often filtered out as useless in conventional data processing.

Spectral decomposition is the use of small or short windows for transforming and displaying frequency/wavenumber spectra (Sheriff, 2005 Encyclopedic Dictionary of Applied Geophysics). In other words, spectral decomposition is the conversion of seismic data into discrete frequencies/wavenumbers: or frequency/wavenumber bands. Spectral decomposition is primarily used for layer thickness determinations and identifying stratigraphic variations. It has also been helpful in identifying DHI (Direct Hydrocarbon Indicator) characteristics and determining fluid types and lithologies.

If you can correlate frequency with any well information that can help to identify the meaning of frequency changes along your reservoir, then you can use Instantaneous frequency or frequency decomposition. If you decide to use Instantaneous frequency you must calibrate it with other geological information because it will change from one area to another and is affected by a lot of geological factors.

Some software enables you to cross-plot the relation between logs and seismic attributes so try to generate cross-plots between frequency and well logs that indicate shale and hydrocarbon to find good meaning for the changes in frequency.

You can use Q-attenuation attributes that mean you study change of frequency with time it may help you to follow change of frequency, Where the frequency loss due to presence of gas drops below the elevated background trend, a high attribute value results supporting attenuation because of gas presence.

http://www.lbl.gov/tt/publications/1663pub7.pdfhttp://www.iosrjournals.org/iosr-jagg/papers/vol1-issue1/B0110510.pdfhttp://www.google.com.tr/url?sa=t&rct=j&q=&esrc=s&source=web&cd=1&cad=rja&uact=8&ved=0CDQQFjAA&url=http%3A%2F%2Fwww.bgp.com.cn%2Fdownload.aspx%3Fid%3D372&ei=LmJyU4vAObKOyQOmwYD4CQ&usg=AFQjCNG1u8D9tuOhtShe9Fu4QHWx6oHJ4A&bvm=bv.66330100,d.bGQ%5C.

SEISMIC ATTRIBUTE AS DHI: CARBONATE RESERVOIR

QUESTION 56

I am performing research on the detection of porous carbonates in 2D data. The porous zone is 1000-2000 m/s lower velocity than the unaltered carbonates above and below (we assume that densities are reduced from 2.6 to 2.3 as well). This porous zone, which is 300 ft. thick, appears as a dim spot with reduced amplitudes and chaotic reflectors that is encased by high amplitudes (bright spots?) at the contact between unaltered rock above and below the porous zone. We are not concerned with hydrocarbons in this prospect, only porosity (hydrocarbons are not present and are not contributing to the anomaly). I have read many of Alistair Brown's papers concerning bright/dim spots but am still a bit confused as to having a combination of dim and bright spots. As I see it, the acoustic impedance contrasts above and below the porous zone are producing the bright spots but the actual rock properties within the porous zone are homogeneous and producing the dim spot. This is not usual when I look at other case studies. Is this because this porous zone is thicker than normal studies looking at bright spots? Am I to assume that the porous zone is imaged as a homogeneous block of porous carbonates since the amplitudes are low within this zone? What are the usual acoustic impedance contrasts between a gas sand and shale? Should I call this a bright spot or a dim spot? Energy based attributes highlight the zone quite well.

Kyle Gray

Geophysicist at Noble Energy

ANSWERS PROVIDED BY INDUSTRY EXPERTS

Mohamed Ouabel

Senior Geoscientist at Petronas

Kyle, Usually the bright spot, dim spot terminology is linked to a seismic amplitude character or signature on the top/base of a porous reservoir filled with hydrocarbon (gas) on a stacked section. This seismic behavior is due in general to the variation of the Vp/Vs ratio, and is linked to the AVO class that your reservoir is presenting. Longitudinal velocity is so sensitive to the presence of gas, rather than Vs, and density also is sensitive to the presence of gas (especially with high porosity).

I mean by that, Class 3 presenting a bright spot on the seismic stack section, class 1 and class 2p present dim spots, but you have to go back to the seismic gather or fluid substitution forward modeling and check which AVO class you are dealing with. Now, let's go back to your model, you are dealing with a porous reservoir with lower impedance compared to the encasing parts (top layer/base layer) with no hydrocarbon. The only parameter which is controlling the variation of your impedance is the porosity (homogeneous model of course), so it's not dim or bright spot, you amplitude variation is due to the porosity variation (impedance variation). So you can, for example, apply a 90° phase shift to your seismic and your amplitude will exhibit the impedance rather than the interface, or you just apply a relative acoustic impedance and you can figure out the extension of the porous zone.

Kyle Gray

Geophysicist at Noble Energy

Ouabel, I have done some AVO modeling, which indicates a far offset anomaly (increasing amplitudes) at the base of the porous zone (not within it), which indicates that there are not unique Vp/Vs ratios associated with the porous zone itself (probably since we do not have hydrocarbons/gas). I have also done elastic inversion with some promising results. As I see it, there are two implications for exploration: 1) far offset anomalies in CIG may indicate unaltered/altered carbonate contact, and 2) inversion can locate zones of low AI (and thus possible increased porosity). I have looked at the Hilbert Transform of the data but am a bit confused as to how this would indicate impedance rather than interface properties? The only way to truly look at impedance (and thus interlayer rather than interface properties) is through inversion. I'd look at the amplitude anomaly as a false positive Class 3 (false positive since it acts like a class three bright spot but there are no hydrocarbons present). I think it would all be easier to explain if the actual porous zone itself produced an offset anomaly rather than the contact. This would indicate interesting changes in Vp/Vs or Poisson's ratio within the porous zone.

Mohamed Ouabel

Senior Geoscientist at Petronas

Kyle, For sure, an AVO anomaly will be shown on the interface, because seismic reflection is matter of reflection coefficient, which is related to interface rather than the layer itself. Did you run the AVO modeling with and without HC, or fluid substitution? Did you do some partial stacks from your seismic data (near, mid and far stack)? Do you have some gathers (or computed super gather) from and off the target zone to check the behavior of the amplitude variation with offset? Concerning the Hilbert transform it will lead you to get the imaginary and the real part of the trace, that will help you to compute some interesting attributes such as reflection strength (good attributes for lithology/AVO class 3), instantaneous frequency, instantaneous phase and etc. If you could present your model with all the parameters as a sketch, then may I can help you more

Kyle Gray

Geophysicist at Noble Energy

Mohamed, I have done some partial stacks but it turns out that the stacking done by the processors best highlights the dimming within the porous zone which occurs at near to mid stacks. I do not have permission to show you the data and I understand that it is hard to comment without seeing it. I work in the mining industry so we do not care about hydrocarbon content and I have thus not done any modeling with different fluid types. This particular problem I am discussing with you is research that I may want to publish as part of my graduate studies (if I can get permission to publish the data). I have run many attributes on this dataset and am familiar with instantaneous attribute theory; instantaneous amplitude and phase are quite interesting. I have a little trouble interpreting the phase data. There are anomalous and highly varying phase values within the porous zone, which I interpret as being due to phase shifting relating to attenuation in the area. This in turn is related to lateral lithologic contrasts. I am not sure if the phase has anything to do with tuning effects in this area; I would not expect that to be the case.

Mohamed Ouabel

Senior Geoscientist at Petronas

Kyle, I see you are interpreting the instantaneous attributes. Did you notice any instantaneous frequency anomaly? If so, that could be related to fractures (as I see you are dealing with altered carbonate zone) because a fracture zone can appear as a low instantaneous frequency which can also help you as indicator of lateral continuity. But as for the specific subject, I believe you are dealing with slope debris flow. Are you dealing with some reworked debris known as calcilutite? If so the elastic impedance as you noticed previously, would help very much. Last question: is your target sited far enough to be away from any biogenic gas accumulation? Otherwise, you may be dealing with some residual biogenic gas that could be in the water. Your case is really interesting, and I would like to know much more but I understand your limit according to your obligation, but keep on.

Kyle Gray

Geophysicist at Noble Energy

Mohamed, Yes, I do see a bit of a frequency low in the area of interest but the response is quite variable for this particular attribute (possibly b/c there are many discontinuous reflectors, some of which are related to noise). I prefer to use an average frequency attribute, which is unrelated to Hilbert/instantaneous attributes (it is based on the mean of the frequency spectrum within the specified time-gate rather than the derivative of instantaneous phase…I think). I then perform a median filter of 5 traces in the lateral direction to get a nice smoothed version of the average frequency. Do you have any experience comparing instantaneous frequency with this average frequency that I just explained? I do not believe anyone has identified calcilutite in this area (this is a new term for me). How would this relate to elastic impedance changes? However, after some review of the area, there are occurrences of carbonaceous material in the district of up to 10 wt%, which is best described as pyrobitumen. Unaltered rocks generally contain <1.5 wt% organic C whereas altered rock can contain up to 10. I do not believe this has been identified in drilling for the particular study area we are speaking of though. This is a very interesting hypothesis though!

Julien Moreau

Basin Analyst

Kyle, I see that you are facing a tricky and interesting phenomenon. I would be very curious to see the image. There are a couple of points I don't understand. For example, the dimming with AVO response over what you think is a 300 ft thick layer. Are we speaking of low frequencies if we are in a single phase? So you have a filter somewhere in your succession. If you don't have gas in your layer, how do you model this kind of AVO except by putting in a very elastic layer (your frame). That would sound very strange to me that weathering creates this kind of behavior.

All that leads to my final question. Which kind of carbonate are you dealing with? Good hard indurated platform style or more soft pelagic ooze more or less lithified? Has it been drilled? And are you sure there are no volcanics or other strange things in the succession?

Kyle Gray

Geophysicist at Noble Energy

Julien, The carbonates in question are postulated to be slope-derived debris-flows with silty limestone and micrite clasts (although structural and hydrothermal breccias may play a role as well). Unaltered limestone below the porous zone is massive micrite to silty limestone. There are a few volcanics in the overburden above but the imaging is decent given the circumstances (working with a PSDM & true relative amplitude preservation). There is drilling in the area (I used VSP data for modeling/inversion). The porous zone is saturated (i.e., below the water table), which could be leading to the unique interface properties we see at the base of the porous zone. Obviously gas would produce a much larger effect but water residing in the 15%+ porosity of the porous zone compared to <5% porosity below could produce some interesting effects on the elastic properties, correct? There is some sort of lateral phase shift going on within the altered/porous zone and a possible attenuation of high frequencies (i.e., porous zone acting as a unique filter). Could you elaborate on your comment on low frequencies and single phase?

Julien Moreau

Basin Analyst

Kyle, I meant at first is that if you have interlayering of volcanics or hydrothermal precipitations you can have bright spots. These can be very elusive and extremely hard to pin down. Most of the intrabasinal volcanism is unknown and awaits precise identification. It seems that you have pretty mature carbonaceous rocks too. Coal can generate anomalous amplitudes and is associated with gas. The point is that you might have HC wandering in the area. In the form of fizz like suggested by Mohamed or others. That remains the most common explanation for AVO effects. If I have it right an AVO anomaly is still a result of an interplay between your frame and your fluid. If your frame is 'mushy' like some unconsolidated sediment or hard and fractured, you basically switch between different classes of AVO. I will have to check my documents tomorrow to look at the different configurations for 3. I might be incorrect but I thought that 3 is for the low hardness (poor framework rock). Finally, if you are sure that even with the phase rotation you are observing a single event and are in the same phase, you can go back to the basic AVO technique. You can pick your base and top and check out the gradient and the intercept in order to estimate S-wave reflectivity and the Poisson ratio. That would tell you much more about the material than trying to fit it into a class designed mainly for standard HC exploration. For your problems with instantaneous frequency, I would not do an average of it across traces since it would be a bit meaningless. It is the real part of a complex number. We ignore that complex part but it is still there in the computation. You can however stack your traces or do a dip median filter before looking at your instantaneous attributes; it de-noises quite well. You are also right, instantaneous attributes are not quantitative attributes. They try to represent an ideal trajectory so they are hard to compare with a standard FFT for example. I think picking your top-bottom event is better if you are confident of its geometry.

Steve Adcock

Senior Geophysicist

I am not sure I see the complication. Bright and dim are just relative terms to indicate absolute amplitude. The harder (less porous) rocks on top and bottom give you high amplitude (bright) events and the low impedance interval contained between them give you lower amplitude events. To see this interval better in reflection space, gain it up. I assume that the polarity of you upper amplitude event is the reverse of your polarity for the base event and you say tuning is not an issue. In this case, I would use instantaneous attributes for examining the top and bottom events. I would be looking for information on the depositional surface upon which this porous unit is deposited and the upper surface which provides the porosity cap. Make sure to examine at least half a dozen time interval slices above and below the interfaces, preferably parallel to the reflection events and before inversion. This is much easier using 3d visualization with opacity and color controls than using slices. I believe one reason Julien leaps ahead to the AVO issue is because the variation of amplitude with incident angle provides us valuable information about the rock physics of the surfaces. Attributes calculated from near vs far incident angle data volumes will give you key information with minimal effort (just take the difference between the near vs far horizon amplitude maps). You're correct in wanting to move to inversion to examine the interval, and for the correct reasons. Do the inversion. Gain the interval amplitudes if necessary, then start doing the analysis. You can use AVO and azimuthal techniques on the porous interval to examine anisotropy effects caused by rock fabric, fractures, fluid/porosity variation, and ductile/brittle lithotypes. I think Julien's description of the AVO effect was clear and essentially correct. However, in my experience it's more useful to think about these issues in terms of actual reflectivity and impedance contrasts due to rock physics, rather than in terms of AVO class. In addition, I always feel that iterating a geological forward model with the seismic to seek the highest correlation is the best way to test multiple scenarios and should address a number of the questions you've raised.

Kyle Gray

Geophysicist at Noble Energy

Thanks for your comments Steve and Julien; I now have a lot to consider when progressing with my research. I have one more question pertaining to the instantaneous phase attribute. As I understand, most people use this attribute for observing discontinuities among reflectors. However, I would rather relate instantaneous phase to Fourier phase in order to better understand possible phase shifts produced by the porous zone, which may act as a unique filter (i.e., deviation from zero phase). I have read that you can estimate the phase in a given area by looking at the phase angle value that occurs at the maximum of the envelope. Accordingly, I have extracted a single trace of instantaneous phase and instantaneous amplitude (i.e., envelope) from my PSDM and plotted them in excel to examine the phase value at max envelope. Does this technique work with real-world data or are there too many assumptions?

Mohamed Shihata

Geoscience Business Development Specialist

Kyle, It is difficult to depend on phase to identify porosity because there are many factors effecting in phases. Real seismic data are not ideal for studying phases in quantitative factors but, you can calibration phase values with real log data with logical calibration or neural network supervisors. That can help identify the real change in your seismic wave and can help to relate porosity or hydrocarbon saturation or any other quantitative characters. You can also use statistics to calibrate your attributes (physical) with porosity. My advice is, if you want to use seismic data for quantitative interpretation, you must try to find correlations between real log data and geological information with your seismic attributes.

http://pangea.stanford.edu/~quany/QSI_Chapter-4.pdf.

http://mmc2.geofisica.unam.mx/cursos/geoest/Articulos/Reservoir%20Caracterization/Quantitative%20Use%20of%20Seismic%20Attributes%20for%20Reservoir%20Characterization.pdf.

ATTRIBUTE MAP

QUESTION 57

I have generated an Extracted Valued attributes map. But the result is completely unexpected, namely the horizon is peak but the result map is trough like, and vice versa. (i.e. if the peak is black and the trough is red, once I generate extracted map for peak "black" the outcome map is "red" which refers to trough) is this normal? What is the usage of extracted valued attributes map?

Mohammed Gumati

Lecturer at Benghazi University

ANSWERS PROVIDED BY INDUSTRY EXPERTS

Mohamed Shihata

Geoscience Business Development Specialist

Mohammed, Point measurements should not be made off of seismic peaks or troughs: In areas of thin beds, the use of point-based attributes (such as amplitude) can be extremely misleading without careful calibration. For example, some data sets have been interpreted in terms of seismic amplitude measured at some time above or below a continuous reflector. The amplitude so measured may result from the vagaries of wavelet interference, or even just a thickening of an intervening bed, and can lead to measurements made at different points on the 'flank' of the reflected wavelet. The resulting attribute map may resemble a geologic feature, but that resemblance may be fortuitous, or caused by a geologic feature other than the one we think we are mapping.

Jerry Coggins

Principal Geophysicist at Coggins Geosciences, Inc

Mohammed, Amplitude measurements on RFC data should always be made on a peak or trough.

COLORED SEISMIC INVERSION

What is colored seismic inversion?

One of the seismic inversion algorithms is called "colored" inversion. It is performed in the frequency domain and the point is to build an operator that converts a seismic trace into a reflectivity trace directly. The question is, what does "colored" mean and why was the technique named in this way?

Colored inversion is designed to approximately match the average spectrum of inverted seismic data with the average spectrum observed impedance (Lancaster and Whitcombe, 2000; DOI:10.1190/1.1815711). The earth's reflectivity can be considered fractal, and the resulting amplitude spectrum favors high frequencies (spectral blueing). If there were no preferred frequency, then you would have a "white spectrum," but as there are some frequencies with more energy, then it is called "colored."

QUESTION 58

Seismic Colored Inversion (SCI) creates volume that is rotated 90 degree with respect to reflection data. Why?

Jamal Ahmad

Deputy Manager Geophysics at Essel Group Middle East

ANSWERS PROVIDED BY INDUSTRY EXPERTS

Asheesh Singh

Because I believe its primary purpose is generate a relative impedance data type. Its then used to identify contrasts in relative acoustic impedance not apparent in the original seismic data i.e. to isolate low and high impedance values & anomalies become immediately apparent.

It is a proven technique that is fast and easy to apply to seismic data, yet produces results that match more sophisticated and compute intensive techniques such as sparse spike inversion. It also helps to match seismic to well log data within the seismic spectrum to reduce noise and signal instability.

Jeffrey Ross

Senior Geophysicist at Whiting Petroleum

All seismic impedance inversions are created for one reason, to convert seismic data displaying changes at impedance INTERFACES to displaying the impedances of the rock volumes above and below the interface. The ability to display bed impedances accurately is by necessity limited to the bandwidth of the seismic data being inverted. There are two ways to do Coloured Inversion, the first is to only use the frequencies contained in the seismic itself, which is referred to as RELATIVE impedance (as per response above). If, however, the interpreter wants to put this relative data into actual numbers that relate to impedances from logs, then a low-frequency background impedance model must be put into the inversion (filling in the missing frequencies from 2-3Hz up to usually around 7-10Hz. This data is interpolated from well control or a well thought out geologic model. Data with background models are considered FULL Acoustic Impedance inversions.

It is exceedingly important to make sure that the data you put in to Coloured Inversion software is verified as being accurately zero phase. Coloured Inversion software is pretty phase dumb as its purpose is to shape the frequency content of the data. It will apply a 90 degree phase shift to any data you put in it and not care. If the data is not originally zero phase, then anything you create using Coloured Inversion will put false energy distributions into the overlying and underlying sediments and the inversion will give bad results that won't tie your wells.

CHAPTER

SUBSURFACE SEISMIC VELOCITY MEASUREMENT 3

CHAPTER OUTLINE

CHECK-SHOT DATA

Surface-recorded seismic data is the largest data set used in exploitation and development of oil and gas reservoirs. However, surface seismic data have one limitation to their usefulness—thc reflection events used to map the seismic sequences and the seismic facies that describe reservoirs and sealing units are measured as functions of seismic travel time, not as functions of depth. To understand reservoir performance, the boundaries of these units need to be mapped in depth (kilometers). Thus, the concept of the velocity check-shot survey has been developed to establish time-depth calibration functions at control wells so that surface-recorded seismic images can be reliably converted to the depth images that are needed to do reservoir volumetric calculations. Check shots also allow for correlations between seismic signatures and rock units and rock type to be established.

Check-shot data are another source of velocity information. It is a type of borehole seismic data designed to measure the seismic travel time from surface to a known depth. In a check-shot survey, the geophone or geophone group is dropped inside the borehole to each formation of interest and fixed at the formation tops. These then receive the direct or downgoing seismic waves generated from the source (dynamite, vibrator, or air gun) located at the surface near the borehole. The geophones receive the direct seismic wave and the travel time of the first arrival or first

Practical Solutions to Integrated Oil and Gas Reservoir Analysis. http://dx.doi.org/10.1016/B978-0-12-805464-2.00003-2

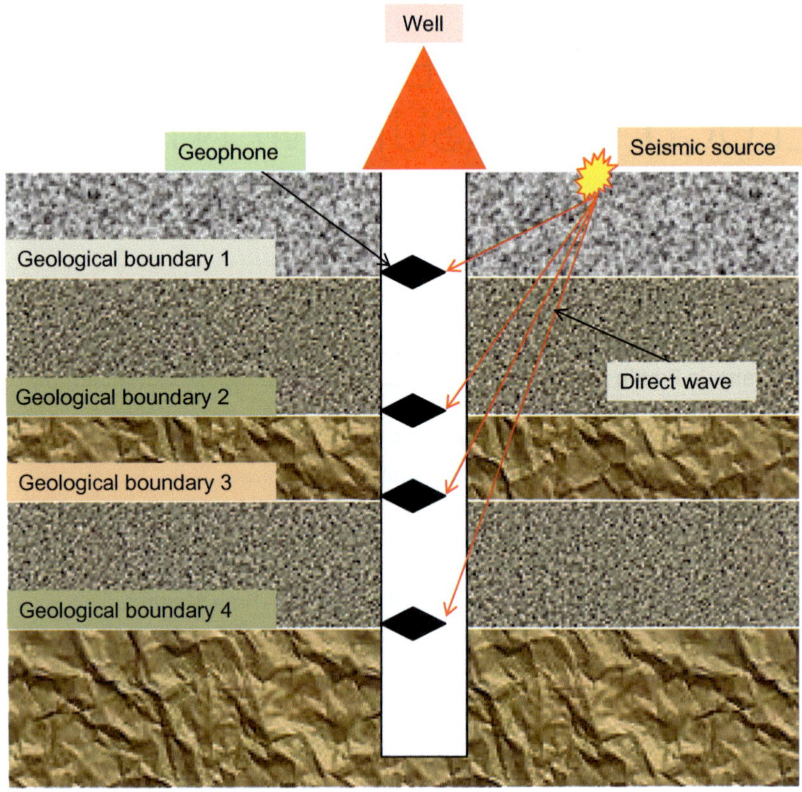

FIG. 3.1

Conceptualized check-shot survey technique.

break is recorded. The geophones are then raised to the next depth level upwards and the source is triggered and the first breaks recorded. This is done until the complete length of the borehole is sampled and a series of arrival times is recorded (Fig. 3.1).

We can calculate the velocities of the medium (check-shot velocity) if we know the distance from the well head to the source, the depth of the geophone and the travel time of the seismic wave using the equation (*Vcht*):

$$V_{cht} = \sqrt{\frac{X^2 + Z^2}{t_{cht}^2}}$$

Remember that unless the source is located directly over the borehole, the check shot measures a diagonal travel path and not a vertical one (*Z*). The check shot first arrival time is actually a function of the angle between the source and the geophone. We must correct for the angle in order to obtain a true average velocity.

The time-depth curve (as shown in Fig. 3.2) is created by using the first arrival time with the corresponding depth (after correcting the slant time to vertical). The curve is used to define the geological formation tops on the seismic line (section) at the well location for seismic mapping. In other words, the check-point survey (CPS) is used not only to correlate the real seismic section with the CPS data and to identify the zone of interest in the seismic section, but also this time-depth data can be used to create a synthetic seismogram by using sonic and density logs of the well.

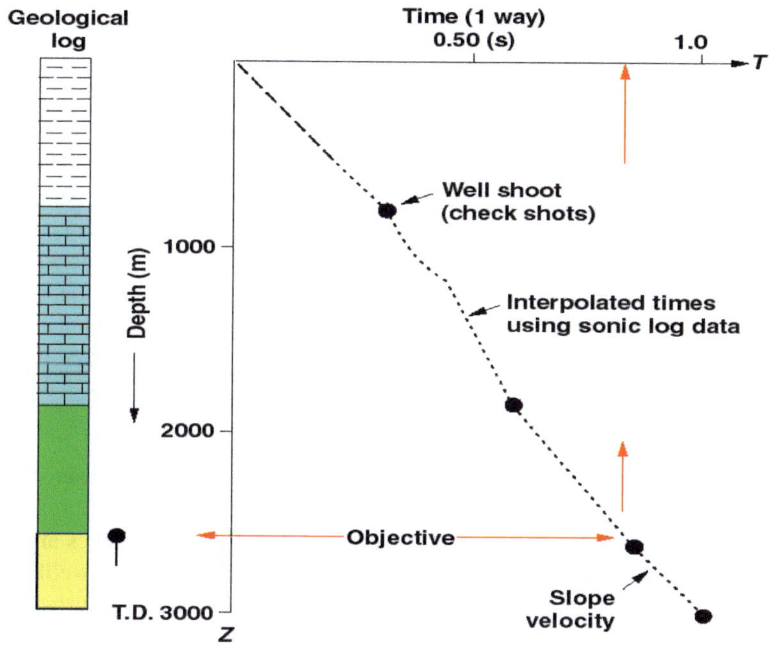

FIG. 3.2

Conceptualized time-depth curve.

QUESTION 38

If check-shot data is missing, how best can you do your well-to-seismic tie?

ANSWERS PROVIDED BY INDUSTRY EXPERTS

Mike Cline

Consulting Geophysicist and President of T/X Resources

Depending on the types, and/or amounts of data that are available, you can:

- Generate synthetic seismograms in nearby wells (if sonic logs aren't available, generate pseudo-sonic logs. This applies for density logs, as well).
- Generate regional velocity maps from other "nearby" velocity surveys (grid/contour all data, being careful to adjust them all to the same datum plane, and correction velocity, if applicable).

- Generate Time-Depth/Velocity functions from other "nearby" velocity surveys (e.g., Slot-nick or Faust methods -- see http://www.akamaiuniversity.us/PJST11_1_563.pdf).
- Generate Time-Depth functions from Dix-corrected seismic processing velocities (works better when calibrated with velocity survey data, if available)
- Generate "Pseudo-velocity" data (can be calculated either by combining well tops, and horizon maps, or with well fault cut, and seismic fault cut data, either by grid/contour maps, or Slot-nick/Faust functions).

Christopher Jackson

Statoil Professor of Basin Analysis at Imperial College London

One qualitative way of doing things is to think about what the rocks encountered in the wells should look like in seismic data. For example, igneous intrusions, salt, and sharp changes from sandstones into mudstones, amongst many of the things, may all be expected to produce reflections. Likewise, thick, relatively homogeneous packages might be expected to be largely reflection free. Fluids, burial depth, composition, etc, can all mess this up, but it can still be a very useful place to start. I use this thinking all the time.

VERTICAL SEISMIC PROFILE DATA

Vertical seismic profile (VSP) is a technique of borehole seismic measurement used for correlation with surface seismic data and for obtaining images of higher resolution than surface seismic images. In VSP measurement, the geophones are dropped in a wellbore and fixed at very short intervals (depth levels) in the wellbore, typically on the order of 15 m (50 ft). The seismic source is placed at the surface near the well. Nowadays, a chain composed of a group of geophones (5, 12, 24, or more) is used to acquire the data (VSP) in a very short time (Fig. 3.3). If you acquired data in this way and used only the first arrival times, it would be the same as a check-shot survey. But this data can be processed to get a seismic section (Fig. 3.4) which can be calibrated and used to define the geological formation tops on the real seismic data at the well location (see Fig. 3.5), which is more accurate than the check-shot method. This technique produces a high-resolution, 2D image that begins at the receivers (geophones) well and extends a short distance (a few tens of meters or a few hundred meters, depending on the source offset distance) toward the source station. In general, VSPs vary in the well configuration, the number and location of sources and geophones, and how they are deployed. Most VSPs use a surface seismic source, which is commonly a geophone or a vibrator on land or an air-gun in offshore or marine environments.

This time section (seismic section generated from the VSP survey) effectively represents the ray paths in the area surrounding the well (borehole), so it has more information than an actual seismic section, because the ray paths are localized. It is important to know if the picked horizons in the seismic section correspond to the lithologic variations in the well (horizontal discontinuities or a fault trajectory). With VSP, the reflections are recorded so you have an image from below the total depth

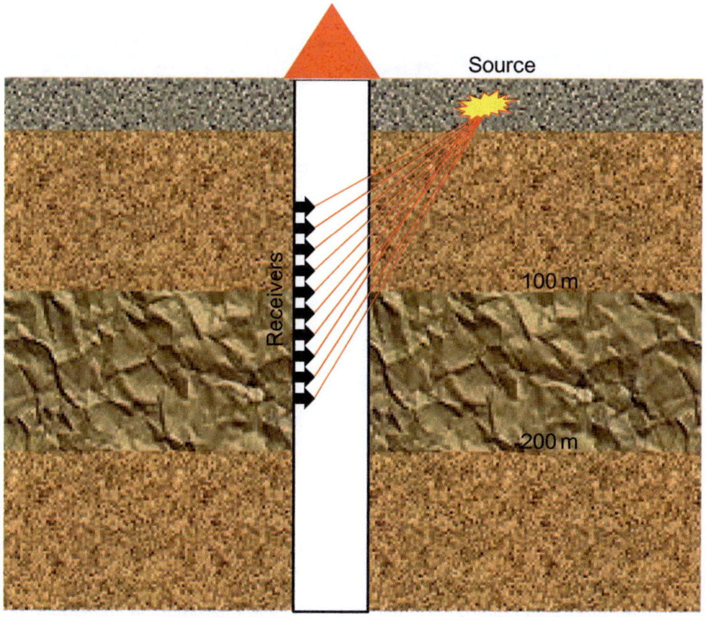

FIG. 3.3

Conceptualized vertical seismic profile (VSP) survey.

FIG. 3.4

VSP seismic section.

Data from An Integrated Multi-Component Processing and Interpretation Framework for 3-D Borehole Seismic Data. https://www.netl.doe.gov/research/oil-and-gas/project-summaries/completed-ep-tech/de-fc26-03nt15418.

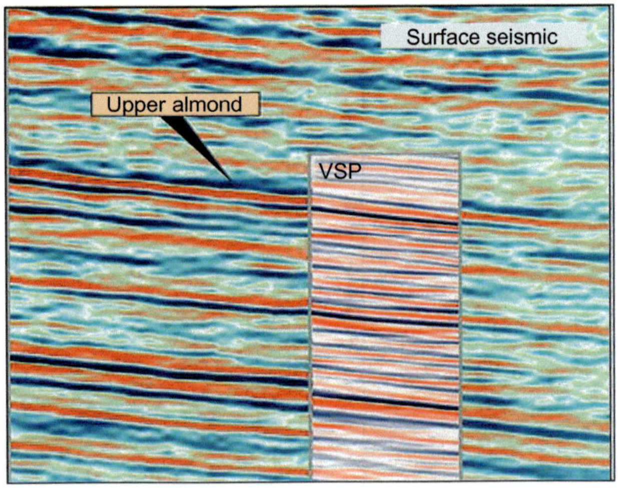

FIG. 3.5

Comparison of a 3D surface seismic image with a 3D VSP image slice at the same location. Note the increased resolution and detail of the 3D VSP image.

Data from An Integrated Multi-Component Processing and Interpretation Framework for 3-D Borehole Seismic Data. https://www.netl.doe.gov/research/oil-and-gas/project-summaries/completed-ep-tech/de-fc26-03nt15418.

of the wellbore, so you can recognize deep targets not investigated by the drilling program. That is, you can see the geological formation below the total depth of the well and more geological information near the well. VSP data is more accurate than check-shot data; hence it is more reliable in interpretation.

Types of VSP Measurements

VSP measurements include:

- Zero offset VSPs: have sources close to the wellbore directly above receivers.
- Offset VSPs: have sources some distance from the receivers in the wellbore.
- Walk-away VSPs: feature a source that is moved to progressively farther offset and receivers held in a fixed location.
- Walk-above VSPs: accommodate the recording geometry of a deviated well, having each receiver in a different lateral position and the source directly above the receiver.
- Multioffset VSPs: involve a source some distance from the numerous receivers in the wellbore.
- Drill-noise or seismic-while-drilling VSPs: use the noise of the drill bit as the source and receivers laid out along the ground.

Difference Between Check-Shot and VSP Data

Acquisition methods for check-shot and VSP data are similar, although in VSP more data is acquired for a larger number of points.

QUESTION 39

What are the differences between check-shot and VSP survey?

ANSWERS PROVIDED BY INDUSTRY EXPERTS

M.S.N. Murty

Consultant Geophysicist

Check shots were taken in the good old days when VSP was not in vogue. As the name itself implies, to get velocity information, shots were taken at random depths and irregular intervals, to generate velocity function which could be used for seismic data processing. It suffered from the defect of averaging velocities from different formations within those intervals.

With VSP, the geophones (in the borehole) can be anchored at formation boundaries which could be known from electrical logs. Thus, formation velocities can be accurately known. The usual geophone interval would be 20-30 meters. Geophone can be anchored at such places after finding rugosity /caving of the well so that it anchors strongly. The resulting average velocity curve would be of great use while processing the seismic data to compare and calibrate. As the geophone is placed very near to the litho-boundary, even a weak reflector can be seen on the up-going wave field when the geophone is moved upwards after each shot. Even multiples can be identified. If there are faults between the source and the well, they too can be seen in the up-going wave-field and throw can be calculated if the data are good. Down-going wave-field would be useful to know the multiple activities and to devise proper de-convolution parameters.

Saad Khurshid

Geophysicist at BGP Pakistan International

CHECK-SHOT

- Check shot surveys measure the first break events.
- In check shot surveys the geophones has larger interval (spacing) in the borehole.
- Check shot records are short in duration.
- Check shot records have fixed gain.
- Check shot surveys use a single receiver recording system.
- Check shot surveys use single source location.
- Check shot surveys have low resolution data.

VSP

- VSPs measure not only the first break but also the reflected events.
- In VSPs, the geophones are at a smaller interval (spacing) in the borehole.

- VSP records are longer in duration and record the full waveform.
- VSP records have variable gain.
- VSPs use multi-receiver digital system.
- VSPs use multiple sources.
- VSPs yield higher resolution data.
- VSP data has smaller sample rate therefore it is more accurate.

Liam Hunt

Senior Geophysicist at CGG

Basically when a check shot is used to get time-depth data, only the first break would be picked to give the Vertical time to depth relationship after geometrical corrections.

VSP's use both the first break to get time-depth data and the remaining waveform from the shot. The waveform contains reflectivity (up-going and down-going) and other information which can be used in synthetic seismic generation and modeling of multiple reflections etc.

Ahmed Hassan

Exploration Assistant Manager at Eni International – Nigeria Agip Oil Company

In addition to all the above differences, a VSP provides a more reliable time-depth curve that matches the event reflection times on surface seismic data and therefore it is useful for depth conversion of seismic time maps.

Fixed-source offset and walk-above (in deviated wells) VSP surveys provide a high resolution reflection image about and away from the borehole environment, which is useful to confirm and assist seismic data interpretation, in particular faults.

A walk-away VSP is useful in measuring velocity anisotropy in the subsurface and with proper processing, may provide P and S reflection images and Vp and Vs profiles.

QUESTION 40

Can VSP be done in open or cased hole? And also what is the use of sonic interval in VSP (geophone spacing)?

Aatef Badshah

Exploration Geophysicist at Pakistan Oilfields Limited

ANSWERS PROVIDED INDUSTRY BY EXPERTS

David J B Dushman

Borehole Seismic Specialist at Chevron

Aatef, VSP's can be performed in both open-hole and cased hole, depending on tool size and casing diameter. It is important however to ensure that the casing is well cemented to the formation, so it is good to run a CBL prior to running the tools. With the large multi level strings it is often preferred to run them in cased hole. The most important consideration is that the geophone has a good bond to the formation.

For the use of sonic interval in VSP, if you are referring to the geophone spacing as defined by velocities from the acoustic log to ensure that special aliasing is not a problem when separating the wave-fields, it can actually cause more problems than it solves. Firstly, the acoustic log has some issues with its measurements which is why we use check-shots to calibrate it.

Secondly, the optimum separation distances can be difficult for the wire-line engineer to stop the tool at;

Thirdly with the multiple tool arrays it is better to have a fixed interval. The standard geophone interval is 15 m or 50 ft, although if you are looking to record higher frequencies (shallower boreholes) then reducing the interval to 25 ft or even 15 ft can be useful, although you do have to take into account the tool length

Basically to optimize your VSP survey it is important to use as much data as you have available in the pre-survey modeling stage to ensure you have a useful result. Most zero offset VSP's are recorded with the 15 m spacing, but when you are looking at recording Walk-away surveys, vertical incidence surveys in deviated wells, offset source surveys, and large 3D VSP surveys, then designing the survey is extremely important in terms of the velocity model, the source type, the depth of the source, the geophone location, the type of geophone used, the number of geophones in the string, the dominant frequency of the geophones, the locking force exerted by the tool and the number of casing strings in the well.

Sergio Filippo Rodriguez Spadavecchia

DCS Borehole Geophysics en Schlumberger Maturin, Venezuela

Aatef, as David mentioned you can run VSP on both open hole and cased hole. I will clarify something. No matter if you have bad or good cement, the fact is, you have free pipe. In this case the casing will ring so much that your Accelerometer will pick that arrival. As casing has high velocity the formation reflections will be affected by casing arrivals.

Neil Peake

US sales manager CGG Geo-consulting Seismic Reservoir Characterization

Aatef, in response to the use of sonic interval in VSP, a useful rule of thumb is to take the slowest expected velocity in the depth range to be sampled by the VSP, and divide that by twice the expected highest frequency that you might obtain. E.g., slowest velocity = 1800 m/s, highest frequency expected = 60 Hz - gives a value of 15 m. This would be the maximum spacing that you could get away with without spatial aliasing (any frequencies above 60 Hz would be aliased). Don't worry about slower velocities outside of the range of the VSP. Note that if you are trying to obtain shear wave information, then the spacing will need to be much shorter, about half. If you are using a multi-level tool with fixed spacing, you can overlap the tool stations to create a spacing/2 set up to improve sampling, though you run the risk of really annoying your seismic engineer in the process.

Eduard Breuer

Borehole Seismic Specialist at Chevron

Aatef, in addition to the great points above, for some surveys such as walk-away surveys, especially in soft formations you do not want to clamp your geophones in an open-hole. You may have a hard time to get them out of the borehole. This is something to keep in mind especially for deep surveys.

Some of the uses of sonic interval in VSP are to:

- Check if the picking of VSP first arrivals are good or not via the Drift Curve.
- We also use it to extract the Time-Depth law, so that we can make the Well-to-Seismic Tie to know exactly the location of well tops (are in depth scale) on a seismic section (in time scale), which allow us to pick the seismic horizons and do some further interpretations.

QUESTION 41

When we have a VSP, it gives a nice Time/Depth relationship and we can get picks for seismic horizons. If we have a VSP, do we really need synthetic seismogram since we have already confirmed our tops?

Farrukh Hussain

Geophysicist at Dewan Petroleum (Pvt.) Limited

ANSWERS PROVIDED BY INDUSTRY EXPERTS

Zaheer Ali Shah

Technical Coach (Geosciences) at Petronas Carigali, Kuala Lumpur, Malaysia

Farrukh, surface seismic, VSP and sonic log have different resolutions and different wave paths. All of them are capable of giving us time-depth curves. They all can show seismic response of a place/point, though, obtained through different routes. Comparing them against each other helps us understand them. Marking a top on a seismic section is just a start of a very detailed process of extracting subsurface geological, petrophysical, and fluid information. How would you know if one of this data set is not correct if you don't have choices for comparison? We should take advantage of all available sources that help us to validate our understandings and avoid short cut approaches. A synthetic seismogram tells you what generates the reflection. A forward and inverse modeling activity tells you about the details hidden in the seismic signal and it also helps you to understand noise in the signals. More so, interpretation of seismic data means relating / mapping seismic response (reflection, absorption, polarity, continuity, or discontinuity, variety of seismic attributes, etc) to subsurface geology if well information is luckily available. Synthetic seismograms help us to understand the details hidden inside the signal (wavelet) in relation to geology (structure / lithology / Petrology) and in relation to different fluid saturation. Understanding of this information is important to carry out your horizon pick.

Tigor Siagian

Geophysicist at KUFPEC Region Venture Indonesia Ltd

Farrukh, by generating a synthetic seismogram you can bring the well information to your seismic not just the pick top formation and get more understanding about the seismic responses related to lithology or HC. Also you can use the corridor stack from VSP final results to correlate the synthetic seismogram.

Teslim Adeleke

Geoscientist at FFA

Farrukh, Yes, we need synthetic seismograms. It is good QC practice to compare the synthetic seismic traces with the original seismic especially to ensure you have done an accurate tie of the well (impedance log) to seismic. It is also crucial to validate the seismic inversion results by convolution with the wavelet to produce synthetic seismic traces which are compared to the original seismic. A good match is an indication of good job.

How to Import VSP Section to Match Your Seismic

QUESTION 42

How can I import a VSP section to the well to match well tops with seismic?

ANSWERS PROVIDED BY INDUSTRY EXPERTS

Jonny Pinto

Senior Geophysical Advisor at Vantage Energy

I am assuming you have the SEG Y for the seismic section. If this is the case bring the SEG Y as another seismic line, then proceed as usual to tie your well, just keep in mind your frequency spectrum would be much higher. If you don't have the SEGY, then based on your best estimate for the frequency range in the section, generate several synthetic seismogram, plot them using the same scale as the section and tie them the old fashion way.

Erik Johnson

Fugro Geoscience Division

I interpreted your question a little different from Jonny. Sounds like maybe you want to put the VSP into your well log software, as another curve? If so, what I have done in the past is import the depth-converted VSP to a seismic program, then export it as an ASCII file. Then in any ASCII editor, you can strip off the SEGY header and replace it with a generic LAS header, and then import the LAS file into your well log file with your other log curves. This sound cumbersome but doable.

Jonny Pinto

Senior Geophysical Advisor at Vantage Energy

If Erik is right, an easier way is to read your SEG Y in Excel, strip off the header, save as ASCII and read it into your interpretation software as ASCII not las.

How to QC VSP Data

QUESTION 43

How can I QC the ZVSP and then decide whether the VSP is good or not? The wellbore is vertical and the source is a vibrator. The objective is to use this VSP in the reprocessing and for a well-to-seismic tie.

Raouf Nasr
Geophysicist

ANSWERS PROVIDED INDUSTRY BY EXPERTS

Zaheer Ali Shah

Technical Coach (Geosciences) at Petronas Carigali, Kuala

Raouf, to have a good VSP, you should take appropriate steps at the following stages such as planning, execution of the acquisition program, processing stage and interpretation stage. If this VSP has to be used in the reprocessing then you have to know the acquisition parameters of the seismic survey you want to tie/reprocess. You must take some VSP shots with the same parameters as of the surface seismic acquisition which I hope is a Vibroseis source too. In a rare case if the surface seismic is not done with the Vibroseis but with dynamite, then please use air-gun as a VSP source. You must consider what type of subsurface you are trying to image. If the target is highly dipping or the well is intersected by a fault, you should do some modeling to know in which direction you should keep the source even if it is a zero offset VSP. I have imaged faults even with zero offset VSP. Remember shallow shots are very important in terms of tying with the seismic. They give you idea of static correction at different datum levels. It is a good time to do some work on this aspect of seismic processing as well, using VSP to help correct static correction. Take some shots with spike mode too.

If you want to do seismic inversion, then it is important that you take appropriate steps to reduce the drift between sonic and VSP. If you want to tie with s-wave sonic then try to have high quality tri-axis acquisition. The source to the well head distance is related to tube wave activities. I would also suggest you save the uncorrelated signal as well. A major use of VSP is to understand the phenomenon of wave propagation, whether a VSP matches or not with the surface seismic. Another important aspect in the VSP is always to check if the zero time reference geophones/accelerometers are working properly at every shot. Sometimes down-hole equipment is working but the surface sensors are not working. This is an important matter in VSP acquisition monitoring. If you can acquire the VSP during the day it will be good as in the night the company men sometimes are not allowed to stop the generators which are main source of noise. At the acquisition phase it is very important to have as quiet as possible an environment around the wellhead to have a clean signal.

Mohamed Badawy

Drilling Supervisor, LUKOIL Iraq

Raouf, as mentioned above, QC extends over the various phases and I will try to add to Zaheer's comments:

PLANNING:

- If you are using Vibroseis then use a broad sweep to take full advantage of the VSP, taking into account that earth filtering response will reduce the frequency content.

- You need to have the tool anchored at the SRD level or as close as possible if the cement behind casing is poor. Some tools as the ability to record high quality direct arrivals in even in multi-casing zones.
- The vertical spacing would have an impact on the spatial resolution, you might want to consider multi-level tools depending on the feasibility, expected Run in Hole (RIH) time and the stability of the open-hole section (if applicable).

EXECUTION:

- One of the common problems are tube-waves which travel along the surface and down the well at the mud velocity. Many solutions exist such as digging a trench which will act as an acoustic barrier, or having the shooting pit along the line of well and one of the rig legs. Another solution would be either adding heavy weight mud ex. barite or reducing the mud column. Both will need the company man to approve.
- Generators would add some noise at 50-60 Hz and a notch filter will cause a hole in your frequency-amplitude spectrum. It is highly recommended to shut it down at least at the target/reservoir zone if not the entire survey or as **Zaheer** mentioned having it run during daytime.
- If there is a lot of wind adding some slack to the wire-line cable after anchoring the down-hole tool would reduce cable noise, too much slack can cause differential sticking of the cable. This is applicable unless you are using a tool which the geophone is fully decoupled from the tool sonde.

PROCESSING:

- Picking is one of the most critical steps as wave-field separation; de-convolution and corridor stack design depend on it entirely.
- If the subsurface is of low dip then you would expect most of the direct arrivals being recorded on the vertical component. If dipping, then you would need to have a careful look at the horizontal components and after both wave-field separation and de-convolution, you might consider either migration or CDP image.
- Environmental corrections need to be applied to the open-hole log suite; however the most important are the acoustic and density logs especially for washout intervals.
- The drift correction needs changing from positive to negative to follow the formation markers and significant acoustic barriers. Minor fine tuning of the picks can reduce the jitter effect.
- Following that the knee points will be selected also at formation markers and significant acoustic barriers.
- Both the synthetic seismogram and corridor stack should match the surface seismic.
- Having a match is most important. Then you can extract a lot of information from the VSP including: Statics correction to be applied on seismic, Q-Factor to compensate for inelastic frequency attenuation and accurate time depth relation for the CDP gather at exact well location.
- 4- Multiple (generators and periodicity) at the well location.

QUESTION 44

What are the main issues on sonic calibration using VSP times?

Sergio Filippo Rodriguez Spadavecchia
DCS Borehole Geophysics en Schlumberger Maturin, Venezuela

ANSWERS PROVIDED BY INDUSTRY EXPERTS

Zaheer Ali Shah
Technical Coach (Geosciences) at Petronas Carigali, Kuala

Sergio, sonic calibration may involve calibrating the component of sonic log and the component of VSP times that are aligned along the same travel path between two shot/receiver intervals.

Sergio Filippo Rodriguez Spadavecchia
DCS Borehole Geophysics en Schlumberger Maturin, Venezuela

Zaheer, the thing is that the sonic log has another scale, and it's a different measurement. I don't understand what you mean by "may involve calibrating the component of sonic log and the component of VSP times that are aligned along the same travel path between two shot/receiver intervals". As you know sonic calibration will be handled using the transit time picking on 3C and selecting the Knees where a lithology change is seen. Can you please be more specific?

Zaheer Ali Shah
Technical Coach (Geosciences) at Petronas Carigali, Kuala

Sergio, I wished to look at the problem, "How can we calibrate sonic log with VSP times between two knee points if the well path and wave path are not exactly the same?" So I have the following suggestions:

Firstly, we assume that the well path between two knee points is not a straight line. My suggestion would be that we re-sample the logs between the two knee points in two orthogonal directions, that is, horizontally and vertically. Then compute the integrated transit time from these two re-sampled logs between the knee points. Let's call these times sonic-h and sonic-v.

Now for VSP times we compute the time interval (delta T) between two VSP levels at the same knees as already chosen above. Then compute horizontal and vertical component of this time between the two knee points. Let's call them VSP-h and VSP-v. And finally compare sonic-h with VSP-h and sonic-v with VSP-v.

Secondly, we assume that the well path between two knee points is almost a straight line, and then we may not need to resample the log in horizontal and vertical directions. We can compute the vertical and horizontal component of the integrated transit time straight away and compare them with VSP-v and VSP-h.

If the interval between two VSP levels is small then we can assume that the well path between two knee points is straight if we calibrate at every VSP interval. But if we choose knees at points which are many VSP levels apart then we may consider the above suggestions.

Sergio Filippo Rodriguez Spadavecchia
DCS Borehole Geophysics en Schlumberger Maturin, Venezuela
Zaheer, the VSP interval is 15 m.

Zaheer Ali Shah
Technical Coach (Geosciences) at Petronas Carigali, Kuala
Sergio, 1m (MD) is a well known VSP interval. However, when you convert 15 (MD) to TVD in a hole with deviation of 35 degree then it can be actually small. This means that you are going to have a high density VSP. This helps me to assume that if you keep your knee interval short, you may assume that well path and wave path are nearly the same. I think that sonic-h and VSP-h may match well as I proposed above. Please Let us know how it went.

Last time I processed a deviated hole VSP was in 2005/2006 in Angola. But it was with a walk-above source. Geoframe handled it well. If I remember correctly, Schlumberger has very good algorithms for TVD correction (re-sampling) of logs. And I hope you will have a very good calibration. If you see a big drift, +/-20 ms between two knees, check your VSP times at them.

Sergio Filippo Rodriguez Spadavecchia
DCS Borehole Geophysics en Schlumberger Maturin, Venezuela
Yes Zaheer, everything went really well. Thanks for the tips!

How to Determine the Static Correction for VSP Data

QUESTION 45

How can I determine the static corrections for the VSP data? Is it possible to pick the first break for the VSP file and use it?

Mahjoob Mansoor
Geoscientist

ANSWERS PROVIDED BY INDUSTRY EXPERTS

Adell Hakim
QC Geophysicist and Client Representative for HBSI
Mahjoob, during the VSP operations you are going to have two elevations references:

- The rotary table of the drilling, rig and the assigned datum plane for your seismic sections.
- The source point locations will be with a known elevation.

The static correction for your source points could be either obtained from the seismic section points, or by running short refraction shots in order to predict the shallow weathered zone layers information. Now you are going to have for each source point its static correction which is related to the pre-assigned datum plane.

Meanwhile, your depth measures are related to the rotary table, you have to refer your depth measures to the same datum plane. That is because there is no static correction for the piece of the geophone cable between the rotary table and the assigned datum plane. It is just a cable, not a ground. So, the depth is corrected as if it is measured from the datum plane. Reading the first arrivals slant time, modify them to vertical time by multiplying with Cos their angle with the vertical. Then apply the static correction for the arrival time from that source point. You can have a symmetry shot point at the other side of the well, and then obtain the corrected average arrival time for both the two symmetrical shots. Obtain the average velocity for that depth, and then from the next depth obtain the interval velocity of the covered section between the two depths, and so on.

Mohamed Badawy

Field supervisor at Schlumberger

Mahjoob, If during acquisition the VSP geophone stations is anchored at SRD, you can possibly use it to calculate the statics correction value provided the direct arrivals are clear enough to reduce uncertainty in the picking.

Zaheer Ali Shah

Technical Coach (Geosciences) at Petronas Carigali, Kuala

Mahjoob, place the down-hole geophone at SRD and take the shot. The true vertically corrected time of this shot is your static correction. The problem is that if SRD is too shallow the vertical components of down-hole geophone may not have a reliable signal record. In such cases of very shallow recording you should check the data on x and y components. Whichever is the best component at shallow level (500 ft and shallower), you may consider to use it for time-depth curve.

VSPs Replacement Velocity

QUESTION 46

If, in VSP processing the calculated replacement velocity is not giving a good tie at reservoir levels (~80 ms. mis-tie), is it advisable to change the replacement velocity and get a good tie?

Harshvardhan

Lead Geophysicist at Cairn India Ltd

ANSWERS PROVIDED BY INDUSTRY EXPERTS

Adell Hakim

QC Geophysicist and Client Representative for HBSI

Harshvardhan, That is for sure. The incorrect estimation of the weathered layers velocities and the replacement velocity, beside their thicknesses generate mis-ties for the known markers. Completely review the static correction values and the elevations to avoid such mis-ties.

Fernando Roxo

Director of Geophysics at PanAmerican do Brazil

Harshvardhan, I think it would be good to have some more information such as:

- Is it a marine or a land well? This can make a whole lot difference in static processing.
- Relative to well head, what are the azimuth of the deviation and the azimuth of source position? This can make the deviation [almost] unimportant, or more critical.
- What is the depth of the reservoir top? What does 80ms represents? If shallow reservoir - Wow! It is a lot! If very deep reservoir, it is still important, but the problem could be just related to a velocity model interpretation.
- When processing the VSP data, is the datum used the same as the surface data? Or was it processed at the SRD level? If so, did you move to the seismic datum later? How did you do that?
- What is the behavior of the mis-tie along the data? Is it a constant or does it vary? If it does vary, is it steadily increasing or decreasing, or is it changing in a semi-random way?
- Was your surface data pre-stack migrated? Time or depth? If depth, how did you converted it back to time? (I am assuming the seismic data is in time scale)
- This is out of curiosity; do you have sonic and density runs in this well? Did you try to do a synthetic seismogram? If so, did you correct the travel-time from sonic to those of VSP?

I think those things can help a lot in understanding your problem.

David J B Dushman

Borehole seismic specialist

Harshvardhan, since the well is deviated it could be worth checking the headers and making sure that the x and y values are in the correct co-ordinate system (feet/meters) that the wellhead location is correct, and that the deviation values are correct relative to the wellhead. Then, check that all your datum values are correct, errors can often happen. A further good point was made as to how the VSP was picked break to break or trough to trough, what vintage is the seismic, along with what is the seismic phase? What is the dominant frequency at the reservoir where there is a mis-tie? How deep is the problem? All these things can have an effect on your results.

Sergio Filippo Rodriguez Spadavecchia

DCS Borehole Geophysics en Schlumberger Maturin, Venezuela

Harshvardhan, let me throw some light in another direction. If you are having a mis-tie in your reservoir -about 80 ms- do you think your velocity model (PSDM) is accurate? Try the following: extract a velocity profile from your PSDM on depth and compare it to the ZVSP velocities. Do they match? If not something is wrong with your PSDM or your VSP. After that you can start calculating your replacement velocity. Does it match with your surface seismic? This is a basic QC that you need to do before changing any velocity value.

Harshvardhan

Lead Geophysicist at Cairn India Ltd

Just to add to my question if the well is around 30 degree deviated and it is a ZVSP with source offset 45 m. How much will anisotropy play a role in getting higher velocities at reservoir level and hence a mis-tie?

Andy Jeffreys

VSP Processing Geophysicist at Upwave Ltd

Harshvardhan, if the mis-tie is constant over the whole VSP interval then it is probably due to either different static corrections being applied to the VSP and the seismic data, or the timing methodology used by the VSP processor (or both). If the tie is good at the top of the VSP, and the mis-tie increases with depth, then it is probably a VSP processing error. With 30 degrees deviation anisotropy may indeed be an issue and this needs to be accounted for in the processing, but potentially more important will be the accurate derivation of the vertical times from the VSP. These are likely to be too shallow at large depths if a simple geometrical correction is used rather than predicting the refracted ray travel paths.

David J B Dushman

Borehole seismic specialist

Harshvardhan, I agree with Andy's comments, to get good VSP times in deviated wells it is normally better to record the data with source placed vertically above the geophone, failing that as **Andy** mentions, the deviation anisotropy is likely to be the issue. When you are looking at ties to the seismic, first place to check for problems is the VSP as this is easy to reprocess and work with, if there are indeed no errors in the VSP then move to check the seismic processing, a constant value of 80 ms is likely to be different static corrections.

Paulus Allo

Geophysicist at Statoil Indonesia

Harshvardhan, 1 msec. mis-tie suggests something fundamental has gone wrong. Have you worked out what is the replacement velocity value needed to correct 80 msecs. mis-tie? If it is something like changing the value from 1500 to 5000, you might have an issue on inconsistent use of units during the processing (in 75meter water depth, error on units between 1500 and 5000 already giving you differences of 70msecs).

Hüseyin Özdemir

Reservoir Geophysics Consultant

Harshvardhan, I would check first if it is a block shift: everything ties after ~ 80 ms shift including top reservoir? If yes, it is probably due to the wrong replacement velocity. However, dipping layers cause the seismic wave path deviate from being approximately vertical. The check-shot times from refracted wave path are smaller than the integrated sonic times. Similar results are obtained when check-shots are acquired over a deviated well even when the layering is horizontal. Here the discussion is confined to the check-shot acquisition using a fixed source position close to the wellhead. In practice, the wellbore are almost never straight vertical. However, deviations of about up to 5 degrees and small doglegs near the termination depth (TD) do not introduce appreciable errors in check-shot times. Projections to true vertical depth (TVD) smooth out most of the discrepancies. Anisotropy is also an issue (see Walsh et al., 2007). However it is unlikely that anisotropy will cause 80 ms mis-tie. In summary, the check-shot data acquired from a source close to the well head of a highly deviated well may not be usable for check-shot calibration. I would like to add the obvious that, a proper well-seismic survey should always have a shot at the top of sonic.

Difference Between Check-Shot and VSP and Depth-Time Model

QUESTION 47

What are the subtle differences between VSP, Check shots, and a Time-Depth model?

Benjamin Egbure

Graduate Student, University of Port Harcourt

ANSWERS PROVIDED BY INDUSTRY EXPERTS

Zaheer Ali Shah

Technical Coach (Geosciences) at Petronas Carigali, Kuala

Benjamin, VSP's and check shots are acquisition techniques (procedures) used during borehole seismic data acquisition.

A depth-time model is among the final results of a VSP or check shot survey. Remember that there are other output/uses of VSP and check shots too. And there are several other ways of making a depth-time model. A check shot survey or a VSP provide a basic depth-time function that can be used to convert time seismic images into depth seismic images and/or to convert depth-indexed logs into time-indexed logs.

The difference between a check shot and a VSP survey is that the check shot is limited to recording just enough data to provide a time-depth function while the VSP is used to obtain not only a depth-time function but also a seismic image at the location of the borehole. The objective of the VSP is to understand the phenomenon of wave propagation in the context of the local geology in as much detail as possible to help interpreters understand/calibrate surface seismic data.

Afeez Popoola

Geophysicist at Schlumberger

In addition to Zaheer answers, the big difference is that check shot just give you a kind of 1D time-depth relation but on the contrary, VSP gives you an actual seismic image around a well and you can extend the area you can see around the borehole by using some more advance acquisition geometry (walk-away, 3D VSP etc).

SEISMIC TIME-TO-DEPTH CONVERSION

Time-to-Depth Conversion

Seismic data are acquired in two-way time, whereas well data are acquired in depth. In order to tie seismic events to well data, seismic time data must be converted to depth. Depth conversion is an important step in the seismic reflection method, which converts the acoustic wave travel time to actual depth, based on the acoustic velocity of the subsurface medium (sediments, rocks, water). Fig. 3.6 shows measurements in time and in depth.

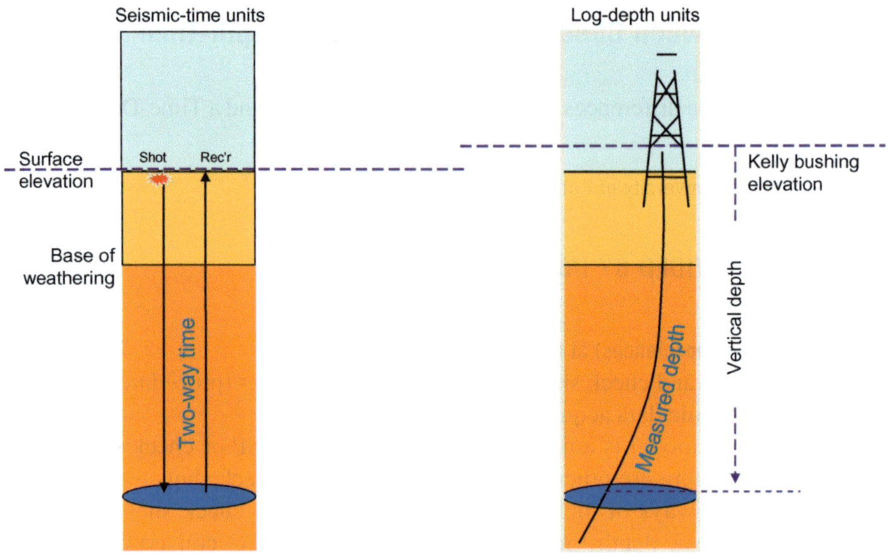

FIG. 3.6

Time and depth measurement.

In depth conversion several sources of information about the subsurface velocity are integrated to derive a three-dimensional velocity model:

- "Well tops", i.e., depths of geological layers encountered in oil and gas wells.
- Velocity measurements made in wellbores.
- Empirical knowledge about the velocities of the rocks in the area investigated.
- Root mean square (RMS) stacking velocity, which is derived from the processing of the seismic reflection data.

Time-depth conversion permits the production of depth and thickness maps that depict subsurface layers based on reflection data. These maps are crucial in hydrocarbon exploration because they permit the volumetric evaluation of gas or oil in place and provide a picture of geologic structure in depth (Fig. 3.7).

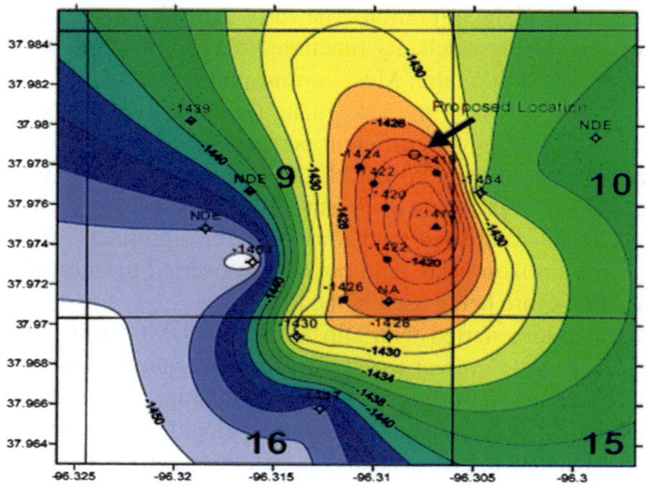

FIG. 3.7

Depth structural map.

Data from http://blackstar231.com/a/recent.php.

QUESTION 48

Is stacking velocity the same as migration velocity? And in tying the seismic data to well data which one should be used to produce a depth map together with the well data?

Raouf Nasr
Geophysicist

ANSWERS PROVIDED BY INDUSTRY EXPERTS

Guy Maslen

Chief Operations Officer at Globe Claritas

Raouf, there are two different things to think about here what the velocity is, and how the migration uses it. "Stacking velocity" is just the velocity that makes the best stack; that makes the reflections within CDP gathers "flat" when the NMO correction is applied. This is not always a good measure of the real subsurface velocities for a number of reasons.

Firstly, the NMO equation is based on plane, horizontal layers. Secondly, the NMO equation doesn't take into account refraction of seismic energy at layer boundaries or within a layer (if the velocity changes) and finally it assumes that the vertical and horizontal velocities are the same, which also might not be the case. All this means that the stacking velocity (Vstack) can be a pretty poor approximation for the true RMS velocity (V rms) and hence the actual interval velocity (Vint) of the rocks.

Migration tends to need some kind of velocity model that represents the geological structure; exactly what this model is tends to depend on the algorithm type, and the desired accuracy of the migration. The early algorithms were constant velocity migrations, or used a single velocity function, which is about as simple as you can get in terms of a velocity model. Most current post-stack time migrations need an V(rms) field; it was common to approximate this with a smoothed V(stack) field, and then test the migration with different percentage scalars to try and compensate for the inaccuracies; tests from 80%-120% were common. People would also make horizon-consistent models, where the V(stack) would be extracted on interpreted horizons, and the results corrected mathematically for dip effects. These would then be smoothed on a horizon basis (and possibly converted to interval velocities and smoothed further) before being used in the migration.

Pre-stack migration, started with Dip Move-out (DMO) and gradually grew in sophistication to initially correct for dip and then for some elements of refraction. This meant that people had to "boot strap" their final migration model, by starting off with a V(stacking) model, and then re-pick the velocities (or residuals) on the pre-stack time migrated data to give a V(rms) model. Pre-stack depth migration took this a step further, and uses an interval model in depth, usually derived from the V(rms) model and additional well information to fully correct for the structural and refraction effects.

The last remaining issue - the mistie between "imaging depth" and "well depth" caused by the differences in horizontal and vertical velocities and is resolved by applying an anisotropic migration which requires additional parameters for the velocity model, usually derived in part from well data. As a final point, if you are using velocities for depth conversion, then you need to be aware of the limitations of the migration algorithm that was used, and which type of velocity you have been supplied.

Fritz Damianus

Multicomponent Geophysicist at CGG

RAOUF, In addition to Guy answer, in depth model building we usually calibrate our velocity model with sonic log which should match in its low frequency pattern, since they are both Vint. We use iterative approach in building depth velocity model, through some user control horizon and tomography inversion.

Guy Maslen

Chief Operations Officer at Globe Claritas

Raouf, it is worth noting that you also have to be a little cautious with the sonic log as a velocity reference. The sonic log tool works at ultra-sound frequencies, which means the delay times that are recorded are a little different from those we see with seismic waves at a much lower frequency. The sonic tool is also not a "point sensor" - the tool has a finite length which means you get distorted readings where you have a series of thin beds of strongly contrast velocities. As a result the sonic log needs to be calibrated against a check-shot (or VSP) to correct for this "drift" before it can provide a link between vertical depth in the well and two-way-travel-time on the seismic section. More critically, perhaps, you can only have an accurate time-depth relationship over the depth range for which you have either check-shots (or VSP) and a sonic log. Outside of this overlap range, you are extrapolating the relationships and so may not have an accurate tie.

Fritz Damianus

Multicomponent Geophysicist at CGG

Guy, I am in agreement that sonic logs measure the velocity at a different resolution than seismic. That's why in depth model building we always see sonic low frequency trend as reference to seismic velocity model. Calibrating seismic velocity depth models with sonic log for QC happens in the depth domain, so there is no need to have check-shot correction (all the modeling is done in depth). In fact, we use check-shots as an additional cross-check of velocity trend for our seismic velocity depth models. I think that is why we call it depth imaging, since everything is done in depth.

Guy Maslen

Chief Operations Officer at Globe Claritas

Fritz, sonic logs use frequencies in the kHz range, at least a couple of orders of magnitude above those used in seismic and hence are of much shorter wavelength, so dispersion is a factor. A significant amount of seismic interpretation is still conducted in two-way-time, which requires the sonic to be calibrated (especially where you have alternating thin high/low velocity beds, as you sometimes find in coaly sequences). Otherwise there is significant "drift" between the well velocities and seismically derived ones. That said, I fully agree if you are working with an anisotropic pre-stack depth migration in depth then this isn't an issue. The thin beds will give a pretty significant anisotropic effect, too. However, if you are converting seismic data (to depth) based on a time migration (pre- or post-stack) then it is something to keep in mind. One other thing to think about here is what we actually mean by "depth." A number of years ago I commented to a very experienced geologist that while seismic processing was all approximations and models, at least we could rely on the depths from wells as being the "truth." His face dropped at that point, and he started to list out all the variations between "logger's depth" and "driller's depth" he had seen (especially on older wells) and the struggles he had faced on occasion getting these into a realistic "measured depth"; in fact he had usually placed more faith in the seismic at well ties than the depths on some occasions. Things have no doubt improved with modern measurement techniques (and algorithms), however if you are working with older wells being aware of the limitations and error bars associated with the "measured depth" values is also useful.

QUESTION 49

How can I use the stacking velocity to make depth conversion?

Nasser Srour

Senior Geophysicist at SUCO / RWE

ANSWERS PROVIDED BY INDUSTRY EXPERTS

Ridho Affandi

Senior Geoscientist at MontD'Or Oil Tungkal Ltd

Nasser, exactly the same way when you use well velocity data, but the stacking velocity need to be calibrated or corrected to well velocity data due to the fact that stacking velocity is slower than the real measured (sonic log) velocity.

Nasser Srour

Senior Geophysicist at SUCO / RWE

Thanks Ridho, actually I have the stacking velocity as a SEG Y cube and convert it to average velocity using Dix conversion and calibrate it with average velocities from the wells but still have miss match ranged between +or- 5 - 15m and more than 50m at blind wells. For me from 5 -15m miss match expected but the value at blind wells is not expected. So what do you think?

Mahmoud Noaman

Geophysicist, Seismic Delivery & WND, Exploration at BP Egypt

Nasser, firstly, you should use all your wells in the calibration process, unless you don't have time depth tables or sonic to calibrate with.

Secondly, you can perform another iteration of calibration but using well markers. In this case it is not necessary to have any velocity information at the well. You start by identifying the well marker in depth in terms of XY and Z, where Z is depth. Project the time value from TWT horizon to the marker point so you have XYZ depth-time value. Calculate the average velocity at this point in space and use it to convert depth to time so you have XYZ depth-time average velocity and force Z to equal time and change its domain definition to time domain. So now it is in time domain, then you calibrate your seismic velocity to the marker velocity by interpolating the correction factor in 3D, doing this for all markers associated with their TWT horizons and then interpolating the CF in a stratigraphic grid should end up tying every single marker to its TWT horizon.

QUESTION 50

How good can an inverted seismic volume be used for time picking and be then used for depth structural conversion?

Debabrata Majumder

Geophysicist

ANSWERS PROVIDED BY INDUSTRY EXPERTS

Hesham Shalaby

Senior Geophysicist at Dragon Oil

Debabrata, yes you can use the inverted seismic volume to interpret the reservoir and then do a depth conversion. For QC, just overlay the horizon on the conventional seismic and look at the difference in the geometry. Assuming the inverted volume has a good quality input, concerning the extracted wavelet and the low velocity model, the interpretation will be more precise than the conventional seismic especially when the reservoir is thin.

Debabrata Majumder

Geophysicist

Hesham, how do you prefer to convert to depth especially when you have lateral velocity variation? How reliable is a velocity model used for converting the inversion model to depth?

Hesham Shalaby

Senior Geophysicist at Dragon Oil

Debabrata, after interpreting the horizons in the inverted volume, you can depth convert them by applying the best and most reliable velocity model you have. If you have the inverted volume in depth, you should check the input velocity model and see how it affects the time. If you see a reliable seismic picture in depth, you can rely on it.

Peter Wang, P.G.

Geophysical Technical Advisor, Paradigm

Debabrata, you can also run the seismic inversion through a FIRST DERIVATIVE operator in the time domain. This will create pseudo reflection events which an auto-tracker will follow. It works Ok.

Debabrata Majumder

Geophysicist

How does the inversion process vary from vendor to vendor, especially in simultaneous elastic inversion? Resolving and characterizing the reservoir are perhaps the next level of challenges.

Hesham Shalaby

Senior Geophysicist at Dragon Oil

Debabrata, the input background model, wavelet extraction window and parameters and the algorithm used vary from one vendor to another. In addition, the acoustic logs editing vary also.

QUESTION 51

There are two wells, 500m apart lying on the same seismic line. One well has a shallower time and greater depth as compared to the other which is shallower but shows more time. So is this possible? Seismic also supports the shallow time for the first well.

Aatef Badshah

Exploration Geophysicist at Pakistan Oilfields Limited

ANSWERS PROVIDED BY INDUSTRY EXPERTS

Kashif Mohammed

Geosolutions Geophysicist at Schlumberger – WesternGeco

Aatef, if the wells are correctly converted to TWT (two-way time), then, I doubt the seismic. Either velocity or statics is causing this problem. In a radius of 500m I expect no such difference in velocity. Please check your seismic velocity. If the values are close enough on the two well locations. If the velocities are close, then check CMP statics. If these values are also close enough you should then also verify the geology, which can cause miss-ties with the wells. If all of above seems ok with seismic, you are left with wells. Please check that both wells have the same datum.

I just put my words up there without looking into you seismic and wells. This all needs proper investigation. You need to dig in to it as you are working on it.

Abdelrahman Samir Abdelsamad

Sr. Geophysicist Advisor at Hellenic Petroleum

Aatef, If you ask simply if it is possible or not, yes it is possible and could happen in complex structures and in some cases of velocity inversion.

Amr Moawad

Geosolutions Geophysicist at Schlumberger

Aatef, I totally agree with Kashif and Abdelrahman. Moreover that you can also check the time depth relationship in your wells from any check-shot or VSP data you have and see how the velocity changes with depth in each well. Also check your geological section in each well because an abrupt change in velocity if it is not related to seismic issue, may be attributed to structural complexity as Abdelrahman said.

Mark Leander

Contract Geophysicist at Maersk Oil

Aatef, the distance between wells (500m) seems rather small to expect reasonable seismic velocity variations explain the difference. Make sure you are completely comfortable with the well correlations and seismic-to-well ties. Again, the wells are pretty close together but, if offshore, is there an abrupt change in water depth? Is there any chance for a fault separating the two wells leading to another way to correlate the wells? Lastly is the seismic you are tying to one single line or a survey? Murphy's Law will put the merge or tie point for separate datasets between your two wells, if that's the case look for differences between the two lines or surveys (datum, velocity picks, and wavelet/polarity). Beside, how much of a mis-tie are we talking about (10's or 100's of msec)? How deep in the section? Can you say where in the world this dataset is from? If your dataset is onshore, check the well elevations and make sure you are properly tying using the datum elevation and replacement velocity applied to the seismic. If none of the above, start thinking about your geology above the horizon that is mis-tying. Could there be an isolated or narrow feature or an abrupt boundary (channel, reef, slump, etc.) that could contain faster or slower material unresolved by the velocity picking? By the way, do horizons above or below the mistie you mention also mis-tie by the same amount?

Mark Leander

Contract Geophysicist at Maersk Oil

Aatef, what I was driving at in my comments above is that often seismic velocities are explicitly picked on a 1 km gid/interval. So for a significant mistie to occur for wells only 500m apart say (to me) that the ambiguity must be driven by a problem with the well correlations/ties, the local geology (unresolved by the velocity model), or data-related issues. Furthermore, you did not supply any metric on the mistie in your question – could it have been within expected uncertainties? Onshore/offshore?

Peter Rowbotham
Geophysics Advisor at Apache Corporation

Aatef, here are a few more suggestions for how this could happen and routes to investigate. Either

(1) Something is wrong with your data
  (a) As commented above, check the well data in your software back against original data (well documents etc., but even these original reports have errors, and the further we go back, there could be less precision and less confidence). What was actually recorded? And what historical interpretation has become to be used as data? Common problems are elevations, deviations, confusion between depths (measured or true vertical; driller's depth or logging depth; relative to KB/RF or datum; seismic datum or ms) and not updating all data items when transferring between them.
  (b) Is the seismic correctly loaded? Would you expect the seismic time image to be of good quality? Again check back through acquisition/processing reports. Is the velocity model used for seismic imaging robust?

Or

(2) The combination of geology and seismic imaging results in the effect. Note that pinch-out/channels could cause the overburden velocity seen by the seismic to be different at the two wells. You did not mention whether you have 2D or 3D seismic. If 2D, is there dip causing out-of-plane reflections to arrive at earlier than anticipated? By the way, in my answer I assume you are talking about seismic reflector horizon time rather than check-shot/VSP time prior to tying to the seismic.

QUESTION 52

Can anyone tell me about how to correct the seismic illusion while doing time-to-depth conversion? Signature wise time-to-depth sections are same. The high frequency effect on a time section is due to higher velocity in deeper section. A thickness of 40m and velocity is 2500 m/sec gives a time (thickness) 16ms. With a velocity of 2000m/s the same unit will be seen as 20ms. If the velocity is 1000m/s the same unit will be seen as 40ms in thickness. This is the illusion.

Akshay Audhkhasi
Geologist, ONGC Videsh Ltd

ANSWERS PROVIDED BY INDUSTRY EXPERTS

Mangat Thapar
President, International Geophysical Company Inc.

Akshay, this is not an illusion. It is the effect of displaying wavelength which is a function of velocity and frequency. Modifying the signal in depth to look like the time section will lose its meaning because you will be modifying wavelength and hence arbitrarily changing the frequency. You could minimize this effect by showing the depth interval covering your zone of interest. I do not recommend the modification you are looking for.

Akshay Audhkhasi

Geologist, ONGC Videsh Ltd

Mangat, can you please elaborate more about what you just explained? And, please tell me about the sentence "You could minimize this effect by showing the depth interval covering your zone of interest".

Mangat Thapar

President, International Geophysical Company Inc.

Akshay, time to depth transformation is also governed by the following relationship:

Wavelength = velocity / Frequency.

As the depth increases, frequency decreases and velocity increases which translates into longer wavelength with increasing depth. This phenomenon is more pronounced in geological environments of older rocks where there is a steep velocity gradient, that is, the velocity increases from very low velocity near the surface to a high velocity at greater depth. In such cases you should see an obvious elongation of the wavelength with depth. However, in the case of younger rocks the velocity does not increase as much, and the elongation of the wavelength is dependent only on decreasing frequency, and the elongation will not be as pronounced as in the case of the higher velocity gradient. Once you have experience in an area, you may be able to relate the wavelet elongation with depth to a known stratigraphic section. About my comment of displaying a small depth interval to minimizing the effect of wavelet elongation, if this phenomenon is distracting to someone, then they can work with a smaller depth interval and not display the whole depth section. The wavelet elongation is only obvious over larger depth intervals as compared to shorter depth intervals. This phenomenon should be used by interpreters to detect anomalous wavelength decrease/increase which could be due to saturation of hydrocarbons. This is why I recommended you not try to modify or eliminate this variation as it can serve as a HC (Hydrocarbon) diagnostic tool.

QUESTION 53

As we all know, time-to-depth conversion can be done using different sources of well velocities (not talking about seismic velocity). So what is the most reliable source of velocity (VSP or/and Check-shot, Sonic logs, Depth markers)? And if possible do I have to correct the sonic log against the Check-shot/VSP to perform the depth conversion, or what?

Raouf Nasr

Geophysicist

ANSWERS PROVIDED BY INDUSTRY EXPERTS

Mark Morrish

President at High-Definition Seismic Corporation

Raouf, the zero offset VSP will give you the most reliable velocities.The sonic tool uses ultrasonic frequencies and only provides data from just around the well bore. The VSP has a similar a frequency range and similar ray-paths to the surface seismic data. Frequency dispersion is one effect that can cause the sonic velocities to be off, but even after compensating for this, there could still be significant differences between VSP derived velocities and sonic velocities. Two examples of this: In the oil sands of Canada, shear wave velocities in the bitumen reservoir can be 50% higher from dipole sonic logs than those obtained from the VSP data. Another documented example occurs when a heavy oil reservoir is being heated. In this situation the P wave velocities from sonic logs might be off by 20%.

The term "Check-shot" is usually used to describe a Zero-Offset VSP that is carried out in an open hole with a tool that only has a few levels (usually 3 to 5). Moving and reshooting continues until the desired interval has been covered, but positioning errors/uncertainty will reduce the accuracy of the data. A project that is referred to as a **"VSP"** is more likely to use a larger array (from dozens of levels to hundreds of levels) which might not even need to be moved. Velocities obtained from a large array will be more reliable than from a small array since the spacing will be accurately known and constant. The idea is to have an array long enough to cover the whole well, with sensors spaced closely enough to avoid spatial aliasing. For the high resolution VSPs in the oil sands we use arrays with receivers at 1m or 2m intervals.

Mohamed Badawy

North & East Africa Business Manager, CGG

Raouf, the best velocity would actually depend on the data quality. For example if the check-shot or VSP have noisy direct arrivals or casing ringing then in this case sonic might be considered better. But you need to pay attention to its quality in terms of borehole rigousity and dispersive effects since the sonic transducer transmits acoustic waves of 10-25 KHz (Formation acoustic properties will alter depending on the frequencies used to measure them). To simplify the difference between VSP and Check-shot the first (VSP) has a closer spacing for the receivers hence a better control on spatial aliasing (but not frequency aliasing which depends on the recording sampling rate). The reduced spatial aliasing would allow better separation between the direct arrival and up going waveforms and hence you get a corridor stack to compare with the seismic. Meanwhile the check-shot survey will record only direct arrivals that will be used to calibrate the sonic/acoustic for washouts and dispersion. Due to the physics of measurement there can be a difference between VSP and Check-shot travel times which called the drift curve. Calibration would require establishing a knee point correction (a linear best fit to the drift curve) the points of which should strictly follow the formation markers and significant acoustic barriers to avoid creating artifact reflectors on the synthetic seismogram. One thing to pay attention to is that deviations in wells will have an impact since the travel path of the waves in acoustic logging is not the same as in the VSP or check shot surveys.

Zaheer Ali Shah

Technical Coach (Geosciences) at Petronas Carigali, Kuala Lumpur, Malaysia

Raouf, it is not just a choice between VSP, sonic log, check-shot or stacking velocities because the depth conversion strategy fully depends upon the complexity of the geology you are dealing with. Depth conversion must also consider the objectives of your interpretation and the amount of accuracy you need or tolerance of error you can afford in your results. It also depends on which data you have in abundance. For example;

- In the, stacking velocity may be the best choice instead of applying a borehole data from a far off well.
- In case of just one borehole data set available (VSP, check-shot or sonic logs) layered cake depth conversion may be best if you have many fault blocks or lateral variation of thicknesses above your target horizon.

Please focus on the geology and use borehole data as a starting point which is accurate near the well. VSP times are similar to seismic times as Mohmed Badawy said. But they should be considered near the well only if you have severe structural or stratigraphic variations above and around the well.

SECTION

Reservoir Characterization

2

CHAPTER

QUANTITATIVE SEISMIC INTERPRETATION

4

CHAPTER OUTLINE

Quantitative seismic interpretation (QI) is the skillful integration of geophysical, geological, petrophysical, reservoir engineering, and production data, in the predictive determination of reservoir properties, fluids and fluid contacts (Fig. 4.1). In addition, it predicts reservoir extent and sand development away from the wellbore. It can also be used to optimize well placement and to support appraisal and field development planning.

PETROPHYSICAL WORKFLOW FOR QUANTITATIVE SEISMIC INTERPRETATION STUDIES

Well log data (Fig. 4.2) give us detailed information at the location of the borehole. The key criteria for the selection of petrophysical data/well logs to be included in a QI seismic study are the availability of the necessary well logs and the need for adequate sampling of depth/lateral variations in rock properties. The well logs should have been recorded over all the reservoir layers and into the overburden (ideally as close to the sea floor as possible for accurate ties with seismic). The key well logs data and collected data may include:

- Gamma ray: needed for sand-shale differentiation and Volume of Shale (Vshale) calculation.

Practical Solutions to Integrated Oil and Gas Reservoir Analysis. http://dx.doi.org/10.1016/B978-0-12-805464-2.00004-4

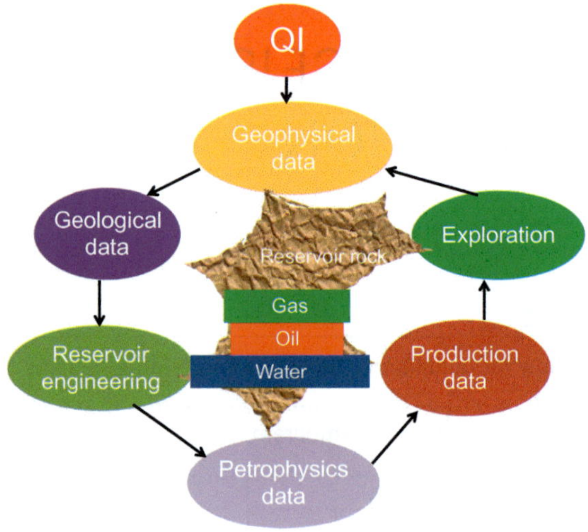

FIG. 4.1

Conceptualized different data used for quantitative seismic interpretation.

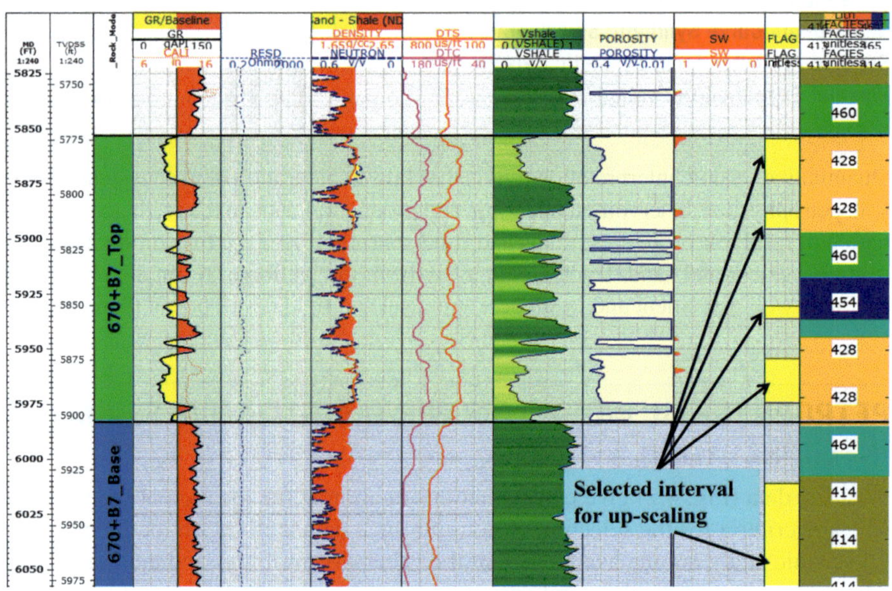

FIG. 4.2

Petrophysical well curves for QI project.

- Caliper: to measure borehole rugosity for data quality control (QC).
- Resistivity (shallow, medium, and deep): for hydrocarbon identification (calculation of water saturation (Sw)).
- Neutron porosity: sand-shale differentiation, Vshale calculation and gas-oil differentiation together with bulk density.

- Bulk density: needed for calculation of porosity and acoustic moduli. Bulk density measures the bulk (average) density of the formation (rock and fluids).
- Density correction: needed for QC of bulk density curve, though density correction is optional.
- Sonic: for interval seismic velocity. The sonic log measures the time it takes for sound energy to travel a specific distance. The sonic log displays the interval transit time (Dt) in units of microseconds per foot (μs/ft). The inverse of Dt is the acoustic velocity.
- Shear sonic (highly desirable): needed to generate shear wave velocity (Vs). Ideally the raw sonic waveforms should be available for QC.
- Water/hydrocarbon saturation and flag curve needed for denoting fluid type (i.e., brine, oil and gas): key input to sand models.
- Porosity: key input to sand models. Where possible, log derived porosity estimates should be calibrated by (stressed) core analysis.
- Pressure data (e.g., repeat formation tester (RFT), reservoir characterization instrument (RCI), modular formation dynamics tester (MDT)): used for gas-oil differentiation, formation pressure, temperature and fluid density and identifying fluid contacts.
- Photoelectric curve: though it is optional, photoelectric curve is needed for QC of bulk density curve.

SYNTHETIC SEISMOGRAM

A synthetic seismogram is an artificial seismic trace created from sonic and density logs used to tie the geology and seismic data collected near the well location. The synthetic seismogram is a direct, one-dimensional model of acoustic energy traveling through the subsurface (Fig. 4.3). The primary well data required to generate a synthetic seismic trace are the sonic curve and, often, the density curve. Check-shot data from the well are also very important.

Reasons Why We Do Synthetic Seismogram

- Synthetic seismograms improve the accuracy of the time-depth-velocity pairs, hence the time to depth conversions.
- The synthetic seismogram improves the overall static and dynamic match of the well to seismic tie.
- The synthetic seismic trace improves the ability to extract the proper wavelet regardless of the seismic section's phase.
- Synthetic seismograms increase the ability to identify multiple generators and multiple periodicity.
- The synthetic seismic trace is critically important for amplitude variation with offset (AVO) and amplitude variation with angle (AVA) analysis, to discriminate between AVO anomaly classes.
- The synthetic seismogram is a source of elastic and acoustic moduli for seismic inversion of both lithologies and fluids.

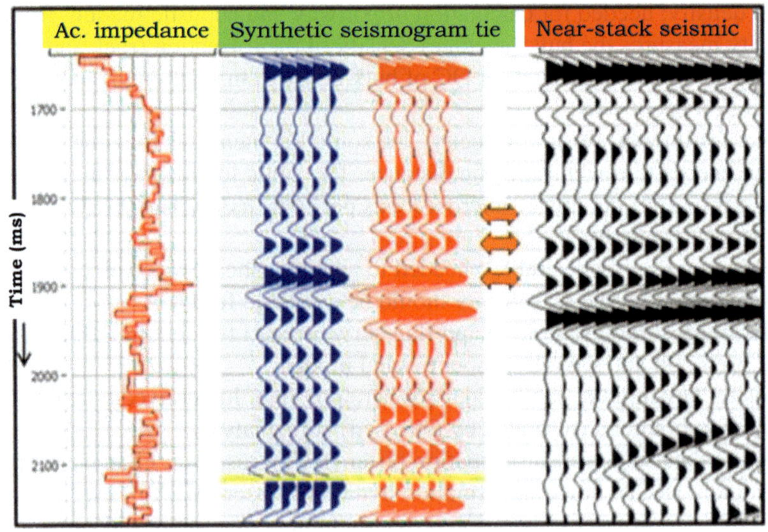

FIG. 4.3

Synthetic seismogram tie. Synthetic seismogram tie for the low-angle (~10°) near-stack. The synthetic seismogram was generated by using the impedance log curve. The correlation seems to be reasonably good.

Courtesy: Arcis SEISMIC SOLUTIONS. http://www.arcis.com/Pages/AngleDependentInversion.asp.

- Seismograms provide information to extract low-frequency acoustic impedance, which is critical to convert the relative acoustic impedance inverted seismic to absolute ones.
- Synthetic seismograms provide a calibration for *Q*-factor to correct for inelastic frequency and amplitude attenuation.

QUESTION 54

Can I generate a synthetic seismogram with only a sonic log?

Diego Timoteo

Exploration Geologist at Hound Exploration

ANSWERS PROVIDED BY INDUSTRY EXPERTS

Jose Antonio Sierra

Sr. Geophysicist

Diego, the answer to your question is "Yes, but ". Assuming that you are making a normal-incidence synthetic seismogram (i.e. not for AVO modeling) you will need to know the density and P-wave velocity of your rocks which ideally comes from their respective logs. However, if you are missing a log, you can infer the properties from other measurements and/or from assumptions based on your own geologic knowledge of the area. If you assume Vp=Acoustic Impedance and make a synthetic using only the sonic, then you are making the assumption that density is relatively constant or that it tracks the sonic log almost identically. This may be a good assumption, but you should try to check whether this holds in your basin based on other wells, laboratory data, etc. If you choose to use Gardner's equation, you are assuming that your rocks behave similar to the ones Gardner used in his paper. The most robust thing to do is to try to understand the rock physics of your basin by seeing how different measurements/properties relate to each other, form your own locally-calibrated Gardner-like equation, and then apply that wherever you think the rocks are similar to the ones you used to build the equation. Now, the amount of effort you put into this mostly depends on how much uncertainty you are willing to accept and what you are planning to use the synthetic for.

Asif Zaidi

Sr. Technical Advisor – seismic Processing and Depth Imaging, Paradigm Geophysical Company

Diego, I absolutely agree with all **Jose**. The effect of density is negligible as compared to velocity for the calculation of reflection coefficient. It does not matter much if we use a straight line (constant density), slant line (reasonably varying density) or Gardner equation (a curve following the sonic trend). The most important caution is never use a bad density log because bad density is an indicator of washouts not density. Always keep an eye on the caliper log, especially in open wells and edit the washout section of the density log accordingly. If more than 50% of the density log is bad ignore it all together.

Bob Avakian

Geophysicist, Instructor, Oklahoma State University Institute of Technology

Diego, "The effect of density is negligible as compared to velocity for the calculation of reflection coefficient" is not always true. I made synthetics for Gulf Oil units worldwide and sometimes had to make a synthetic from just the density to assess the effect on the seismic trace.

Jerry Coggins

Principal Geophysicist at Coggins Geosciences, Inc

Diego, I would add that one needs to use care in neglecting density. One tertiary basin comes to mind where the sand-shale velocity contrasts were small, and density was an important component in the synthetic. An interesting side to this is the use of other logs for synthetics. In places where the gamma ray log is a good sand -shale discriminator, the gamma log can be converted to time and filtered for use as a synthetic. (Typically, it would tie to integrated seismic traces.) I saw this used effectively in the Gulf of Mexico when sonic and density were not available.

Asif Zaidi

Sr. Technical Advisor – seismic Processing and Depth Imaging, Paradigm Geophysical Company

Thanks Jerry. I agree with you. This is a very important addition. Very well mentioned, there is no replacement for experience.

Diego Timoteo

Exploration Geologist at Hound Exploration

Well, actually I only have a sonic log for this reason I am going to generate 2 synthetic seismograms. In the first case I will consider a density as a constant (value=1) and in the second case I will use the Gardner equation to generate a pseudo-density log and later I will compare the results. I am sure that I need to calibrate the Gardner equation, however, this procedure seems to be more real than assuming a constant density along the sedimentary succession of my well (only Cenozoic, Eocene to Quaternary, silici-clastic units).

Asif Zaidi

Sr. Technical Advisor – seismic Processing and Depth Imaging, Paradigm Geophysical Company

I believe that there will not be any problem if the velocity contrast across the formations is reasonably good as mentioned by Jerry. If there are not then you have to rely on pseudo density to play its role.

Andrew Forbes McIntosh

Senior Exploration Geophysicist Specialist

Diego, not sure what kind of rocks you are dealing with but if all you are doing is trying to tie the well bore data to surface seismic, I would not worry about not having a density log. Velocity probably varies from 10000-18000 feet per sec. Density probably varies from 2.2 to 2.8. Impedance is velocity scaled by density. The small change in the density scalar which you cannot account for because you do not have a density log is going to have very little effect on the position of any interface that corresponds to a much greater velocity change. You might miss a few subtle interfaces but your overall tie will be just as good and just as convincing.

Hesham Shalaby

Senior Geophysicist at Dragon Oil

I would add that in case of the presence of salt the density log will reduce the amplitude strength of the synthetic generated as salt has low density. However, the density log enhances the synthetic response to low velocity gas charged sands.

Diego Timoteo

Exploration Geologist at Hound Exploration

Let me share the results of the test results (After the test). I performed the test, generating a synthetic seismogram with a "constant density (value=1)" and a "pseudo-density (from calibrated sonic log)" (Fig. 4.4).

Analyzing the results:

- Graphically both synthetics are very similar (pdf).
- The acoustic impedances obviously differ because the density values are different for each case (pdf).
- The comparison of the RC (reflection coefficient) values obtained for each case is very close, showing little difference (pdf).

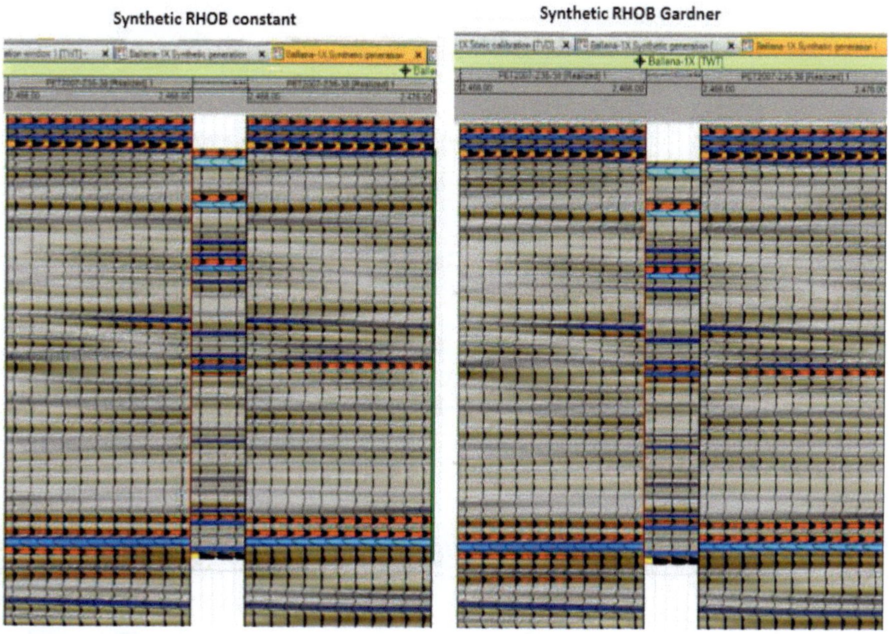

FIG. 4.4

Test result of synthetic generated with in situ density log.

Preliminary conclusions

- Apparently both cases have the same result.
- The synthetic generated with constant density is only valid for tying the well to seismic, but the resulting impedances are incorrect and therefore the reflectivity is not scaling.
- The synthetic generated with Gardner's density correctly tied the well to seismic and in addition the impedance is correct and reflectivity is scaling.

Well-to-Seismic Tie

Well-ties are where and when one compares seismic data at a well location with well log data from the well. This allows us to relate horizon tops identified in a well with specific reflections on the seismic section. We use sonic and density curves to generate a synthetic seismic trace. The synthetic trace is compared to the real seismic data collected near the well location. This is shown in Fig. 4.5.

Objectives of Well-to-Seismic Tie

- *Correlation of formation tops and seismic reflector*: In the exploration stage, we would like to tie well data, (e.g., the top of a stratigraphic horizon) to

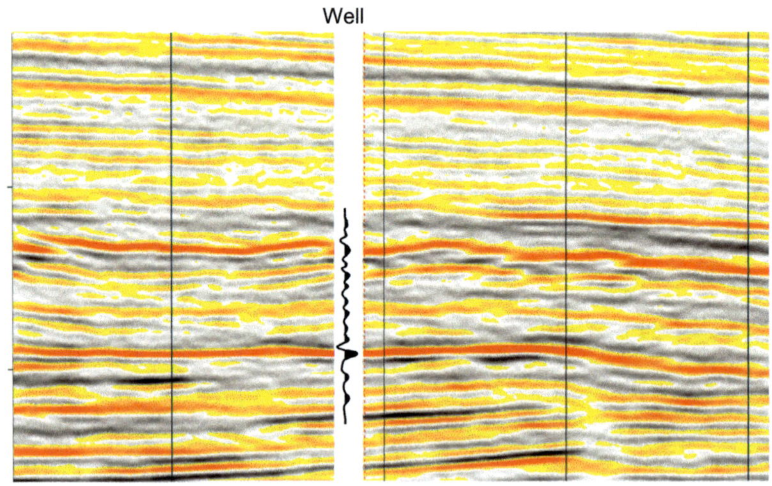

FIG. 4.5

An example of well-to-seismic tie.

Courtesy: ExxonMobil.

within half a cycle of a specific waveform on the seismic section. This requires good data quality in both the seismic section and the logs.

- *Derivation of wavelet for seismic inversion*: In the exploitation stage (development and production), we need to not only know the seismic event within half a cycle, but we need the shapes of the real and modeled seismic traces to be quite similar. For this, we need very good seismic data quality. If we obtain a good character (shape) tie between the real and synthetic traces, then we would be able to extract various seismic attributes (measures of the seismic wavelets) to predict rock and fluid properties with some certainty. We may also be able to use the inversion technique to transform the seismic data into an estimate of the rock properties in cross-section views or as a 3D volume.
- *For regional mapping*: If we are doing regional mapping, e.g., mapping a significant erosional unconformity or a flooding surface, then our tie does not need to be very precise, within 1 or 2 seismic cycles (peaks or troughs)—and the seismic data quality does not have to be very good.
- *For calibration of seismic processing parameters*: As processing parameters can often affect what we see on the seismic section, it is necessary to check these parameters using the synthetic data.

Uncertainties in Well-to-Seismic Tie

- The signature of the seismic imposes some uncertainties in well ties. Seismic processing corrects for spherical spreading and absorption to obtain a stable response across an entire survey.

- Most well ties include generation of a synthetic seismogram from measured impedance contrasts along the borehole. The signature of the synthetic should match the seismic to enable comparison around the borehole.
- The comparison between the synthetic seismic section and the real seismic section may be affected by noise, e.g., logs and synthetics can be impaired by borehole conditions or the seismic section contaminated with multiples.
- Seismic data may be distorted by a complex overburden and must be migrated into an accurate image at the correct location. Errors in (migration) velocities may lead to lateral and vertical uncertainties in well tie.

QUESTION 55

What will be the best way to tie the seismic data to a well which crosses a thrust fault and has a repeated section? The well correlation includes tops from hanging-wall and foot-wall. Assume there is no check-shot data acquired in this well and we need to tie well tops above and below the fault with the corresponding seismic reflections. Neighboring wells have no faults or repeated sections. The well-tops are a good tie with hanging-wall seismic reflections but for the deeper part of the well the correlations are off.

Leonardo Velasquez

Interpretation Geophysicist

ANSWERS PROVIDED BY INDUSTRY EXPERTS

Mike Cline

Consulting Geophysicist

Leonardo, here are three suggestions, going from easiest, to most difficult. You could generate a synthetic seismogram in the faulted well, and use that for the tie. If there isn't a sonic and density log in that well then you would need to create pseudo-logs:

- Pseudo-sonic created from the neutron log works very well, or you can use the deep resistivity logs.
- Pseudo-density data can be calculated from the pseudo-sonic log.
- Or, you could also do some layer-cake velocity modeling, using the interval velocities from the un-faulted well, and carry that over to the faulted well, varying the tops/thicknesses, such as you would do for a simple structural model. Often there are velocity-related, false time structures (pull-ups, or push-downs) created under faults that exhibit significant velocity/structural complications.
- Finally, you could convert the time seismic data to depth, or reprocess it in the depth domain, but this would obviously involve a bit more effort than the two earlier suggestions.

Leonardo Velasquez

Interpretation Geophysicist

Thank you Mr. Cline. To start I would try to follow your second recommendation. Yes, there is a pull-up effect in the seismic reflectors under the thrust fault, but it still appears above the expected structural level according to the interpretation in the footwall of neighboring wells. But I think varying the tops/thickness could work.

QUESTION 56

I have an issue with a well-tie, a North-Sea well. The stratigraphy down to BCU is relatively flat, but dips increase below BCU. Down to and at BCU the well synthetics match the seismic, while below the BUC, there is a mismatch/shift, with events on the synthetic coming in shallower (sooner) than the corresponding seismic events. Stretching the well does not give a good fit, both because there is still some difference in synthetic and seismic time, and also, doing so alters check-shot based velocity profile. I find, however, that comparing synthetics to a seismic trace 6 traces up-dip from the well gives a much better match. I have tried to find relevant literature on this topic, but what I find is very limited, especially since I am not an expert on seismic processing and imaging. White & Simm (2003) states that "For various reasons the best match often does not occur at the well location since velocity typically increases with depth. Time migration commonly moves the best match location up-dip from the well. The effect of imperfect migration velocities is less predictable and it is even possible that the well location is suspect." Does this imply that the problem is most likely in the seismic section I am working on? The well was drilled in 1982-83, and has a deviated well path with maximum inclination of 70 degrees. I have tried to tie the well to two different seismic 3D datasets, with the same result. Is there a way to solve this? Or should we just leave it as it is and be aware of the problem?

Connie Christiansen

Senior Exploration Geophysicist at Pure E&P Norway AS

ANSWERS PROVIDED BY INDUSTRY EXPERTS

Rob Simm

Principal Geophysical Advisor at Origo Exploration UK Ltd

Connie, the first issue I would look at is the effect of the deviation of the well. Are the check-shots you have from a rig source? If so, there is a large uncertainty in the relative timing of logs and seismic that is very difficult to resolve. Of course the other problem is VTI. If you have other wells at different deviation angles (or some other data indicating likely VTI parameters) then it is possible to estimate the correction needed to verticalise the velocities (particularly in the shales) by a comparison of log responses. If so, then I would try and tie the well by using the timing from the integrated (corrected) sonic and if not then I would suggest that you could spend a great deal chasing your tail.

John Chamberlain

Consultant Geophysicist

Connie, I would say that it is very likely a combination of depth-time conversion, imaging and possibly log response. As Rob Simm says, if the check-shots are from a rig-source VSP then it is very difficult to reconcile these with the seismic times. Regarding the mispositioning, it sounds like you have time-processed data. In my experience, this can be mis-positioned by up to 100m or so even in relatively simple overburden. This is even more possible once you get into dipping sequences with velocity contrasts (it is quite easy to estimate the positioning error using an image-ray trace technique, which can be done in Excel). It is interesting that you see the same offset with two datasets. Another source of error would be the deviation survey; for a well of this age where uncertainty in vertical and horizontal positioning (especially orthogonal to the trajectory) could be some 10s of meters. Given these uncertainties, 6 traces (I assume this means 150m) is not a very large error and I would be tempted to do what you say - note it and move on. Incidentally, I would much prefer this to introducing arbitrary stretch and squeeze; this is an issue for tying depth maps as well. People often seem to assume that there is no lateral positioning error and then force a vertical tie.

Duane Pankhurst

Principal Geophysicist at Edge Geophysics, LLC

Connie, Well-tying can be as much of an art as it is a science. However, in your case I would suspect the problem is with the logged seismic data (sonic), particularly in the highly deviated portion of the well. It was in West Africa where many wells were drilled where TTI (tilted transverse isotropy) was first understood and defined. Many of the deviated wells had poor well-ties, even after reprocessing the waveform data. One of the elements had to do with bedding at a high angle to the wellbore causing issues that processing could not undo. Also, there was a suspicion that the sonic tools were not truly centered.

The two elements combined led to erroneous sonic data, particularly in the highly deviated portions of the well. Current well-logging tools are designed to take the off-center condition into account and correct for it. I would trust the check-shot data and in Excel turn it into Vint (Interval Velocity) and splice the upper half to your sonic. This has two benefits. You reduce your uncertainty in your depth to time initial step. Many times, I need a very short (10-20ms) bulk shift using this technique. The other benefit is accurately calculating turning waves through the overburden. Furthermore, you can use the check-shot to verify what sections of the sonic are suspect. If after you have convinced yourself that things are reasonable (survey loaded correctly, etc.) and you feel the well is mis-located one or two CDP's (25m spacing), you should not worry too much. A typical issue I see often is practitioners who generate a zero offset synthetic and attempt to tie to the full stack. The full stack can look nothing like a zero offset due to things like Class 2p AVO. To be rigorous, one should tie to the gathers or at the least a near trace stack.

Amir Abbas Babasafari

Sr. Reservoir Geophysicist at Tehran Energy Consultants

Connie, The following methods are recommended:

- During check-shot correction, use relative change option instead of time-depth relation to see if sonic log is changed in an incorrect manner. If so, try to do well-to-seismic ties without check shot corrections and just using defined Depth and Time value at each horizon for this purpose.

- I think you can use VSP seismic data if available (Walk away or Walk above based on your inclination) to solve this problem by comparing seismic and VSP matches and also Synthetic and VSP ties to see if the reason for your difficulty is the well log or seismic data set.
- If there is a time lapse between well logging and acquired seismic data, you can also test fluid substitution for synthetic seismogram generation and then try to match it with seismic.
- On the other hand I propose to try well-to-seismic tie using near offset stack data to find any probable processing issue.

Nayyer Islam

Geoscientist

Connie, as velocity is one of the inputs in migration, an incorrect value can move the reflectors up or down. So it could be an imaging issue. If the mismatch is only at one horizon, you may want to consider possible tuning that can cause a localized mismatch. If there is a time gap between the two data sets, you may want to check for any likely changes in the overburden associated with production or any other activity. You may also want to consider the anisotropy effect. A few models with different scenarios may help you nail down the issue. There are many factors that may cause such issues. It is hard to pick one without modeling.

Peter Rowbotham

Geophysics Advisor at Apache Corporation

Connie, perhaps the imaging (DMO & migration) has not been done correctly in moving reflections to their true subsurface location. Points to investigate would be the velocity field and in particular the velocity interface around BCU. What drove the velocity field – imaging, that is, what were the focusing objectives? Is this biased by a particularly strong reflector? Of course it would be nice to know whether (anisotropic) PSDM would improve the situation.

The well should be relatively Ok as far as positioning goes. Yes the check-shot velocity could well be interrogating the rock at different angles form the wire-line velocity therefore introducing an anisotropic effect (e.g. Hornby or Rowbotham publications from 1999-2002). I am not sure that is particularly relevant for the well-seismic tie but what are you using to make your well AI log? If this is from sonic, perhaps you should test an anisotropic correction factor in the shales. Where is your time-depth curve coming from? If from sonic rather than check-shot then perhaps an anisotropic correction is needed again.

You say check-shot velocity is too high because of deviation. If this is rig sourced survey, then having a 70 degree deviation in the lower section will increase your interrogation angle a bit. I suggest you ray-trace to investigate. If it is walk-above source VSP, then this should not be such a big issue.

To answer your final set of question: being aware and leaving things as it is could be acceptable depending on the purpose of your work. But ensure that it is well communicated for the next person/team who pick this up. Can it be solved? Potentially, maybe through better imaging or by checks on well data it could be solved.

Arslan Tashmukhambetov

Geophysicist at LLOG Exploration Company

Connie, which software are you using? Working with time data, some software packages require you start tying data from top to bottom, saving tied check-shot for every tie point and re-extracting the trace. This is, because some programs shift positions of deviated wells.

Phil Lock

Consultant Geoscientist/Reservoir Modeller at PETREL Phil Pty Ltd

Connie, the one facet I would look at very closely is the stratigraphy. You mention that below the BCU, the data is dipping and this is where your misties start appearing. Some aspects that should be considered are: Age of data and acquisition technology as well as the stratigraphic dip of the geology. Note that the VSP is a relatively new technology. Two things that spring to mind are that the drill bit will migrate (move or walk) up-dip. Perhaps check your driller's logs also to see what the drillers are saying. Also, ray path or optics theory would suggest shorter ray-paths will arrive sooner hence again the stratigraphy would influence your VSP outcome. So, you can see how the geology will indeed affect the ray-paths. And you haven't yet reviewed just what additional distortions will be introduced when you take a simplistic ray path model and post process the data and finally migrate it and then try and tie it to well. In essence, consider the geological setting, look at the original drilling report, compare the drilling report to the VSP report (in case there is any error made). Armed with that information, then look at your Vint (interval velocity) and remember that migration relies on velocity assumptions.

STRETCH/SQUEEZE TO IMPROVE WELL SYNTHETICS TO SEISMIC MATCH

QUESTION 57

I receive the maximum CC during well-seismic correlation. To effect a tie may stretch and shift the synthetic seismogram, which results adjusting the current velocity or time-depth log. But, what percent of initial velocity is the most we should accept?

Masoud Aali

Seismic Interpreter

ANSWERS PROVIDED BY INDUSTRY EXPERTS

Maher Maklad

President at Signal Estimation Technology Inc

Masoud, The sources of mismatch requiring stretch/squeeze to improve well synthetics to seismic ties include:

- Difference in sonic and seismic velocities - This should be addressed using time-depth curves from VSP or check-shots to correct sonic velocities. A better approach is to model the relationship between sonic and seismic velocities using cluster-guided regression to generate a model for each lithology using a regional data base.

- Presence of multiples or converted waves in the seismic. The matching algorithm should be able to detect such events; otherwise the wavelet estimated using well reflectivity and seismic data will attempt to model them as well and be accurate.
- Modeling errors. Those resulting from ignoring the truncation effect and working with evenly sampled reflectivity, to mention a few. We should strive to address such modeling errors in the process of matching. How much change in well log velocity should be allowed to improve the match through stretch and squeeze? This will depend on the available data.
- Sonic plus VSP or check shots: very small changes should be allowed. Remember the other factors that may cause the mismatch.
- Only a sonic log: up to 5% after original match using an initial wavelet designed according to the band-width of the seismic (Ormsby or wavelet estimated from seismic). The optimal constant time shift and phase shift should be found first to improve the alignment of the reflectivity with the seismic.
- No well information: Rely on wavelet estimation, amplitude and phase from seismic data alone without white reflectivity assumption. We had excellent results using this wavelet in de-convolution and inversion.
- Finally, CC, being a gross measure may not be the best indicator of fidelity and one must rely on examining the statistical properties of the residuals such as their spectra to assess the quality of the match (See F. J. Ancombe, American Statistician, vol. 27, 1973, pp. 17-21).

MORE TECHNICAL ANALYSIS OF QUESTION 57 BY INDUSTRY EXPERTS

Erik Johnson

Fugro Geoscience Division

Masoud, well, in ideal conditions there shouldn't be more than a 1% or 2% difference in velocity between that measured by the sonic log tool and the seismic velocities. Of course, these measurements are not perfect. Borehole conditions affect log velocities, while picking accuracy and structure affect seismic velocities, etc. In my experience, I could usually get a good CC and QC if I limited myself to a 3% difference, with 5% being an upper limit. If I could not get a good tie within those bounds, it was time to ask a petrophysicist or rock physics expert to make a synthetic sonic, I could compare to the measured sonic to look for possible log errors that could be corrected to give a better tie. Depending on the experience of different professionals, you are likely to get a lot of different opinions on this question based upon: hard rock versus soft, deep versus shallow wells, age differences between the well logs and the seismic shoot, and a host of other factors.

Andrew Wilson

Team Leader, Reservoir Geophysics and QI at BG Group

Masoud, how many wells are you working with and do you have good quality VSP, or maybe a check-shot, in at least one of your near-vertical wells? Dispersion means that sonic velocities are faster than seismic velocities so you expect some drift between the time-depth integrated from sonic and the time-depth measured by a VSP. When you run the drift curve interpretation you should get an idea of how big this difference is. After drift curve interpretation you will have an updated time-depth curve that should then only need a very small stretch to tie to seismic (although of course, real data is never perfect). If you do a well tie in a well which does not have a VSP/check-shot survey you can then reasonably expect to need a stretch that is similar to the drift you saw in your wells that did have check-shots.

Michael Doty

Principal at Doty Geoscience

Masoud, I agree with Andrew that a good VSP is the best way to come up with a drift curve. As Erik points out, the amount of stretch can vary depending on lithology, age, depth, and burial history. In the Western Canadian Sedimentary Basin I have found 5 to 7% stretch to be fairly common.

Andrew Wilson

Team Leader, Reservoir Geophysics and QI at BG Group

Michael, that was a useful and interesting observation. Please can I check to make sure I understood? When you say 5% to 7% stretch, did you mean that much of a change during drift curve interpretation, or that much stretch needed after drift interpretation? Were you working on with wells that did not have a VSP/check-shots?

Michael Doty

Principal at Doty Geoscience

Andrew, the 5-7% stretch is on wells that don't have a check-shot or VSP and is an average from surface to TD.

Sbonelo Ngcongo

Geophysicist at Petro SA

Masoud, I have performed seismic well ties by interactively doing the sonic log calibration and synthetic generation at the same time. I prefer the integrated seismic well tie approach because it combines the sonic calibration and synthetic generation as one workflow. I do not think there is a set limit of stretch with real life data. The most important thing is that the calibrated sonic log should not create false events on the synthetics, so you should be careful when creating your knee points. It also depends on how much drift is there between seismic times and sonic times, or seismic velocities and sonic velocities. We know drift results from all sort of problems, starting with borehole condition effects (i.e. negative drift) and so on, sonic velocities are usually slower than seismic velocities. Positive drift is usual from dispersion effects, as it was explained by Andrew. Petrel software came with a solution where you corrected for drifts while performing a seismic-to-well tie. Loosely speaking, one is searching for best calibrated sonic log that generates synthetics that best tie seismic trace, while honoring check-shot or VSP data.

Andrew Wilson

Team Leader, Reservoir Geophysics and QI at BG Group

Sbonelo, yes I strongly agree with you. Avoid knee points that create false events and/or change the amplitude of real events. If you do drift curve interpretation by fitting your drift curve with a series of straight lines between "knee points" then each change in slope of the drift curve creates a bulk shift in the drift-corrected sonic. I avoid using knee points by running check-shot QC and edit before going into well ties and then using a smooth line to fit my drift curve. Do you make your synthetic using the original sonic or the drift corrected log?

Sbonelo Ngcongo

Geophysicist at Petro SA

Andrew, I use a drift corrected sonic log to generate synthetics.

I overlay the corrected sonic log over the original sonic log to make sure that I do not create false velocity anomalies on the corrected sonic log. Yes, between the two knees, the slope of the line segment joining those knees is the average drift that needs to be applied on the original sonic log in that interval. When a slope is positive, that means average drift is positive in that interval, therefore sonic times are less than check-shot times or the sonic velocities are faster.

I create the keen points at changes in the sonic log or lithology or at the major changes in acoustic impedance. I manually vary the knee point locations on the drift plot panel, while observing a tie: it is a trial an error method.

Andrew Wilson

Team Leader, Reservoir Geophysics and QI at BG Group

Sbonelo, I agree that if you are going to use knee points and the drift-compensated sonic then it is good to be careful in the ways that you mention, including only placing the knee points at the location of strong seismic events. Of course, if you were to place a knee point at a low amplitude event then the impact of the jump change in the sonic might make a significant change to the event on the synthetic. For example, a low-amplitude trough could be changed to a peak. With a strong event we can hope the change is not big enough to have a noticeable impact and so we can still use the synthetic to make a "structural" well tie. However for "QI" well ties, I prefer to avoid the problem and use the original sonic with the drift-corrected time-depth relation.

KNEE PICKING IN SONIC CALIBRATION

QUESTION 58

I want to know about Knee picking in sonic calibration. What is the concept and are the steps in doing this?

ANSWERS PROVIDED BY INDUSTRY EXPERTS

Ajay Badachhape, PG

Sr. Reservoir Geophysicist, Huston Consulting, Global Geophysical Services

I have heard and used the term "knee-points" for sonic calibration in time-to-depth conversion. These are the checkpoint/VSP data points and additional ones that are often (manually, with software designed to do this) added to align the synthetics with the seismic. They are also called "tie-points" and several other similar terms. Note that the calibrated sonic would be used only for time-depth conversion, and the original sonic should be used for computation of Vp, Zp, etc.

Steve Adcock

Geophysical Consultant at dGB Earth Sciences

We used to call these (Knee picking in sonic calibration) pseudo-check shots in the old Cogniseis DLPS software days. You are using interpreted seismic well-tie points to help you calibrate to depth. The concept and scientific steps to do this are for you to start by selecting some obvious lithology breaks which should generate a seismic reflection. Test it out with the new calibrated synthetic tie. Iterate, adding points one or two at a time, but don't add points past your confidence level that the geologic marker is a seismic event. Avoid overdoing this as the process gets unstable. You can clean up minor time mis-ties when you do your deterministic inversion later.

Grant Hanson

Consultant-Seismic Data Processing Software Testing, DISYS

"Knees" is a common term in picking seismic velocities for various stages of processing. They are generally picked where linear trends in velocity change. The term should refer to the same thing as picking points to calibrate seismic to sonic logs with all the warnings Steve provided about "overdoing" it.

DEPTH-DEPTH CURVE SYNTHETIC SEISMOGRAM

QUESTION 59

Is it possible to make a synthetic seismogram with seismic data in depth domain to generate a depth-depth curve (DD) and tie the well data and seismic data in the depth domain?

Ngel Jimenez Carballo

Ingeniero en Geociencias

ANSWERS PROVIDED BY INDUSTRY EXPERTS

Ifeanyi Ifeonu

Executive G&G Exploration at Petronas

Ngel, Definitely, most interpretation platforms allow one to select the domain of the seismic data; either time or depth. In depth domain you don't need a TZ curve. They assume a one to one, time to depth ratio. However at the end of your interpretation in depth, you calibrate the depth where you have wells and then create an error map (this can range from positive to negative values), after which you add it back to your final depth map - Some refer this process to backward interpolation.

Nestor I. Cortes U.

Geophysical Advisor

Ngel, yes, it's possible. You can generate synthetic seismograms in the depth domain. The principal objective is to define precisions of your depth or time migrations seismic data, as explained by Ifeanyi. Today's software has the tools to do seismograms in the depth or time domain. This depth-depth table allows you to correlate events in seismic data time or depth domains, with the well logs.

Peter Rowobotham

Geophysical Advisor at Apache Corporation

Ngel, One strong word of caution - synthetic seismograms are generated by convolution of the reflectivity sequence with a wavelet. The wavelet is a time domain object as this is what the seismic experiment is. If you do convolution in the depth domain, you need a 'depth wavelet' but this wavelet will be velocity dependent. For example, if you are convolving with a spike at the interface between a low velocity and a high velocity, a zero phase wavelet in time will be depth-squeezed in the low velocity (at the back of the wavelet) and depth-stretched in the high velocity (at the front of the wavelet). From this you can see that convolving a reflectivity sequence with a constant wavelet in time is equivalent to convolving with a wavelet that changes at every depth sample, depending on the vertical velocity profile.

Jeffrey Ross

Senior Geophysicist at Whiting Petroleum

Ngel, I agree with Peter Rowobotham 100%. You are being 'foolish' if you make a 'depth' synthetic seismogram with a constant wavelet. This would ignore the effects of Q filters (i.e. the progressive loss of high frequencies with depth). If you wished to make a synthetic that ties your depth section, make the synthetic in the time domain then depth stretch it with the same velocity function you used to depth correct your seismic section. I always caution that just because you can find some software somewhere that allows you to do something, doesn't mean it is a good/valid thing to do.

Ifeanyi Ifeonu

Executive G&G Exploration at Petronas

The software recognizes the rudimentary pitfalls pointed out here and allows anyone the leeway to import or generate wavelets (statistical, analytical, etc) from wherever. The software are very robust and have been tested even for specialized studies.

Rajeshwaran Dandapani

Director Geophysics at Telesto Energy

Ngel, Synthetic seismogram is convolution of a wavelet with reflectivity series. A wavelet cannot be defined in a robust manner in the depth domain as it is defined in time domain. The wavelet will become highly variable as we move down in depth as the velocity function is varying. Hence in my view a synthetic seismogram should not be attempted in depth domain.

SEISMIC ATTRIBUTES FROM SYNTHETIC SEISMIC DATA

QUESTION 60

How useful are seismic attributes when dealing with synthetic seismic data, in particular a Marmousi 2 synthetic dataset? Are they as good as seismic attributes for real data for estimating rock properties?

Hussain Alkhalifah

Exploration Geologist at Saudi Aramco

ANSWERS PROVIDED BY INDUSTRY EXPERTS

Nicholas Cooke

Geophysicist at Maersk Oil

Hussain, Seismic attributes from synthetics are a very good way to calibrate the attribute response to something more meaningful (e.g. some rock physics property). The advantage that synthetics provide (Marmousi or otherwise) is that the interpreter has all the control over the inputs. Attributes are straightforward mathematical processes so that the interpreter can 'play around' to assess what the qualitative or quantitative values of the attributes are when applied to real data. My only recommendation would be to either use some statistical wavelet or normalize the synthetic and real data and give some consideration to Q such that the synthetic results are given more of a chance to have some meaningful numerical value.

Mohamed Shihata

Geoscience Business Development Specialist

Hussain, I completely agree with Mr. Nicolas. Synthetics can help you to identify the ideal response for seismic attributes for different geological features, it can also help you to generate your test cases and control all the items that can affect seismic attribute responses. You must take care: synthetics are not equal to nature but can give you a key to predict complex parameters in your real case study.

Hussain Alkhalifah

Exploration Geologist at Saudi Aramco

Mohamed, Thank you for your comment. When I create seismic attributes from synthetics that should be better in terms of how much noise there is because it can be noise free as compared to real data which as you said can give better relations. But I am worried there can be major differences between the relation of physical properties and the attributes generated from synthetic and real data when dealing with attributes affected by frequency/wavelet shape. Would you please comment on this?

Hussain Alkhalifah

Exploration Geologist at Saudi Aramco

Nicholas, Thank you for this valuable comment. You have mentioned it is important to consider Q factor which some synthetic models don't as they are either acoustic or elastic. The Marmousi 2 model is elastic but does not consider real effects such as attenuation or anisotropy. Do you think it is possible to use a fully processed 2D synthetic line and make it more realistic with respect to attenuation or this has to be done earlier (before processing or before shooting the synthetic gathers)?

Nicholas Cooke

Geophysicist at Maersk Oil

Hussain, You could either use some function/filter derived from real data and use that to mimic Q on your synthetics or use a software package that does the job (I think NORSAR does this). Generally though, when I do synthetics it is to better understand wavelet effects and geology so Q may not be necessary as I am mostly interested in different interference patterns or seismic characteristics.

Don Adams

Chief Geophysicist at Saddleback Geosolutions

Hussain, The modeling should be representative of the seismic response of the geology and wavelet effects. The attributes from the model will be correct given the level of model detail and the seismic effects modeled. The biggest differences between modeled attribute response and observed attribute response may be caused by:

1. How interference between reflectors is handled
2. Changes in how velocities in the beds are represented (e.g. constant, increasing or decreasing impedance) will affect some attributes.

Both will affect spectral decomposition and phase attributes. In an exploration study it would be best to build a model where you have seismic and well control. Change the modeling parameters or type of model until you match the seismic or understand why it does not match. Then you can build scenario models based on best matching model. Ideally you would model gathers and then process them and stack using the processing flow from your the observed data.

Mohamed Shihata

Geoscience Business Development Specialist

Thanks Mr Don, Nickolas, and Mr. Hussain for your valuable comments. My advice is to start your model using arbitrary lines between nearest wells to help you to calibrate your models to your real seismic data. That can help you to solve your geological problems especially if they result from conflict over frequencies for not interested geobodies or result from noise conflict.The benefit of 2D modeling is in testing and calibrating seismic attribute responses with real results from well information. There is a lot of software help available which allows you to enter many parameters so you can create a model close to your seismic. Modeling can also help you to remove some effects due to noise cancelation, band frequencies change, phase shifting, and so on. You compare model seismic attributes with real data to determine a best workflow to follow analyzing seismic attributes or in choosing the most diagnostic attributes for your target.

Another important use of synthetic 2D modeling between wells is to compare the seismic attribute responses in your known reservoir to other shallow or deep prospects after you change velocity and density of your model. Also you can use software to help you change oil character by Model Fluid Substitutions.

Abdulnaser Abousetta

Geophysical Advisor at Tatweer Petroleum

Hussain, I think one has to choose the right modeling algorithm (i.e. elastic or acoustic, full or truncated wave equation) and the domain (i.e. 2D or 3D, Pre or post stack) capable of modeling the phenomena that your specific attribute is attempting to reveal.

Hussain Alkhalifah

Exploration Geologist at Saudi Aramco

Dear Abdulnaser, I was thinking that a visco-elastic model is needed to model for attenuation but found certain popular synthetic models are less than that. As I said earlier Marmousi 2 is elastic and Marmousi 1 is acoustic.

Turgut Ozdenvar

Seismic Application Developer at Wave Equation

Hussain, software for modeling visco-acoustic media (P-wave velocity, density, and Q) or visco-elastic media (P-wave & S-wave velocities, density, Qp and Qs) is rare in commercial platforms, especially for full wave-equation based solutions. For a Marmousi 2 model which is a large size model even if it is 2D model, you need a high-performance computing platform to get your CSGs overnight. Check your with R&D department as to where you might find help among geophysical gurus. Otherwise let me know if I could be help.

WAVELET EXTRACTION/DERIVATION

A wavelet is a relatively small sequence of seismic amplitude values at a constant sample interval in time, which represents a seismic impulse response that propagates through the earth. The wavelet is convolved with a reflection coefficient series to generate a synthetic seismogram. A wavelet represents the seismic pulse generated by dynamite, air gun, Vibroseis, or other source of seismic energy – as detected on return from the earth. The geophone or hydrophone "hears" a series of overlapping echoes, each echo being a return of the seismic pulse. If you yell "Hi!" across a canyon, you will likely hear "Hi!" in return – diminished in volume but recognizable. If you yell "Hi!" in a vacant stairwell, you may hear a series of sounds, each being an echo of "Hi!" but difficult to recognize, after the first one or two returns, because of overlap.

When creating a synthetic, you would like to have the same wavelet as applied by the processor to the seismic trace. Convolution with the logged impedance would then presumably do the best job of approximating the seismic trace. To obtain this wavelet, you may extract one from the seismic data itself, or you can use one of the theoretical wavelets.

A wavelet is numerically represented as a series of numbers separated by a constant time increment. The reflection coefficients derived from the logs are represented in time as provided by the time-depth chart. The wavelet is then replicated centered on each impedance boundary. The wavelets overlap, much like the echoes in the stairwell and the overlapping values are summed to approximate the seismic trace.

Various methodologies can be used for wavelet extraction, as follows:

- Extract from seismic traces (frequency matching) – extracts an average frequency spectrum over a specified time range of selected traces, and then computes a zero-phase wavelet that has the same spectrum.
- Extract from seismic traces (Walden-White) – computes a wavelet that acts as a matching filter between seismic data near the borehole and the reflection coefficient series calculated from well-log data located at the borehole.
- Extract from seismic traces (Wiener-Levinson) – compares the reflection coefficients from the well log with the sample values in the extracted trace. The extraction computes a wavelet that will most closely reproduce the extracted trace when the extracted wavelet is convolved with the reflection coefficients using a least-squares algorithm. The extraction result is a mixed-phase wavelet that is neither a zero-phase nor a minimum-phase wavelet.

QUESTION 61

During wavelet extraction using seismic and well data, how does an incorrect time-depth relationship impact the phase of the extracted wavelet?

Rajesh Soni

Seismic Interpreter at TOTAL E&P

ANSWERS PROVIDED BY INDUSTRY EXPERTS

Dhananjay Kumar

Geophysicist at BP

Rajesh, In terms of wavelet, you are after amplitude spectrum (amplitude-frequency relationship) and phase (think of side-lobes and interpretation linkage to geology/wells). If your T-D relationship is poor, you can still get an amplitude spectrum from seismic data alone if you know (approximately) the zone of interest on the seismic data (so called statistical wavelet). If your well-tie (T-D) is poor, you get poor/unstable phase estimation at the well. Run this at multiple wells to get a sense of variability. I consider good well-tie (in 100+ time samples) a prerequisite for phase estimation. Alternatively, if you can find the phase of the seismic data by other means, apply an appropriate phase rotation to make the seismic zero-phase and then estimate a zero-phase statistical wavelet from the seismic data alone. People apply various tricks to estimate phase of seismic data in the absence of well data. One way is to find a strong reflector on the seismic near target zone. If you know the geology you expect to give a strong reflection, you should be able to guess whether the seismic phase is zero at that reflector by looking at the shape of waveform. If it is not, then rotate seismic by various angles and observe the shape of that strong reflector till you are happy. For example, if reflection is due to increase in impedance, one should get a strong peak with symmetrical side-lobes for a zero-phase (SEG polarity) data. However, a wrong time-depth relation prevents a good match between seismic data and well data such as synthetic seismogram and corridor stack from VSP data. Before extracting the wavelet you need to make sure that a good match can be seen between both data sets. The best thing to do to avoid this problem is to run a VSP in the well. A VSP has high resolution and provides more correct time-depth relationship than a synthetic.

Gunnar Holmes

Sr. QI Geophysicist

Rajesh, wavelet extraction has to be done in time, while the well data is in depth. The linkage between the two is subject to a number of errors, both large and small. Check-shots are required to get the bulk shift established, but its measurements are rarely well matched in frequencies to the 3D seismic, and especially the well log data, leaving significant inaccuracies in the TZ pairs. Problems with too a short interval in the sonic logs, plus logging errors and seismic imaging issues, make finding the best TZ an iterative process. If most of these errors are adjusted for and a detailed well-to-seismic tie can be made, then a wavelet can be derived that will be a close estimate and that will show some phase information. But how precise is the phase calculation? Doesn't the phase vary with depth and lateral location? Too often the well-to-seismic ties are too poorly done or suffer from missing data so do not give phase results of any precision. So instead we should do what gives us an accurate result. I would rather use a statistical wavelet derivation approach around the well from the best imaged seismic, then check that with a synthetic seismogram correlation. You can then perturb the tie and analyze the phase variance to look for major deviations from expectations. Or you can invert the seismic data with a phase assumption and phase corrections.

Marcin Przywara

Sr. Geophysicist at PGS Reservoir Ltd.

Rajesh, it also depends on the magnitude of time-depth table errors and on the methodology of extraction. There is software (Wavelet Extractor of Gunning & Glinsky) which works in Bayesian framework, where you can allow for some variance in your time-depth tables during extraction process (how much you will allow your check-shots to move, will depend on the probability distribution assigned to T/D pairs). It is however user-unfriendly software (unless somebody developed some wrapper for original scripts, which could handle numbers of possible exceptions). As a more sensible approach, you need to make few assumptions, and derive wavelet which is simply practical for your purposes, but not necessarily the "correct" one - whatever "correct wavelet" means in a non-stationary world. Assuming the phase is stationary in the interval of interest you get rid of non-stationary effects (trickier when the length of extraction needs to be significant, i.e. while working on broadband data). The second assumption many practitioners make is that the phase spectrum is constant. The truth is most likely far from it. But to get phase spectrum of the embedded wavelet, in general, is not trivial task and depends on the excellent quality and reasonable amount of the data. With those two assumptions, the approach suggested by Gunnar, to extract statistical wavelet and assign constant phase shift on the basis of synthetic/seismic trace comparison seems to be characterized by the best complication/value factor.

Doug Knowles

Geophysicist at Santos Ltd

Rajesh, there are many things at play when extracting a deterministic wavelet using a sonic log which has been corrected for drift, bulk-shifted and then used to produce a time-depth curve. For one thing every time you modify your drift curve you will get a differing sampling of the sonic and density logs and then slightly different reflection coefficients. In my experience, the deterministic wavelet will have quite different time-lag, frequency and phase characteristics if the time-depth relationship is non-optimal. Drift of the sonic while potentially linear for geologic zones, might be quite variable over the larger window from which you are extracting your wavelet. I agree that a corridor stack from a zero-offset VSP is a very useful way to get on the right track with your time-depth curve. You will have to band-pass filter the VSP to match the seismic more closely, due to the Q-effect from the two-way ray-path in surface seismic (versus the one-way VSP product). Tying your synthetics (or VSP) to near-offset stacks is a good way to avoid stretch and AVO effects.

Hamed Amini

Geophysicist at Senergy Energy Services

Rajesh, Yes, the wrong time-depth law has an impact on the phase of the extracted wavelet.

The reflectivity series for k interfaces is:
R(w) = Sigma[r_i * exp(-i(w*t_i)]; i = 1:k
w: frequency
r_i: reflectivity series for each interface
t_i: TWT (two-way-time) for each interface

Now, if the time-depth table is wrong, it means that t_i for some of the interfaces is not right, to compensate for that you need to do stretch and squeeze:

R(w) = Sigma[r_i * exp(-i(w*(t_i+-dt))]
dt: twt shift for stretch and squeeze
and you can consider the above equation in a form of
R(w) = Sigma[r_i * exp(-i(w*(t_i+theta_i))]
the t_i can be considered as a phase distortion.

The effect of this depends on the sequence and amplitude of the reflectivities that are shifted.

Gautam Sen

Advise/Consultant in Exploration at Independent Oil & Gas Professional

Rajesh, Hamed Amini, has quantified the phase distortion. A simple explanation is as follows. Assuming linearity seismic is a convolution of the wavelet and reflection coefficient series. If you are off in your time-depth relationship and then use RC values which are from a different segment, your wavelet will have distortion. The Best way of doing wavelet extraction is to target strong reflectors like sea bottom etc and use Q-deconvolution to take care of absorption of high frequencies to get meaningful wavelet, incase stationary assumptions cannot be made.

Jerry Coggins

Principal Geophysicist at Coggins Geosciences, Inc

Rajesh, I was taught to adjust the time-depth curve to get the best tie before extracting a wavelet. This is usually necessary; inversion software does this for wavelet extraction. The VSP wavelet can be quite different from the best effective wavelet derived from a well tie, because it has AVO, stretch, for example. It does matter what you are doing with the extracted wavelet. Often the derived wavelets for inversion are broader band than the true wavelet, because sparse spike uses a tight wavelet with small side lobes to minimize tuning effects. Your window, the data and, wavelet can have significant effects on the answer you get.

Belal Khalil

Geophysicist at BP

Rajesh, wavelet extraction in many applications is based on a correlation process between a synthetic trace created from a well and the seismic data. The initial domain conversion of the well data from depth to time in order to allow for a correlation with the seismic data (which is in time) will affect the estimation of the seismic wavelet within a specified correlation window. But your question was, will it affect the "phase" of the wavelet in particular. My opinion is no; the way the well-driven synthetic trace is posted on the seismic data will not affect the "phase" of the extracted wavelet, however it will affect the "time lag" required by which the synthetic is shifted to match the seismic data.

QUESTION 62

I have been calibrating a set of wells first using the zero-phase Ricker wavelet and then using an extracted wavelet from the seismic within the vicinity of each well. So, each well had its extracted wavelet. The calibrations with the Ricker came out good at each well. The calibrations with the extracted wavelets came out quite well but the phase of the different wavelets varied greatly from well to well. What are the causes and implications of the varying phase of the extracted wavelets?

Joshua Ndugga

Geophysicist at Total E&P

ANSWERS PROVIDED BY INDUSTRY EXPERTS

Milos Savic

Consultant Geophysicist at Cairn Energy

Joshua, your data is obviously close to zero-phase (since Ricker wavelets seems to work at every well). With deterministically derived wavelets, the phase difference from well to well is related to possible small errors in timing (bulk shifts due to inaccurate check shots). Further, if structure changes from well to well the primary assumption of the "layer cake" structure for assumed 1D convolutional model varies from well to well. This in turn leads fluctuating phases. You should check if there are any bulk shifts and correct for them. Further, you should try to constrain the amount of allowed phase rotation when deriving wavelets. If after that you still see big differences from well to well just stick to your initial Ricker.

Tom Morahan

Geophysicist

Joshua, I agree with Milos. Deterministic wavelet extraction is difficult because small timing differences equate to large phase differences.

Edward Lalande

Sr. Geophysicist

Joshua, the shift of the wavelets phase is likely due to near surface effects. Do you have a weathered layer that is varying in thickness over the area of interest? From what you say, the deconvolution operator has done a poor job of producing a zero-phase wavelet. You may want to go back to your processor, if the seismic processor can revisit the processing sequence, to see if he could propose a different approach.

Gwendal Salmon

G&G Technical Consultant at Halliburton

Joshua, depending on the geology and faults, both phase and frequency change with depth along the borehole. If you are doing a relatively long synthetic, it is in fact recommended you re-extract new wavelets periodically (e.g. every 1000ms or even less), instead of using only one wavelet with a single phase rotation and frequency for the whole interval. You will then have several wavelets used along your synthetic, with a progressive merging of the two adjacent wavelets in the overlapping zones. The quality of the seismic where you extract the wavelet and the quality of logs used to compute the synthetic will also have a great impact in your final calibration.

Shahjahan Laghari

Reservoir Geophysicist, United Energy Pakistan

Gwendal, don't you think using multiple wavelets from different time windows for constructing a synthetic is not a recommended approach, since it will results in good calibration but you are merging wavelets from different zones and forcing a good tie?

Gwendal Salmon

G&G Technical Consultant at Halliburton

Shahjahan, your seismic wave frequency content and phase will be changing with depth, and this does affect the final seismic section that you see and use to correlate with your synthetic. Using varying time wavelets, you should get a more realistic approach. I have seen this procedure applied in different companies, and it is highly recommended.

Shahjahan Laghari

Reservoir Geophysicist, United Energy Pakistan

Gwendal, I submit that using multiple wavelets in order to get a perfect tie is the same thing as using a modeled wavelet. Also the Time-Depth relationship you get at the end will not be a true picture of subsurface and will cause problems in velocity modeling. Hence it is always recommended you try to extract a single wavelet and calibrate your borehole and seismic data with it.

Peter Rowbotham

Lead Geophysicist at AGR Petroleum Services

Joshua, there have been publications and workflows on this by Elio Poggiagliolmi of Entec (then Western, then Schlumberger) and others published and marketed as SAWP - space adaptive wavelet processing which attempts to find a solution to zero-phase deconvolution by estimating the effective wavelet in every single trace of the seismic data.

I remember trying to unravel this with a student back in 1997 (test Total's archiving system by searching for Angus Hamilton - An implementation of Space Adaptive Wavelet Processing on Google). The method assumes that the minimum phase component of the local wavelet is laterally variant and the non-minimum phase component is laterally invariant. The reasoning behind this is that propagation in the earth is a minimum phase process and will depend on geology, whereas any acquisition or processing impact on the wavelet will probably be constant phase and identical for every trace (ignoring non-linear processing steps line NMO).

I guess the message to take away is that whenever we make the assumption of a constant wavelet in our seismic, we are making a big assumption. And assuming zero-phase is a second big assumption on top of that.

CHAPTER

ZERO-PHASING OF SEISMIC DATA

5

CHAPTER OUTLINE

By zero-phasing the interpreter is attempting to make every reflection on the seismic section into a symmetrical zero-phase wavelet. One reason this is done is to secure a better tie between the seismic section and synthetic seismograms. This is especially useful when accurate correlations are of great importance, such as when dealing with thin or closely spaced target horizons. For example, let us say that the filter (wavelet) used during wavelet extraction/derivation appears to have +142-degree phase shift and +2.4 ms time shift. By applying the inverse of this (−142-degree phase shift and −2.4 ms time shift filter) to the seismic data we hope to achieve a fairly good

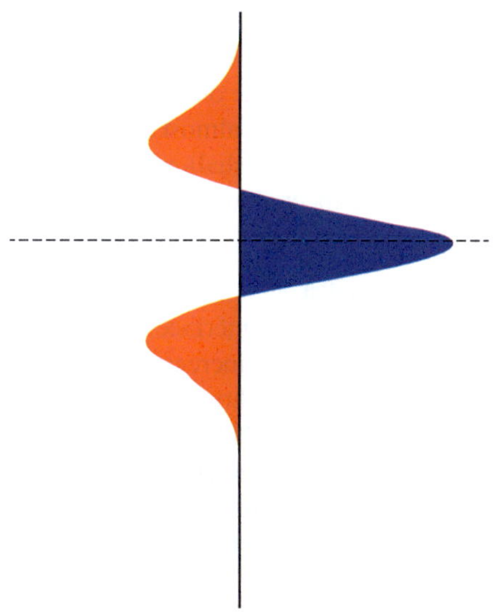

FIG. 5.1

Zero-phase wavelet.

Practical Solutions to Integrated Oil and Gas Reservoir Analysis. http://dx.doi.org/10.1016/B978-0-12-805464-2.00005-6

zero-phasing of our seismic data. Although the data should have been already zero-phased by proper seismic processing, certain adaptive seismic processing steps, (e.g., deabsorption), can distort the phase of the data, especially if the filters applied were derived from traces with significant amounts of noise in them. This application of bringing the seismic data phase into agreement with the synthetic should be verified with more than one well, and should be expected to change as a function of depth on the seismic data (Fig. 5.1).

QUESTION 61

How do you perform phase analysis on 2D seismic lines? In seismic stratigraphic interpretation the interpreter really needs to know whether the data are processed to zero-phase and whether the dominant loop in the resulting seismic wavelet is delayed.

Diego Timoteo

Exploration Geologist

ANSWERS PROVIDED BY INDUSTRY EXPERTS

Rocky Detomo

Research Lead Reservoir Monitoring at Shell International Exploration and Production

Diego, For 2D data, you do exactly the same thing, but you need to account for cross-line dip. If the 2D line is shot along dip, I would expect that this process would work fine. However, if it is acquired along strike, it may be very difficult to perform this analysis. Usually, we look for an unambiguous marker reflection that can be identified clearly on the log, and on the seismic, and simply tie them at that level, and then work from there.

In Exploration, we usually look for something we "know"! Examples are looking for hard reflections at the seafloor on marine data, or in deepwater, we can sometimes find an inversion or salinity layer to calibrate our seismic reflections. Onshore, it can be more difficult, but top of salt reflections are almost always "hard" and maximum flooding surfaces are usually hard, at least in the shallower depths (<15,000ft). Additionally, AVO responses of some of these expected interfaces are usually pretty clear.

Andrew Forbes McIntosh

Senior Exploration Geophysicist Specialist

Diego, The easiest way is to find a major interface is to look in areas where you know from the local geology that the velocity is increasing or decreasing and an area that is isolated from interference from interfaces above or below (i.e. a half space type model). What phase is the wavelet at that interface? Hopefully you have more than one of these type interfaces so that you can check for agreement over a large depth range.

Here is a great link to get you well on your way to dealing with this problem. http://subsurfwiki.org/wiki/Phase_determination

Mike Cline

Consultant Geophysicist and President of T/X Resources

Diego, Below are list of some of the available techniques that can be used for phase analysis. I think it would be a good reference for those not familiar with the current technologies.

So far, the list would include:

- Synthetic seismograms
- Envelope/instantaneous phase technique (R. Roden's single event technique)
- Areal mapping of R. Roden's technique
- Amplitude squared/Skewness
- High frequency inversion
- AVO/AVA crossplots
- Manual inspection of seismic reflections (at the large impedance boundaries, similar to Andrews's suggestion above).

Hesham Shalaby

Senior Geophysicist at Dragon Oil

If your area is offshore, investigate the seabed reflection. It will be the easiest reflector to extract the phase and polarity. You may also try wavelet extraction at a good horizontal reflector.

Diego Timoteo

Exploration Geologist

(Study area data: only 2D seismic grid and 1 well with check-shot)

I calibrated a sonic and density logs and I loaded a time-depth relationship (from check-shot) I extracted a zero-phase wavelet and finally I generated a synthetic seismogram. Then I tied the synthetic with the 2D seismic and identified a shift between them.

I have the following observations about this:

(1) The 2D seismic line tied with the synthetic is not zero-phase processed (correct me if this is wrong).

(2) For seismic stratigraphic interpretation (in exploration) is it necessary the correct phase of the seismic?

Tom Morahan

Geophysicist at Petrotechnical Resources of Alaska

In the marine environment it is common to have to lag a check shot corrected synthetic seismogram from a vertical well down by 30-40ms in order to tie seismic traces with the same datum. On land it can be an up or down shift because of differences in replacement velocities and statics. Make sure that your check shot time and depths are corrected to vertical and listed relative to seismic datum. I have seen listings where times are relative to MSL, but depths are relative to KB. Or, make sure that the software that you are using properly handles datum variability. Once you have a good tie, you might be able to improve it with slight stretches or squeezes, being careful not to alter interval velocities more than a percent or two. Properly processed seismic data should be close to constant phase, but not necessarily zero phase, unless special efforts have been made to correct it as such. One approach is to cross-correlate the synthetic with real seismic traces to determine the best constant phase rotation to bring the seismic close to zero-phase. For confidence, I like to see a correlation coefficient of at least 0.5 using at least a 500ms long window. Be aware that different analysis and processing programs may use different assumptions regarding phase rotation, that is, you might determine using an analysis application that a +40 degree rotation is required but find that you need to apply a -40 (or +320) degree rotation using a different processing application. There should be a standard, but apparently there is not. It is advantageous to have zero phase data for seismic stratigraphic interpretation, but not an absolute necessity if the work is non-quantitative. The advantage is that peaks and troughs should correspond closely to geological interfaces, usually bedding surfaces. Also, isolated high amplitude features can be easily identified as either relatively high impedance or relatively low impedance lithologic units.

PHASE OF SEISMIC DATA

An increase in acoustic impedance across a reflecting boundary corresponds to a positive reflection coefficient and should, theoretically, produce a perfect zero-phase reflection. Globally, we have two polarity conventions:

- Normal polarity for zero-phased sections plots a positive reflection coefficient as positive peak (rightward deflection of wiggle by convention). This corresponds to the S.E.G. convention (Figs. 5.2 and 5.3).

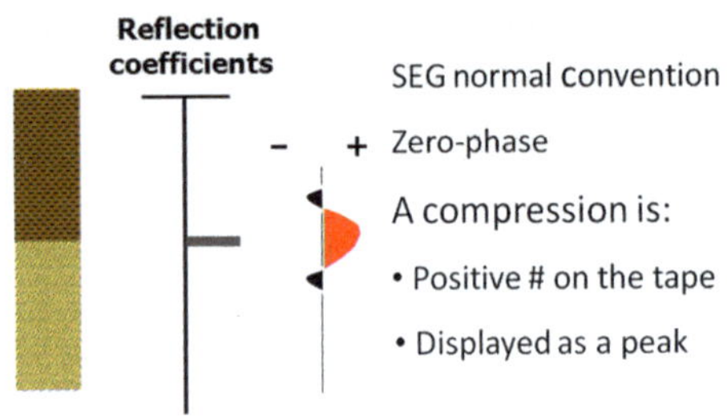

FIG. 5.2

SEG normal convention for zero-phase wavelet.

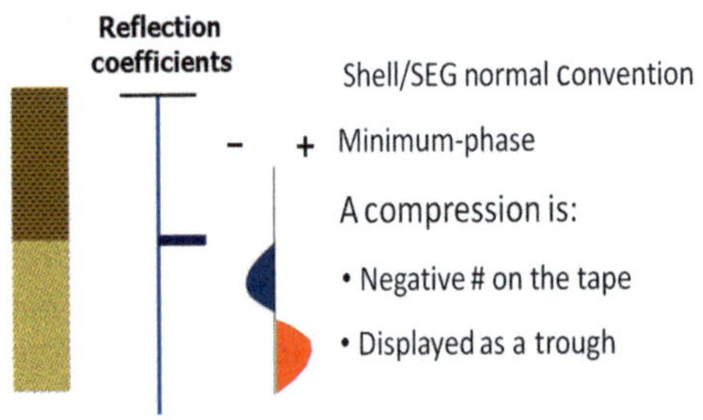

FIG. 5.3

SEG/shell normal convention for minimum-phase wavelet.

- Reverse polarity for zero-phased sections is the opposite of the normal polarity convention. This is the Shell convention (Fig. 5.4).

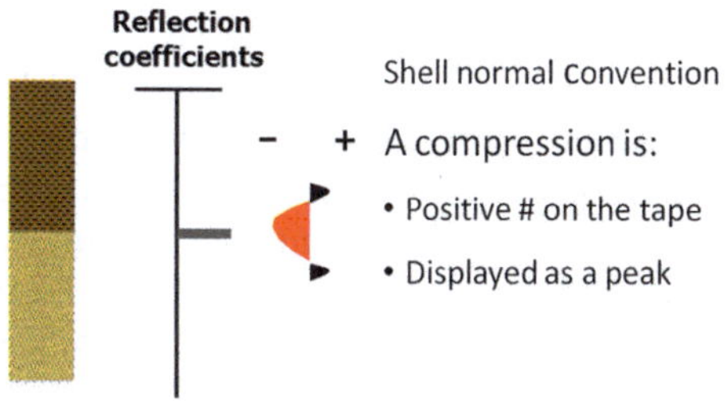

FIG. 5.4

Polarity reversal for zero-phase wavelet.

How to Determine the Phase of Seismic Data

QUESTION 62

How can i identify the phase of the seismic wavelet in land and offshore seismic data? (Fig. 5.5).

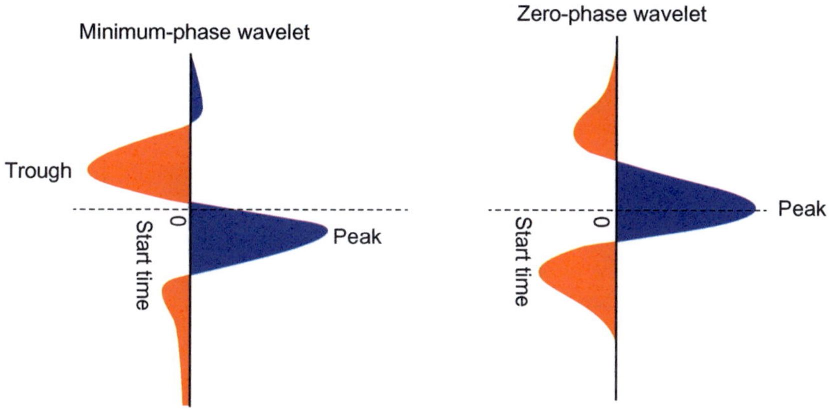

FIG. 5.5

Minimum and zero-phase wavelet.

ANSWERS PROVIDED BY INDUSTRY EXPERTS

Brian Schulte

Geophysical Specialist Quantitative Geophysics - Talisman Energy

In seismic data processing we generally just assume that the acquisition data is minimum phase and after deconvolution it is zero phase but, we are generally unsure of the phase of the data input to the processing. You need to tie the seismic data to a VSP or synthetics created by well data. If you have a VSP then utilize the VSP data since it will tie the seismic a lot better. In other word, phase is identified using VSP data.

A synthetic should be generated from sonic and density logs to create the acoustic impedance response and convolved with a wavelet such as the Ricker wavelet. If you do not have a density log you can utilize equations to derive density from the Vp such as Gardner's equation. To do the phase tie, we should look at the near-stack seismic data and not look at the full-stack seismic data. That's because the near stack is actually the acoustic impedance but most interpreters only have the full stack. The full stack can have issues due to changes in amplitude with offset. To model the far offsets we need to have the dipole sonic or Vs information. There is software within the interpretation packages that does a cross correlation between the VSP/synthetic and the seismic, to tell you how much to rotate the data and the time shift to get the best match between the seismic and the VSP/synthetic.

Mohand Cherif Ammour

Geophysicist at BP-SH-Statoil Join Venture

Basically you need to extract the wavelet from your seismic data that gives the best match with a synthetic seismogram. The phase of the wavelet is the amount of phase rotation with reversed sign one will apply to the seismic data to turn it to zero-phase. For instance, in the Landmark Openworks package you use SynTool to generate a synthetic and extract the wavelet. You can phase rotate the data using PostStack. The figure below shows rotated wavelet (Fig. 5.6).

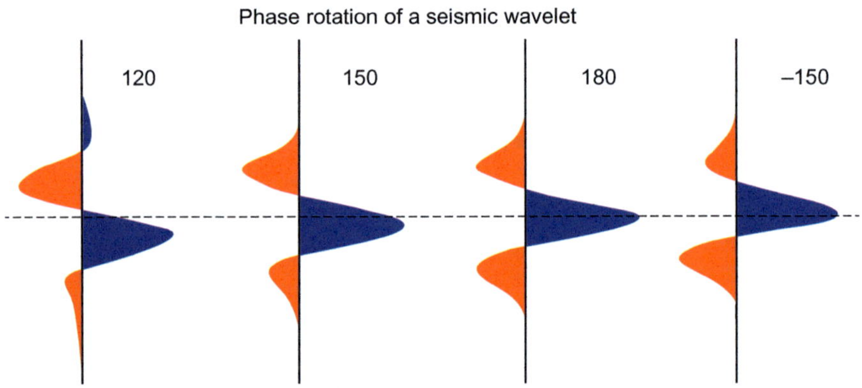

FIG. 5.6

Phase rotation of seismic wavelet.

Thomas Klopf

Senior Geophysicist at KUFPEC

If there is still uncertainty about the polarity of the seismic, some interpreters look at the water bottom reflection or at a likely shallow gas anomaly on the data to confirm the polarity of the data.

Brian Schulte

Geophysical Specialist Quantitative Geophysics - Talisman Energy

The water bottom reflection is a good guide to the phase of the data when it is deep water data but we need to use the synthetic for land data. Remember we are dealing with reflection data which means the reflector represents the interface between two layers. We then use the synthetic with the geological tops to tie the geology into the seismic. Without this tie we cannot tell which reflectors are which. It is true we can extract the wavelet from the seismic to convolve with the reflection coefficient series. The shape of the wavelet we extract can tell us the quality of the seismic. When we extract the wavelet we can see if the seismic has reverberations in it which may be due to multiples. We need to also look at the quality of the well log curves when we do the tie. If the tie is poor and the well curves are ratty, we may need to do well editing to accomplish the synthetic tie. I generally look at the gamma ray, sonic and density to help me with making my ties. I want to tie at least 5 layers to feel comfortable (Fig. 5.7).

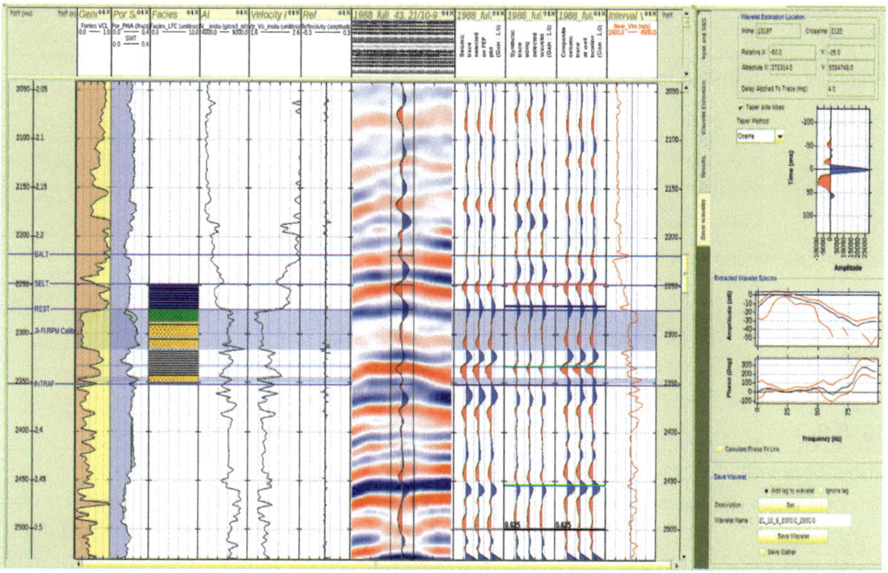

FIG. 5.7

Synthetic seismic and well-to-seismic tie.

Courtesy: Ikonscience, https://www.ikonscience.com/software/rokdoc.

Thomas Klopf

Senior Geophysicist at KUFPEC

There may be a strongly reflecting, regional lithologic change in the subsurface, such as a change from shale to carbonate, that may also be used to give a confirmation of the polarity. Of course, shallow, low impedance gas sand can have a strong reflection off the top of the sand and a reversed polarity reflection off of the base of the sand. The key is the initial polarity at the top of the gas sand which should be the opposite of a hard reflector like a carbonate. The gas sand polarity works much better if the sediments are only weakly cemented (Figs. 5.8–5.10).

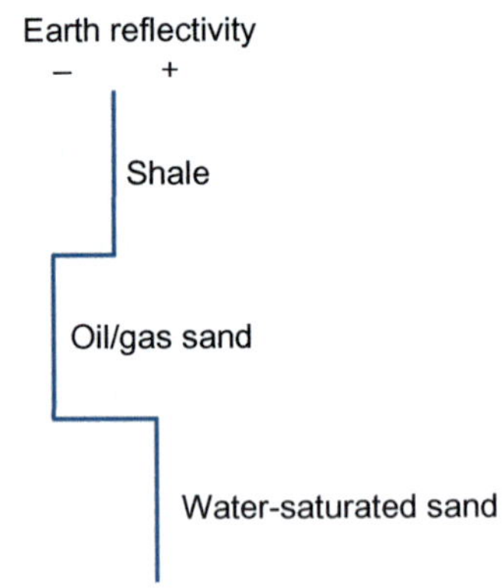

FIG. 5.8

Reflection response of polarity reversal.

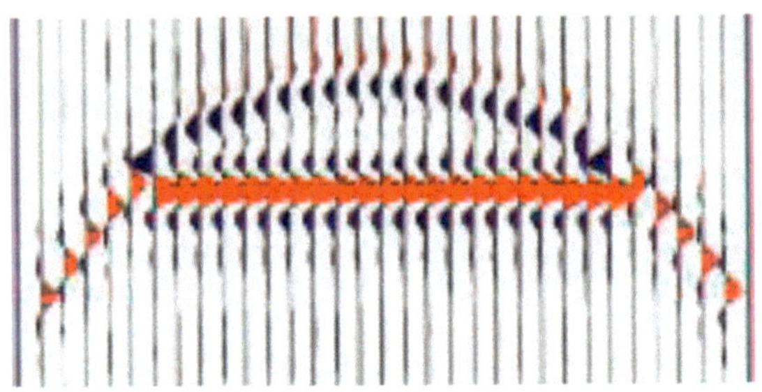

FIG. 5.9

Model of polarity reversal response.

Data from Bacon, M., Simm, R., Redshaw, T., 2003. 3-D Seismic Interpretation. Cambridge University Press, 21 August.

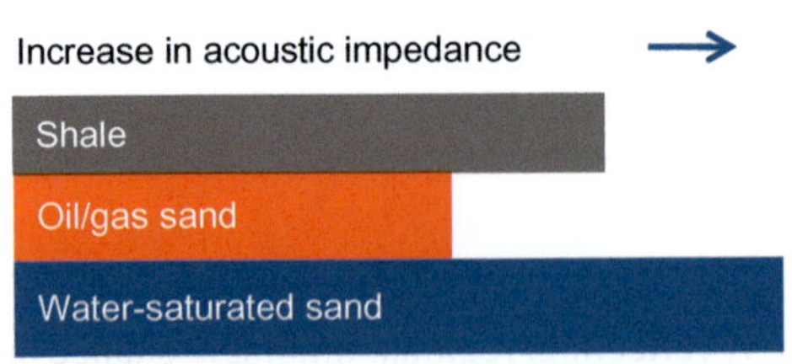

FIG. 5.10

Acoustic impedance response of polarity reversal.

Rajeshwaran Dandapani

Senior Geoscientist at PETRONAS

I think that almost all aspects of estimating the phase/polarity of the seismic data have been covered exhaustively by the experts. Apart from those mentioned there is one particular reflection, the contact between gas and water (Flat-Spot as shown in Figs. 5.11 and 5.12), which can clearly bring out the polarity of the data provided you are lucky enough to have one in your area.

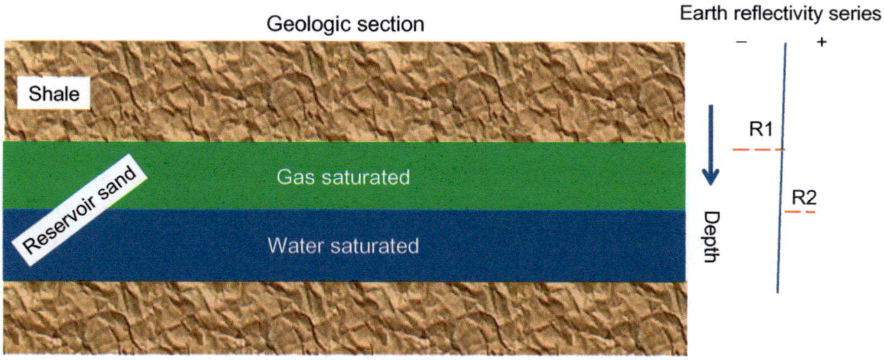

FIG. 5.11

Depicts gas-water contact and the effect on the earth reflectivity response.

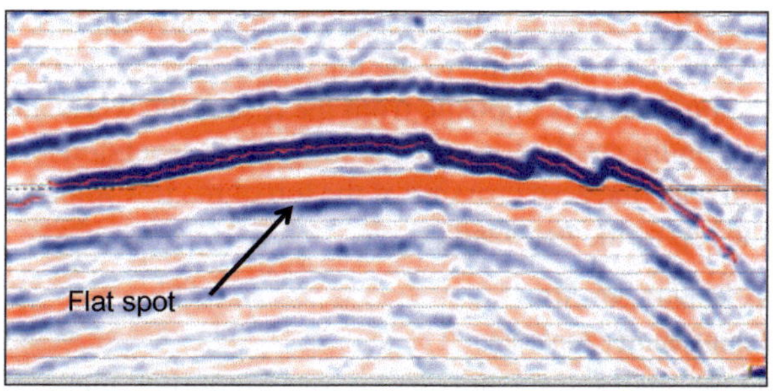

FIG. 5.12

Flat spot on seismic section.

QUESTION 63

In marine seismic acquisition it's quite common to use the water bottom reflection as a polarity check. But what can we use as a polarity check on land?

ANSWER PROVIDED BY INDUSTRY EXPERT

Rajeshwaran Dandapani

Senior Geoscientist at PETRONAS

On land you can look at gas-water, gas-oil, or oil-water contact reflections to determine the polarity. If you have a well with recorded sonic and density logs then you can do a tie to establish the polarity and the phase of the seismic data for your zone of interest.

QUESTION 64

Why is the phase of seismic data important in acquisition and processing?

ANSWERS PROVIDED BY INDUSTRY EXPERTS

Brian Schulte

Geophysical Specialist Quantitative Geophysics - Talisman Energy

Phase is not that important in acquisition and in processing, if a source signature is available. Then, you can rotate the data to minimum phase before deconvolution. This is due to the fact that in deconvolution we do autocorrelation and we do not carry the phase of the data but assume it. When it comes to interpretation and inversion we need to have a zero phase wavelet. This is why we spend time tying the well data to the seismic data. A zero-phase wavelet means that the apex of the peak or trough is at the interface causing the reflection. Our principles of AVO are based upon the concept of zero phase data.

Thomas Klopf

Senior Geophysicist at KUFPEC

Brian is correct when he says that zero-phase could have the center of the maximum peak or trough at the reflection interface. The SEG convention is for the center of the maximum trough corresponding to the top of gas sand with low impedance. The usual Australian convention is opposite from the rest of the world and uses the maximum peak corresponding to the top of a gas sand. After inversion is applied to the 3D seismic reflection data, the zero crossing at the top of the low impedance event would correspond to the top of the gas sand.

Edgar Leonardo Galvan Aguilar

Geophysicist at Schlumberger Tech. Co

The polarity and phase of the seismic data is very important matter to resolve before the seismic interpretation. This analysis should be carried out during the construction of the synthetic seismograms at the beginning of the study. Extracting the deterministic wavelets at borehole locations allows you to understand the geological implication in the seismic trace and you can collect needed information from your seismic: frequency, phase, polarity and seismic resolution. Knowing the polarity and phase of the seismic allows you to identify prospective seismic anomalies.

Harvityanto Muhammad Arief

Sr. Geophysicist

Based on my experience, I usually do an analysis of synthetic seismograms to see the changes of seismic phase in the target reservoir. In reality, there will be changes in the wave phase at certain depths because of different lithologies. You will find the seismic phase changing vertically although the header contained pre-determined phase information. As the first process in seismic interpretation, I always go back to the quote: "Well log data have a better vertical resolution, seismic data have a better lateral resolution". Even if you have a VSP corridor stacked in that well, you have to link it to synthetic seismogram from the well.

QUESTION 65

Is there a direct clue that could make it possible to identify whether data is zero phase or minimum phase and, also what are the advantages of zero phase data?

ANSWERS PROVIDED BY INDUSTRY EXPERTS

Rajeshwaran Dandapani

Senior Geoscientist at PETRONAS

If you notice the water bottom reflector or a contact reflection carefully in case of perfect zero phase data there would be only one strong phase i.e. either a peak or a trough depending on the polarity convention used.

In the case of a minimum phase data you will get both the phases highlighted i.e. you would see a peak over a trough or a trough over a peak reflection pattern depending on the polarity convention used. The above explanation itself tells you the advantage of the zero phase data in exactly identifying the interface between two formations.

If you track a peak or a trough on zero phase data it exactly corresponds to the interface. It is, either at the top or bottom of a formation and it closely matches the tops derived from well logs. But in the case of a minimum phase data a peak or a trough it will not match with the tops obtained from well logs and you need to apply a time shift to match it.

Note that I used the offshore example because the water bottom is a very good reflector. On land if you have a gas water contact you can use that. If you do not have any other clue and you have any well data then you can do the seismic to well tie to establish whether the seismic data is zero phase or minimum phase and, you can also determine the phase angle of the seismic data.

QUESTION 66

Why do seismic interpreters prefer working with zero-phase seismic data over minimum-phase data?

Aatef Badshah

Exploration Geophysicist at Pakistan Oilfields Limited

ANSWERS PROVIDED BY INDUSTRY EXPERTS

Asif Zaidi

Paradigm Technical support Geoscientist, Houston

Zero phase seismic data gives a strong peak or trough at the water bottom depending on the polarity convention used. If we track a peak or trough on zero phase data it exactly corresponds to an interface, i.e. either a top or bottom of a formation and it matches the tops derived from the well logs very well. In the case of minimum phase data if we track a peak or trough it does not match with the tops obtained from well logs and we need to shift the data to match it. That is why interpreters prefer to use zero phase data (Fig. 5.13).

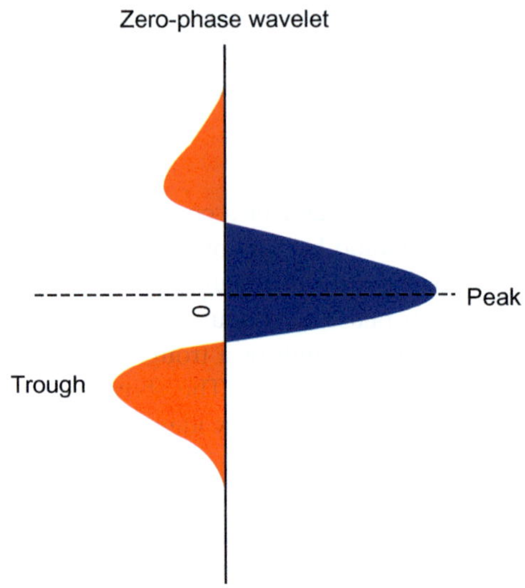

FIG. 5.13

Peak and trough of zero-phase data.

Kamal Omosanya

Sr. Geophysicist at Rockseis

A zero-phase wavelet is symmetrical and the majority of energy is coincident with its peak. This is great advantage for seismic stratigraphic interpretations as the centre of the wavelet is coincident with the geological interface causing the reflections. In addition, it is less distorted as compared to minimum phase. You should know that zero-phase wavelet is non-realistic as you cannot have the maximum energy at zero time. Imagine you have the maximum energy at the same moment you shoot your source. However, the data is phase shifted from minimum phase to zero-phase during processing.

Andrew Forbes McIntosh

Senior Exploration Geophysicist Specialist

Zero-phase wavelets offer increased resolution (shorter time duration than their minimum phase equivalents) and are easier to interpret (the half space interface response is a peak or a trough) for sharp changes in acoustic properties.

Zero-phase data has higher amplitude than the equivalent minimum phase, thus greater signal-to-noise ratio), better reflector continuity, easily identified stratigraphic relationship and thus better subsurface imaging.

Hesham Shalaby

Senior Geophysicist at Dragon Oil

When calculating thin reservoir thickness, the data has to be zero-phased. The zero phase wavelet has a better signal to noise ratio.

The seismic input to the colored inversion has to be zero-phased as well.

QUESTION 67

Generally, we assume that pre-processed seismic data has minimum-phase characteristics prior to deconvolution. As far as I know, this requirement is crucial for constructing a stable Wiener Optimum filter which transforms the minimum-phase into zero-phased data. Unfortunately, reality can be quite sobering and it is not unusual to see phase variations along a seismic reflector or within a seismic trace. Moreover, seismic phase may also change during the processing workflow (e.g. stacking). So, if we treat the input data for the deconvolution process as minimum-phase (even though if this is clearly not the case) and derive a zero-phasing filter from an approximated minimum-phase source wavelet we would get an incorrectly phase-shifted output, right? I, assume, that this phase-shifted output won't be a big issue if a density and sonic log from the wellbore log is available since we can perform a well-to-seismic tie. But how do we real with deconvolved data which is clearly not zero-phased if no well log is available? (Peaks (or troughs) that not align with the reflector anymore and are probably displaced by several milliseconds from the corresponding zero-phase position).

Martin Vögele

Junior Reservoir Geophysicist

ANSWERS PROVIDED BY INDUSTRY EXPERTS

Guy Maslen

Chief Operations Officer at GLOBE Claritas

Martin, "We assume that pre-processed seismic data has a minimum-phase characteristic prior to deconvolution". I suspect that is only the case for an explosive (or highly impulsive) source on land. And with airgun arrays the phase of the source is usually far from minimum phase with the source and receiver ghosts complicating things. For example, on most marine surveys we deal with a single (modeled) source signature for the whole survey. However, in practice there are variations in pressure, depth, volume and firing times (within specification, including swapping in spares) which start to make this "single signature" an approximation too. There are various ways in which we can try and measure or extract a wavelet; however these have a different set of assumptions and approximations underpinning them. At best in a processing sequence we can try and preserve the approximation to true minimum (or zero) phase that we have obtained at the de-signature stage; this tends to be harder for zero phase as thing like NMO stretch come into play (and can add a 20+ degree phase rotation at target depths). So, without well control and calibration I would suggest that it is very difficult to be confident that the seismic is even with +/- one "loop" compared to the geology without well control. If you add in structural uncertainty (from imaging and/ or statics) this could increase significantly. I have certainly seen this cause problems with some fine-scale stratigraphic interpretation work, where tracking the right loop mattered. In that case there was some limited well control but it was still very tricky.

Edwin Quadt

Seismic Processing Delivery Manager at Shell Global Solution

Martin, without a well synthetic this is extremely hard. The only data where you can get a reasonable idea about how zero-phase the wavelet is, is on deep water data where the seafloor reflection should ideally be a zero-phase (symmetrical) wavelet. But that only works for the seafloor reflection, and deeper data might still suffer from phase issues. However, in my experience we are usually out by 20-30 degree with the zero-phasing if we do compare with a well synthetic. Such a phase shift does change the character of the wavelet somewhat, but peaks and troughs do not move that much. Anything more than that and you are looking at more serious shifts. However, if you do not have a well, a more serious problem is the time-to-depth conversion which will introduce a lot more uncertainty than the possible milliseconds shift of the zero-phasing.

Gerardo Clemente Martínez

Interprete de atributos sísmicos en PEMEX

Martin, with reference to a synthetic trace from well information you can measure the phase in a time gate with some precision by comparison with the reference. Without a well what you can do is focus on some interfaces whose reflection are fairly sure of, like the reflection off the water bottom, top salt, or an interface between clastics and carbonates. But in all cases you most make assumptions. Higher order statistics can be used to estimates phase overtime intervals but provides just an estimation. Remember that starting with the acquisition of the seismic data, there are a lot of factors that we assume are correct, but we never know.

David Reynolds

Seismic Processing Consultant

There is a good paper written by a colleague of mine that specifically addresses this challenge.

Mark S. Foster in Geophysics V 62, No. 1, pgs 252-261 (Jan-Feb, 1997) wrote an article "Zero-phasing seismic data without wells in offshore West Africa: Reducing uncertainty and variability of the wavelet".

Of course this is specifically for marine streamer data, not land data, but it is a good approach to this problem. Mark uses 2 approaches to estimate the wavelet - physical modeling and statistical evaluation. It is a nice read and serves as a recipe to follow in other areas. In my experience, the biggest change in phase is with time (vertically) and not so much along the line or around the 3D survey (laterally). For this reason, application of some Q model is most helpful if you are expecting to compare prospects from top to bottom. After all, our customers always say "I want it correctly zero phased everywhere" (and this is a difficult request achieve).

How to Determine the Phase of Thin Reservoir Bed

QUESTION 68

Can I predict the phase of the seismic data if I have very thin sand bodies and their thickness is less than the minimum resolvable thickness? (Fig. 5.14).

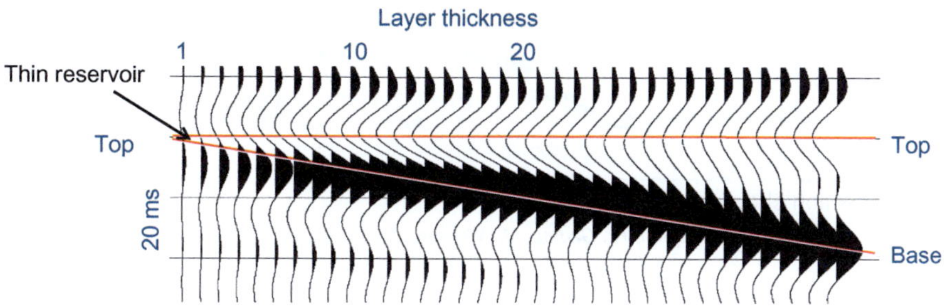

FIG. 5.14

Wedge model.

ANSWERS PROVIDED BY INDUSTRY EXPERTS

Rajeshwaran Dandapani

Senior Geoscientist at PETRONAS

To determine the phase of the seismic data you need to look at an interface. If the sand is very thin then the bottom reflection (base) will interfere with the top reflection. If the thickness is half the wavelet (wavelength/2) you will not get any reflection but if it is one fourth of the wavelet (wavelength/4) then you would get a very bright reflection. If the data is exactly zero phase then the tuning amplitude would be the maximum. As you rotate the phase the amplitude decreases. You can do this experiment by taking a wedge model and playing with the phase of the wavelet. You can plot the resulting amplitude value at tuning thickness for various phases and you can arrive at a relationship between the phase and the reflected amplitude at tuning thickness. If the thickness is far below tuning thickness then the amplitude diminishes because it cannot act as a reflector anymore (say if the thickness is less than one eight of the wavelet (wavelength/8)). The amplitude phase relationship is still valid but the top and bottom reflections vary by such a minute phase difference that it is very difficult to detect. As well both the top and bottom reflections will be very weak and it will be about the same as the wave being getting transmitted without being reflected (Fig. 5.15).

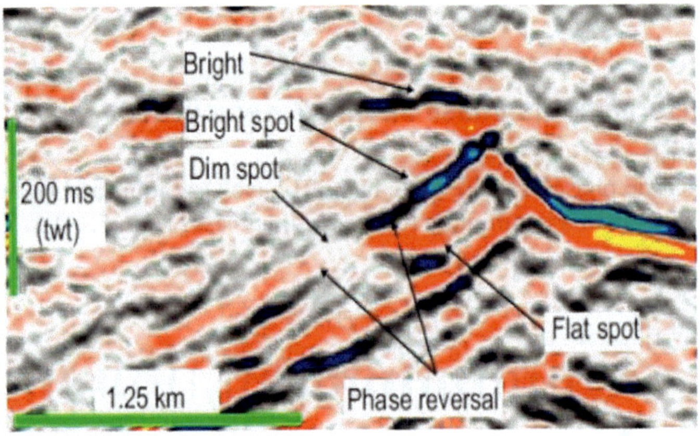

FIG. 5.15

Seismic section showing phase change.

Source: Løseth, H., Gading, M., Wensaas, L., 2009. Mar. Pet. Geol. 26 (7), 1304–1319.

Gautam Sen

Advice/Consultant in Exploration at Independent Oil & Gas Professional

I would just like to add that it is better if you use a 90 deg phase wavelet for detecting reservoirs of thickness less than lambda/4.

Rajeshwaran Dandapani

Senior Geoscientist at PETRONAS

Yes I fully agree with **Gautam Sen** that the visual depiction of a thin layer is better on a 90 degree phase rotated data than a perfect zero phase data. This is because the entire layer appears as one color embedded within two zero crossings instead of a trough peak pair depicting the top and bottom as in the case of a zero phase wavelet. Nevertheless, the apparent thickness measured on seismic data in both cases will be greater than the real thickness when the thickness of the reservoir is less than one fourth of the wavelength. We can address this problem by building some empirical relationships by cross-plotting amplitude versus real thickness using the wedge-model. Inversion can actually help in detuning to some extent and show us the true thicknesses of layers below lambda/4.

Brian Schulte

Geophysical Specialist Quantitative Geophysics - Talisman Energy

If we have zero phase data and we do a 90 degree rotation we are now making the wavelet equal to the layer and the zero crossings becomes the top and base of the event (Fig. 5.16 below). In essence this is what we do when we do an inversion.

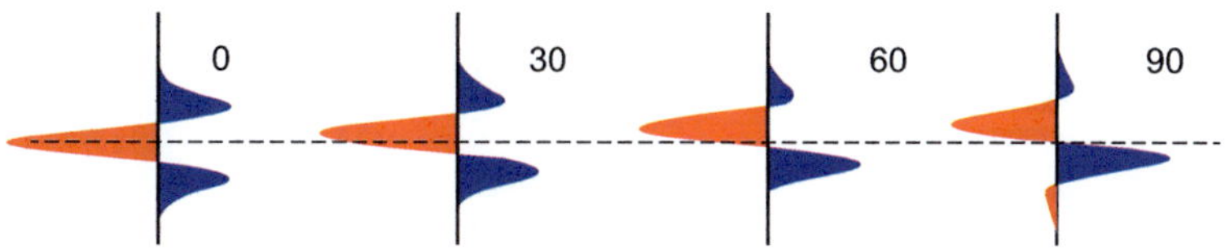

FIG. 5.16

Ninety-degree phase shift applied to zero-phase wavelet.

The purpose of the inversion is to attempt to remove the effect of the wavelet and isolate a rock property such as AI (acoustic impedance) or SI (shear impedance) or density. When we do the integration to do the inversion we are in essence rotating the reflector 90 degrees. We are also trying to remove the tuning effects we see in the data when we do the inversion and we should be able to resolve thinner beds. When we begin to discuss the offset dependent tuning in the gathers then we need to be careful. **Downton** wrote some interesting papers on doing inversion without applying NMO to the data. We need to understand why we do things and how can one process help us to get better results. The comments about tuning above and the 1/4 wavelength are good comments. Building a wedge model always helps and should be done to understand your data better.

Rajeshwaran Dandapani

Senior Geoscientist at PETRONAS

I do agree fully with you **Brian** that inversion also does a 90 degree phase rotation but it also does something extra. In fact every inversion algorithm starts with a Utopian intention to decipher the actual acoustic property distribution of mother earth bereft of the wavelet which is so dear to it. Unfortunately we fall short of achieving this Utopian objective due to rapid variations within the wavelet in 3 dimensions. This makes it difficult to catch hold of it and remove its affect completely. Inversion is nothing but a glorified version of amplitude preserving deterministic deconvolution with appropriate geological information transformed to acoustic properties and input into the starting earth model. The stronger the wavelet gets deconvolved the greater the vertical resolution we can achieve through inversion. In some cases where the overburden is well behaved, the wavelet is very stable and the reservoir is also uniform. I have seen true thicknesses measured even below lambda/8. As you have rightly pointed out the full stack data should never be used for any kind of inversion. If we have the luxury of limited angle stacks we should exploit that information to the full. Acoustic Impedance (IA) inversions should always focus on near angle stacks since it gives a better depiction of lithology (Fig. 5.17).

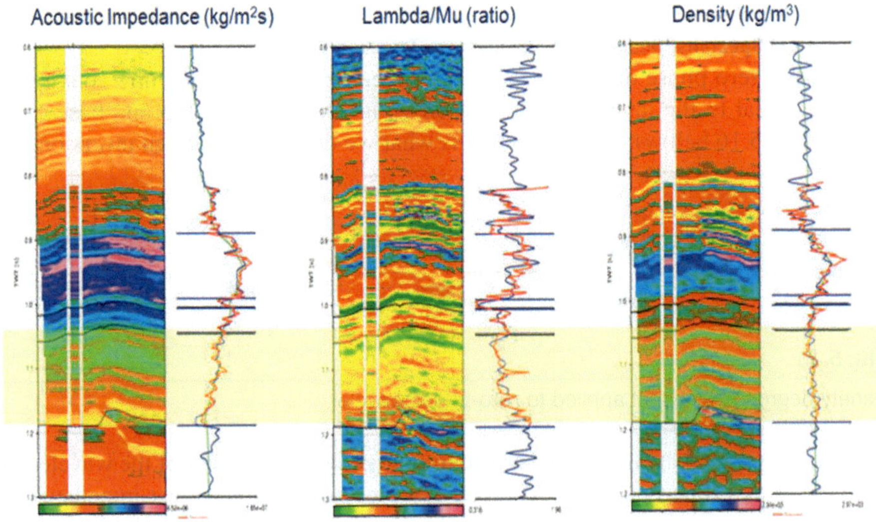

FIG. 5.17

Comparison of well logs (*red curves*) and prestack inversion traces (*blue curves*) at a vertical well for acoustic impedance (*left*), Lambda/Mu (*middle*), and density (*right*). The reservoir zone is indicated in *yellow*. The correlation between inversion results and well logs is high, which gives confidence in the postinversion work carried out.

Data from Leiceaga, G.G., Norton, M., Le Calvez, J., Henery, F., 2013. Predicting hydrocarbon production capacity in an unconventional reservoir from seismic derived elastic properties. Courtesy: Canadian Society of Exploration Geophysicists.

Elastic Impedance (EI) inversion should always focus on the far angle stacks to decipher the fluid-fill and also to exploit the Poisson's ratio difference between shale and sand if it is to accurately delineate the lithology. Nowadays we are using many limited angle stacks simultaneously to invert for Vp, Vs and Rho, where it is difficult to acquire a good quality data. All this is fine as long as there is a good impedance contrast between the reservoir and the non-reservoir. If the contrast is negligible then the inversion based on reflection coefficient (RC) fails miserably. In these cases the Paradigm group has come up with a novel idea of doing inversion based on the waveform shape rather than RC which yields reasonably good results where the impedance contrasts are very low. This is because waveform shapes are immune to the amplitude level (effects).

CHAPTER

AVO ANALYSIS AND INTERPRETATION

6

CHAPTER OUTLINE

UNDERSTANDING AVO

Many geoscientists think of amplitude-variation-with-offset (AVO) as a tool to predict fluid content, primarily gas. But AVO can be used for much more. AVO is a prestack attribute and is a function of rock properties and acquisition geometry. Since geometric effects are relatively predictable, the AVO response is a complex function of the physical properties of the overlying and underlying rock at an interface. If we assume that the properties of the rocks above a certain level are constant, then the AVO response at any level is a function of changes within the underlying rock. These changes may be a result of the change in fluid content, or the changes can just as easily be because of a variation in the physical properties of the rock itself. The changes may be stratigraphic (change in rock type/depositional environment), porosity related, or even influenced by existing fractures. Possibilities include gas-filled reservoirs, porous channel deposits, a high porosity zone, say within a reef, or areas of increased fracture density in the reservoir. Thus, defining an AVO anomaly is just the first step in deciding whether one has a viable exploration or development target.

Generally, AVO analysis allows us to use well log data to model the seismic response across a proposed lithological interface as a function of incidence angle. As these models (see Fig. 6.1) can be used to predict the possibility of a specific reservoir fluid (oil or gas), they form an important part of exploration and development projects.

Practical Solutions to Integrated Oil and Gas Reservoir Analysis. http://dx.doi.org/10.1016/B978-0-12-805464-2.00006-8

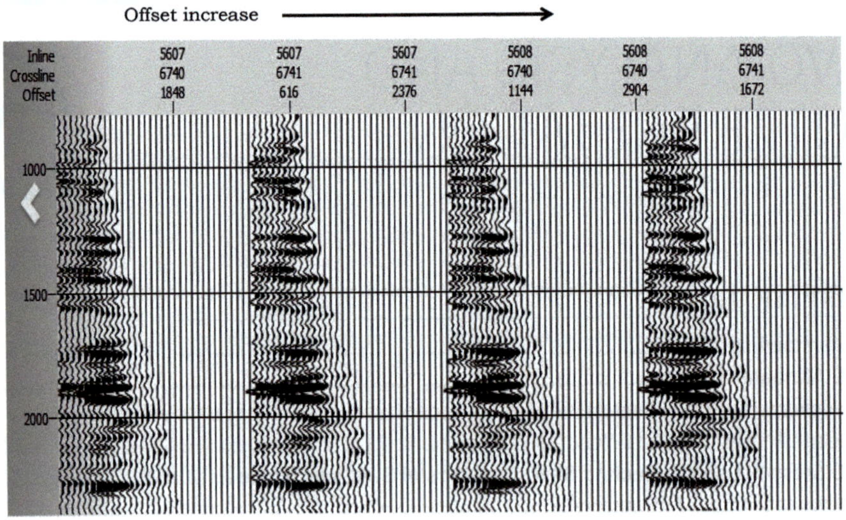

FIG. 6.1

PSTM gathers show AVO effect.

Courtesy: AVO Detect.

For relatively small angles of incidence, usually <30 degrees, Shuey (1985) showed that the compressional-wave reflection coefficient can be approximated by an equation of the form

$$R(\theta) = A + B\sin^2\theta$$

where

$R(\theta)$ is the reflection coefficient at the incidence angle θ.
A is the reflection coefficient intercept (or normal incidence reflectivity at zero offset).
B is the gradient of the amplitudes against the sine of the angle of incidence squared or reflection coefficient slope at normal incidence. B describes reflections behavior at intermediate offsets.
(θ) is angle of incidence.

QUESTION 69

Why is there sudden increase in amplitude at higher angles while modeling amplitude variation with angle (AVA) curve using Zoeppritz equation? Is it due to refraction energy? (Fig. 6.2).

Rajesh Soni

Lead QI Geophysicist- Reservoir Development at Cairn India Ltd

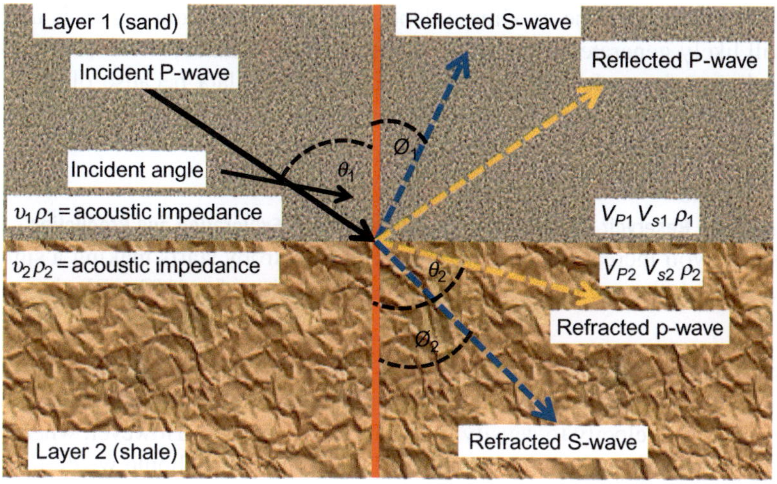

FIG. 6.2

Reflection coefficient at an angle and mode conversion.

ANSWERS PROVIDED BY INDUSTRY EXPERTS

J. Antonio Sierra

Geophysicist at EOG Resources

Rajesh, Most likely, yes, it is due to refraction energy. You are approaching the critical angle, that is, total internal reflection, and your reflection coefficients are approaching 1. If you think it is an artifact of the calculation, consider blocking your logs to avoid dramatic variations in velocity over very small intervals. Also, check out the "CREWES Zoeppritz Explorer" to get a feel for how reflectivity should change with incidence angle.

Huw James

Principal Consultant Geosciences at Integral Geo Services

Rajesh, You need to say which modeling software you are using and what it is trying to model. For reflectivity at a plane interface we commonly have:

R(theta) = R(0) + G sin**2(theta)+F(tan**2(theta)-sin**2(theta)) where theta is the angle of incidence. F & G depend on the density and P & S wave velocities, and the changes in these at the interface. You can see that as theta increases tan (theta) is likely to get much larger than sin(theta) so there is the possibility of very large positive or negative reflectivities as theta approaches +-Pi/2. I doubt that your software models refractions but in the real world these occur and are frequently very strong. Please read your software description to see if it tries to model refractions.

In addition, it is common for models used for AVA analysis to be unrealistic. If a vertical trace through your model is not a virtual match to a log trace from the surface to a little deeper than the maximum depth of your model, your model is unrealistic and will likely generate results that are not matched in nature. Often, AVA modeled data only matches a few of the reflectors in the real data because of this. It may sound obvious but modelers frequently oversimplify the subsurface to make their task easier but then they reduce its utility.

Rocky Detomo

Research Lead Reservoir Monitoring at Shell International Exploration and Production

Rajesh, As you increase angle, the impact of anisotropy starts making a significant impact on both the angle and the amplitude calculations for AVO analysis.

QUESTION 70

When we do not have density data, the Gardner equation is one of the good ways to transfer P-wave velocity to density in ISOTROPIC medium. However, when the medium becomes anisotropic, for example VTI medium, what should we do for density data? Use pha?

Huizhong Yan

Teaching Assistant at University of Houston

ANSWERS PROVIDED BY INDUSTRY EXPERTS

Vic Lamanuzzi

Geophysicist at Noble Energy

Yan, Didn't Gardner use velocity data from vertical boreholes to empirically build his velocity-density relationship? If so, if you want to use Gardner, perhaps you should use the best estimate of the vertical velocity component. If your velocities are coming from significantly deviated boreholes such that anisotropy has an impact, you may need to back out the anisotropy effect. Otherwise, the higher "horizontal component" velocities will presumably give erroneously high density values using Gardner. Unless it makes sense to you, that density should generally be higher in VTI media than in isotropic media (which is not intuitively obvious to me).

John Chamberlain

Consultant Geophysicist

Yan, Vic is correct; the Gardner equation is a purely empirical relationship between vertical velocity and density, and quite a poor one as you see whenever you crossplot the two measurements. It helps to limit the prediction to a particular lithology in a small depth range using parameters established in a nearby well, but the scatter is usually high, especially in less consolidated sediments. So in principle I would do as Vic, suggested, but I disagree with Huizhong that it is ever a 'good' way, especially if you use the 'default' parameters. But often it may be the only way (though petrophysicists have more sophisticated methods).

Gautam Sen

Advise/Consultant in Exploration at Independent Oil & Gas Professional

Yan, Scatter in Gardener's relation between velocity and density is comparable to anisotropy effects in p wave velocity in a majority of cases. Thus it is best to use this equation in vertical holes only.

AVO ATTRIBUTES

The following are AVO attributes:

- The reflection coefficient intercept (or normal incidence reflectivity at zero-offset), A
- The reflection coefficient gradient or reflection coefficient slope at normal incidence, B
- P-wave normal incidence reflectivity, RP, which is equivalent to intercept A
- S-wave normal incidence reflectivity, RS
- Fluid line
- Fluid factor $(A+B)$
- Poisson reflectivity

In AVO analysis, using AVO attributes, such as intercept A and gradient B, is advantageous because:

1) These attributes (intercept and gradient) relate directly to the angle-dependent seismic data.
2) It is easier to predict the effect that a change in A (intercept) or B (gradient) will have on a common-depth-point (CDP) gathers or angle stacks.

Therefore, understanding the impact of changes in reservoir properties on A and B provides insight when interpreting the seismic-amplitude response.

QUESTION 71

What is the relation between seismic AVO attributes and seismic lithology?

Kwasi Obeng

Geophysicist at CGC-Ghana

ANSWER PROVIDED BY INDUSTRY EXPERT

Guy Maslen

Chief Operations Officer at GLOBE Claritas

The AVO response is related to the physical properties of the rocks (such as porosity, pore fluid, compaction, stress, lithology, permeability and geological age) at the interface and is usually modeled/inverted using an approximation to the Zoeppritz

Equations. AVO adds information that will help identify lithology or at least provide additional parameters that can help to confirm the initial lithological interpretation; cross-plotting of the AVO attributes and calibration against well data is a key technique to help with this.

How to Compute AVO Attributes (Intercept and Gradient) From Seismic Gather and the Data Needed

QUESTION 72

How can the intercept and gradient be computed from seismic gathers? And if possible, how can the intercept and gradient be plotted on a seismic map? (Fig. 6.3).

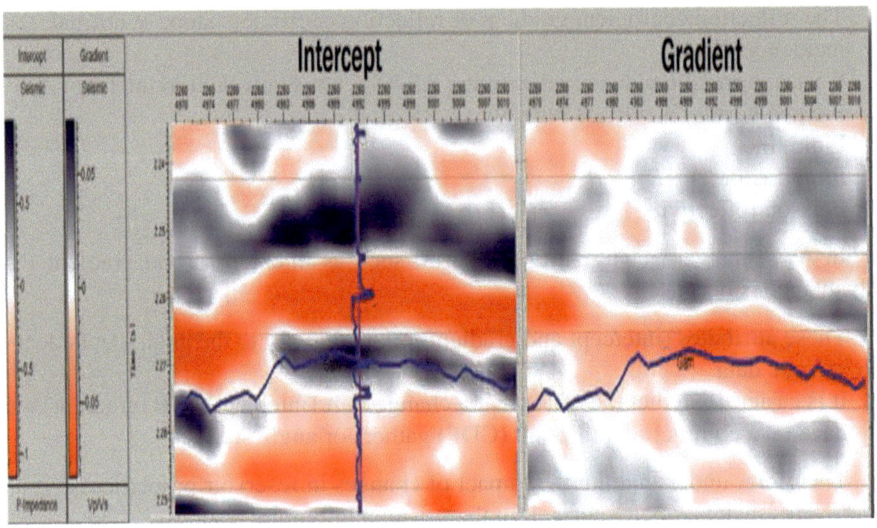

FIG. 6.3

Intercept and gradient sections.

Raouf Nasr
Geophysicist

ANSWER PROVIDED BY INDUSTRY EXPERT

Brian Schulte
Geophysical Specialist Quantitative Geophysics - Talisman Energy

Raouf, There are many programs that actually calculate the intercept and the gradient as attributes within the different processing and rock properties software. However, to calculate the intercept and gradient you need angle stacks. But since your common-depth point (CDP) gathers are in offset, you are going to require a velocity field and software to convert offset to angle stacks. These are standard operations in software like Hampson-Russell. After having these angle stacks you will need to decide which model you want to work with:

Zoeppritz (shuey, Aki/Richards),
Shuey with 2 terms,
Shuey with 3 terms, etc.

After choosing your model you will have to perform some kind of optimization process to find the coefficients of the model (e.g. to do a least squares fit between the amplitude and the square of the sine of the reflection angle). For instance, if you choose the 2-term Shuey model, the linear coefficient will be the Intercept or the amplitude value when the square of the sine of the angle is zero or more simply, when the angle is zero (that's why you can make an approximation with the amplitude values of the near stack). Then the angular coefficient will be the gradient.

Or if you chose to use Zoeppritz's (shuey, Aki/Richards) approximations, you can calculate reflection coefficients using P and S-wave velocities and, densities plus angles. The final step is to calculate intercept and gradient attributes using least square fitting, b=Ax. X includes intercept and gradient, b has reflection coefficients, and A indicates angles. Note that near offset data is considered to be intercept and the gradient is the difference between far and near offset. The figure below shows near and far offset data (Fig. 6.4).

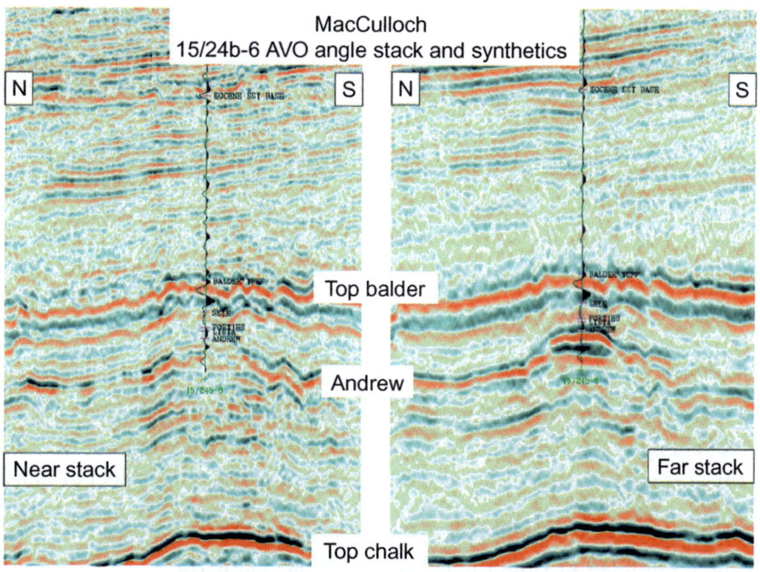

FIG. 6.4

Near and far angle stacks over the MacCulloch field.

Data from http://csegrecorder.com/articles/view/multi-disciplinary-geoscience-the-brenda-north-sea-development, https://www.cgg.com.

However, if you start to compute intercept and gradient from angle stacks, it is easier. You can just take the difference between the far and near stacks and that is a crude estimate of gradient (strictly speaking, you need to scale by the difference between the squares of the sine of the reflection angles). There is also an analytic expression for least squares fits using 3 angle stacks. If you wish to utilize intercept and gradient for mapping then the most powerful tool you can use is the cross plot of the intercept and gradient where intercept is on the x-axis and gradient is on the y-axis. You can then distinguish the background or shale line, identify anomalous events, draw polygons around them and back project them into the seismic. You need the right software to do it but the cross plot is the most powerful tool we have (Fig. 6.5).

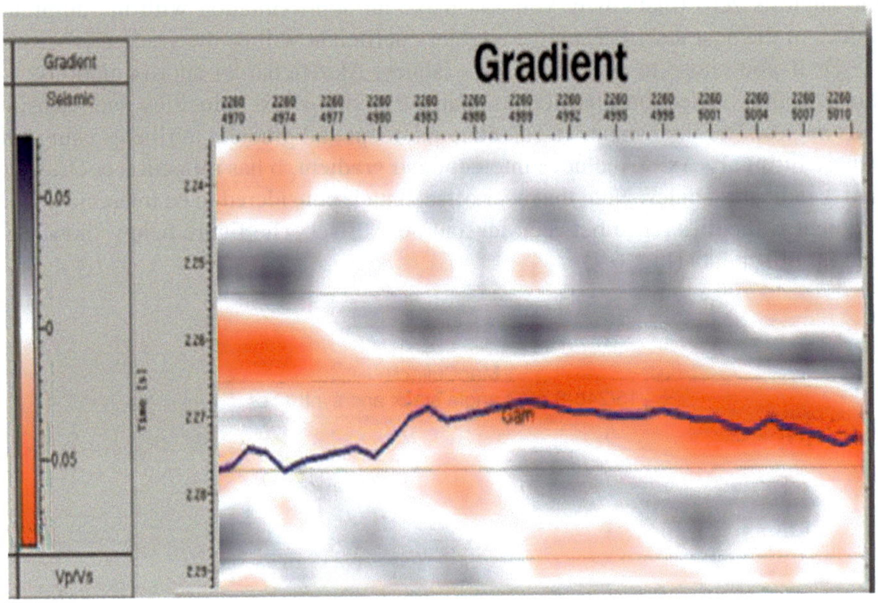

FIG. 6.5

Gradient section.

Note that:

- Near offset component is dependent only on P-Impedance Contrast.
- Middle changes are caused by only Vp/Vs ratio, so Vs is important here.
- Far offset is only P-wave velocity and angle of incidence, no effect of density here

Mike Stone

GEOSCIENCE TECHNICIAN AT LUKOIL OVERSEAS OFFSHORE PROJECTS

Raouf, Generally you first calculate angle stacks. Then do some sort of Zoeppritz approximation (Shuey, Aki/Richards) to get an intercept and gradient volume. Then take your time to map and do amplitude extractions on these volumes. If you're still at the reconnaissance stage, you can even skip doing the Zoeppritz approximation and just use the angle stacks directly. Use the near angle stack as an approximation of the intercept, and the do a (far-near comparison) as a quick and dirty approximation of the gradient.

QUESTION 73

How can we interpret the AVO gradient? Do we need to expect the same event continuity in a gradient section as a stack? Does a gradient section resemble a stack in terms of event continuity?

Jyothi Prativadi

Senior Processing Geophysicist At Reliance

ANSWER PROVIDED BY INDUSTRY EXPERTS

John Chamberlain

Consultant Geophysicist and

Brian Schulte

Geophysical Specialist Quantitative Geophysics - Talisman Energy

Jyothi, Essentially the gradient is a scaled version of the far-near difference (assuming we are within the Shuey limit), so it should look like reflectivity and have the same continuity (as long as there is some increase or decrease of amplitude with offset). But the gradient is a very useful attribute on its own as it depends on both the impedance and Vp/Vs ratio changes across each interface. The gradient is seen more as a stepping stone to other attributes such as fluid and lithology stacks. The strength of the gradient is in cross plots (see Fig. 6.6 below). If we cross plot intercept

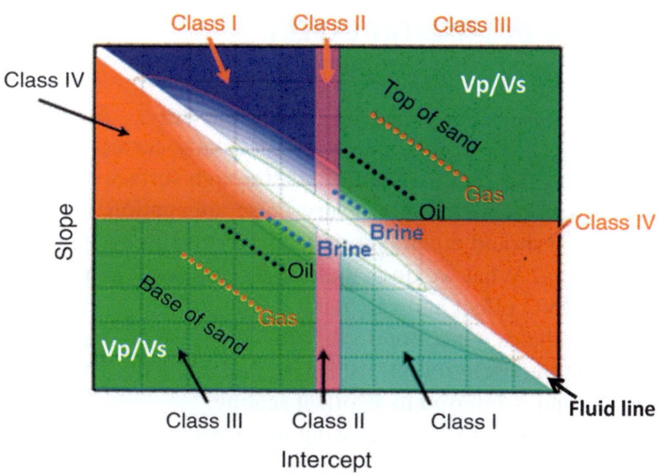

FIG. 6.6

A crossplot.

versus gradient we actually illuminate possible AVO behavior and we can define a background which is the wet sand and shale line. Deviation from the mud-line is in essence the fluid factor.

Though, the gradient is a least-square fit between the amplitude and the square of the sine of the reflection angle it can be noisy. We can actually utilize the correlation factor to control the quality of the gradient and throw out any gradient values with a correlation below 0.8. We also still have tuning present in these reflective attributes. With Intercept and Gradient you also have a two term fit or 3 term fit with the 3rd term being the curvature but sometimes that gives you a better fit. Note that gradient can be noisy because it is the slope of the rate of change of amplitude versus offset. It is a best fit line and the controlling factors of the gradient are:

1) Residual NMO;
2) Third term move out;
3) Offset scaling.

The gradient is sometimes a magnitude different from the intercept and we scale the intercept to match the gradient, forcing a Vp/Vs ratio of 2. If we do this we can use intercept + Poisson Ratio as the fluid factor which is also the Hilterman and Verm Poisson Ratio. We can also plot the Poisson Ratio versus intercept.

QUESTION 74

What is "offset scaling"?

Elsa Makiona

Geophysicist At Total E&P

ANSWER PROVIDED BY INDUSTRY EXPERT

Brian Schulte

Geophysical Specialist Quantitative Geophysics - Talisman Energy

Elsa, When we do our seismic processing we actually correct for frequency with deconvolution. We try to correct for amplitude through surface consistent amplitude correction which we should run after every noise attenuation, and spherical divergence. Our migration should maintain true amplitude but do we really have the right scaling on the gathers? What we do is that we go to a quiet area in the seismic data where we have no AVO effect and calculate scalars for the "background" where we have wet rocks and shales. Once we do this we define the background and anything that is anomalous to the background will be an AVO anomaly. This is generally done in the offset cubes and is referred to as offset scaling.

Note: If you think about AVO processing we should do nothing to affect the amplitudes in processing, but in reality amplitude in the gathers may not be perfect. This causes our background to be rotated and can cause some AVO's to rotate to the background trend.

QUESTION 75

Which gradient fit will be good for class 1 sands? Two terms or, three term Aki and Richards?

Jyothi Prativadi

Senior Processing Geophysicist At Reliance

ANSWER PROVIDED BY INDUSTRY EXPERT

Brian Schulte

Geophysical Specialist Quantitative Geophysics - Talisman Energy

Jyothi, The Class 1 AVO response is a peak getting smaller in amplitude until, at further offsets, it flips polarity and becomes a trough (figure below). Trim, is going to struggle with this because of the polarity flip. Semblance will not see it since semblance is looking for maximum energy and you will get a dim spot. Swan's RVI or Swan's velocities may work because they reduce the error in the gradient. Three term Aki Richards may work because it is picking up what is happening on the far offset. This is just my opinion and how I "think" this out (Fig. 6.7).

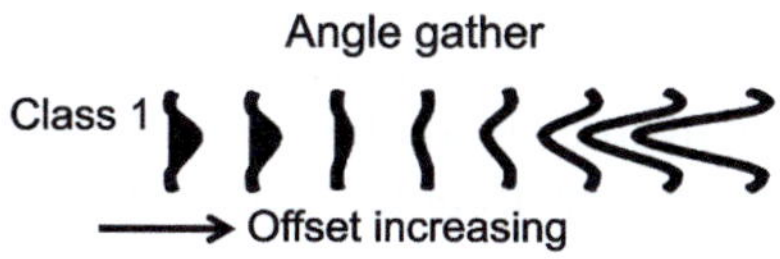

FIG. 6.7

Class 1 AVO seismic response.

Note

For Class 3 and class 2 AVO's we tend to have a negative gradient. Class 1 AVO and class 4 tend to be positive. Class 3 AVO's tends to have a low gradient while class 2 and 1 have higher gradients.

AVO CROSSPLOT AND REFLECTION CHARACTERISTICS OF SAND AND SHALE

In a crossplot domain, reflections are characterized by their contrasts in acoustic impedance and Vp/Vs ratios relative to the fluid-line trend. The crossplot allows us to analyze both intercept and gradient simultaneously. Fig. 6.8 shows an intercept and gradient crossplot.

The slope of the fluid-line depends on the background Vp/Vs. The fluid-line slope is −1 if the Vp/Vs ratio = 2 and rotates counterclockwise as the Vp/Vs ratio increases. The fluid-line is a useful concept because reflections from shales and brine sands that have little contrast in Vp/Vs ratio tend to fall on or near the fluid-line trend.

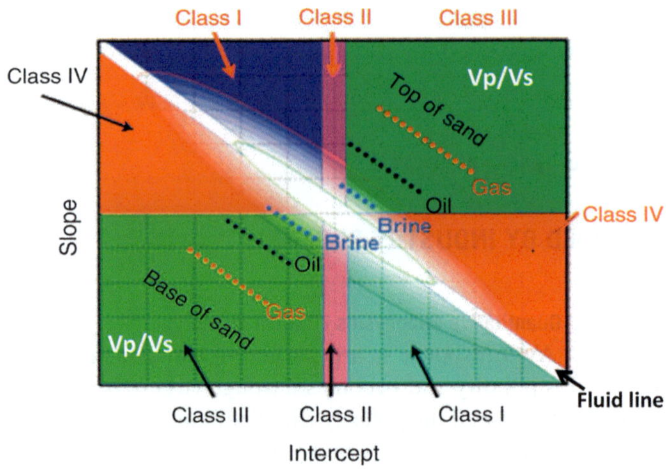

FIG. 6.8

A crossplot.

Reflections from hydrocarbon-bearing sands usually do not fall near the fluid-line trend. Because gas or light hydrocarbons often cause an abrupt decrease in the Vp/Vs ratio of a porous sand, reflections from the tops of hydrocarbon-bearing sands fall below the fluid-line.

Similarly, an abrupt increase in Vp/Vs ratio at the base of hydrocarbon-bearing sand places the gradient-intercept pair on a trend above the fluid-line. Therefore, displacement from the fluid-line can distinguish hydrocarbon-bearing sands from wet sands and shales. If there is a significant Vp/Vs contrast between sands and shales, this analysis can also be used to predict lithology in clastic sediments.

FACTORS THAT AFFECT AVO ANALYSIS

The following factors are very important in AVO/AVA studies:

- Fluid type (gas or oil)
- Lithology (sand or carbonate)
- Vertical resolution (tuning and target's thickness)
- Correct Vp and Vs values to compute the mud rock trend that fits the area
- Seismic quality (*S/N*, amplitude preservation, NMO condition, multiple, migration artifacts, phase and frequency content at target's depth)

Note also that it is important to do a feasibility study to assess net pay discrimination in log resolution and consequently in seismic resolution. If it is positive, you can do AVO/AVA modeling, AVO attribute analysis, and intercept and gradient crossplots to find oil sand zones (maybe class II). Also, you can test elastic impedance inversion. Furthermore, you can find fluid effect by using spectral decomposition at lower iso-frequency components.

In addition, AVO may not work well for carbonates. However, when there is gas present, lambda rho-mu rho could be helpful when carrying out AVO analysis in carbonate reservoirs.

QUESTION 76

Is AVO analysis applied in oil sands?

ANSWER PROVIDED BY INDUSTRY EXPERT

Anonymous

Reservoir Geophysicist

Carrying out AVO analysis depends on the region. Therefore, good knowledge of the regional reservoir lithology and fluid content is a major factor in determining the need for conducting AVO analysis. AVO analysis in oil sand will help if there is at least a small amount of gas associated with the oil. A purely oil sand reservoir will not show an AVO response. However, if you're not sure if AVO analysis will work in your area, do some fluid substitution modeling from well logs and compare oil sands to wet sands to oil with gas sands. More so, AVO analysis also depends on the API of the hydrocarbon of the reservoir. The heavier the oil (lower API) the more difficult it becomes to use AVO analysis. The contrast in properties between sand and shale decreases as the API decrease. Note that the lighter the oil, the higher the API (America Petroleum Institute unit of Specific Gravity, S.G). Generally, light oil has greater than 300 API; medium oil has between 200-300 API, while heavy oil has below 200 API.

AVO CLASSIFICATION

The AVO classification is based on reflection from the top of four classes of gas sands identified by Rutherford and Williams (1989) and Castagna and Swan (1997).

Four Classes of Gas Sands

The four classes of AVO responses of gas sands are classified based on their acoustic impedance contrast with the surrounding shales. Fig. 6.9 depicts the AVO response of reflection from the tops of four classes of gas sands.

The four classes are aligned on a trend in the figure. Batzle et al. (1995) show that the Vp/Vs contrast depends on the type of pore fluid. Therefore, the AVO response of the four classes of gas sands must fall on the trend that corresponds to the Vp/Vp contrast for gas sands. Their position on the gas sand trend depends on their acoustic impedance contrast with the surrounding shales.

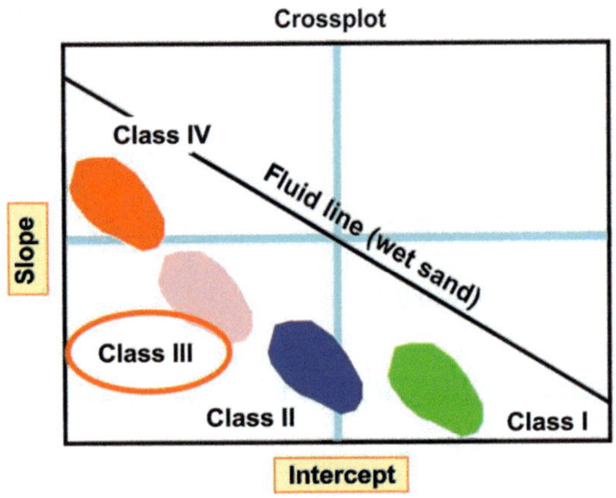

FIG. 6.9

AVO crossplot showing the four classes of gas sand.

Class 1 AVO response

The class 1 AVO response shows a peak on the near offset trace becoming smaller in amplitude across the section on the proposed lithologic interface, until at longer offsets it flips polarity and becomes a trough. This is shown in Fig. 6.10.

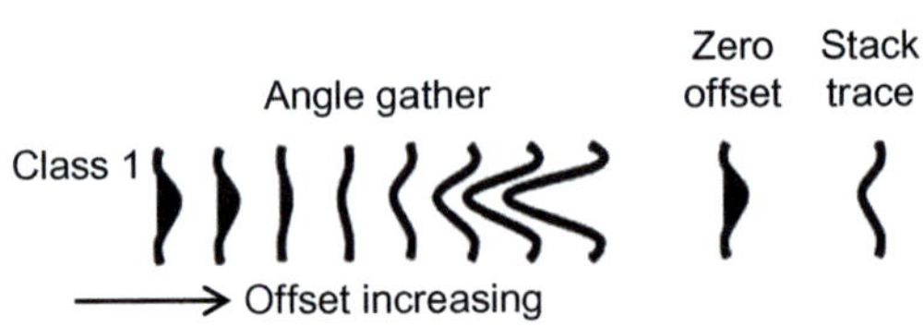

FIG. 6.10

Class 1 AVO response.

A class 1 gas sand has higher acoustic impedance contrast than the encasing shale. A reflection from the top of the class 1 gas sand must lie below the fluid-line trend, to the right of the slope axis (Fig. 6.12). Therefore, the reflection from the top of class 1 gas sand is positive at normal incidence, but its amplitude decreases with increasing offset faster than reflections that fall on the fluid-line trend (Fig. 6.11).

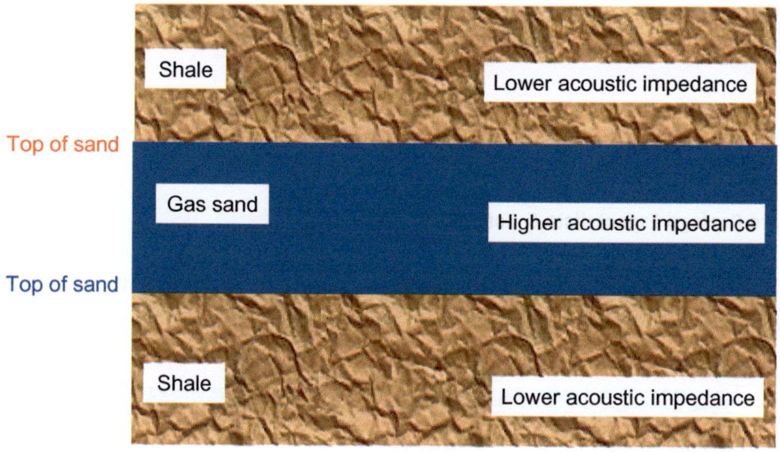

FIG. 6.11

Conceptualized class 1 AVO gas sands.

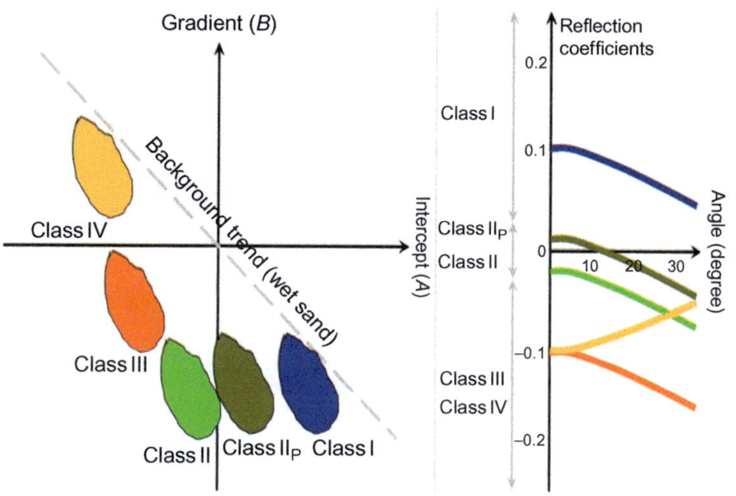

FIG. 6.12

AVO sand trend for the different class of gas sands. Class 1 sand is indicated as blue.

Courtesy: Jason Geoscience work bench, https://www.jason.cgg.com.

QUESTION 77

What can cause changes in the polarity (within the same survey) of the same seismic horizon from peak to trough or the reverse? What is the reason for this because I picked a horizon as trough. Then when I tried to pick it on another line the software did not allow me to pick it as a trough but only as peak?

Laouini Ghassen
Master En Geophysique

ANSWERS PROVIDED BY INDUSTRY EXPERTS

Alfredo Sánchez González
Ingeniero Geologo Especialista & Consulting

Laouini, If you are working in a siliciclastic environment, a possible cause can be associated with sand-shale boundaries. Hence, the polarity will be related to the contrast in impedance between sand and shale. This contrast will vary with depth (Fig. 6.13).

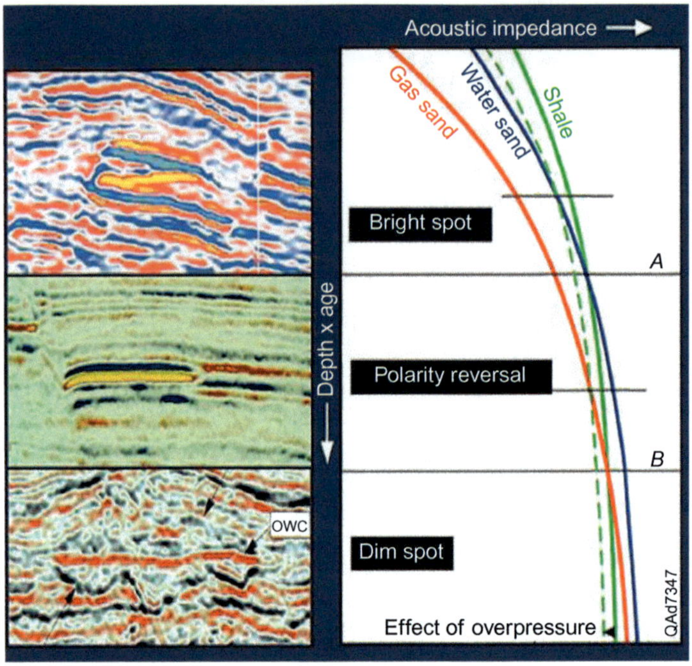

FIG. 6.13

Normal compaction curve.

Courtesy: Dim spots in seismic images as a hydrocarbon indicators by Alistar R. Brown.

Usually, relatively soft sands are found at relatively shallow depths where the sands are unconsolidated. At greater depths, the sands become consolidated and cemented whereas the shales are mainly affected by mechanical compaction. Hence, cemented sandstones are normally found to be relatively hard events on the seismic. There will be a corresponding cross-over in acoustic impedance of sands and shales as we go from shallow, soft sands to the deep, hard sandstones. However, the depth trends can be much more complex. Shallow sands can be relatively hard compared with surrounding shales, whereas deep cemented sandstones can be relatively soft compared with surrounding shales.

Muhammad Iqbal Hajana

Sr. Geophysicist At Weatherford Petroleum Consultants As

Laouini, Alfredo has explained everything about the polarity reversal. I would like to add one point, sometimes there are minor faults below seismic resolution that cannot be resolved seismically. These faults can cause strata to move in such a way that the peak from the hanging wall merges with the trough from the foot wall. So be careful when you are interpreting polarity reversals. Careful interpretation and knowledge about seismic polarity with stratigraphy of the under investigation area will help to you to understand polarity reversal.

QUESTION 78

A recently drilled well encountered a class 1 high impedance gas sand. The synthetics created using logs from the location shows typical class 1 gas sand behavior. We just don't understand the behavior of the actual seismic. We see a very unusual seismic response for class1 gas sands in offset domain: the amplitude is increasing from near to mid range offset and then decreasing from mid to far. The angles ranges are 0 -12 degrees near, 12-24 degrees mid and 24-36 degrees far. Even the raw gathers (prior to normal move out correction) also show similar response. What could be causing this observed effect?

ANSWERS FROM INDUSTRY EXPERTS

Rocky Detomo

Research Lead Reservoir Monitoring at Shell International Exploration and Production

The most likely causes of this observed effect are either:

- Interference/tuning effects between the real seismic wavelet and impedance contrasts of buried mediums. Because the bandwidth of real seismic data changes with offset, effects like thickness, AVO, and anisotropy become distorted by interference effects.
- Multiples often have a strong AVO dependence and their interference with primary reflections can give rise to some very unusual looking AVO responses. This can be confirmed with a VSP.

Peter Harris

Quality Project Manager At Exxonmobil

Interference effects can produce all kinds of apparent AVO responses and may even reverse the polarity of a "top reservoir" reflection. Simple classification only works in real life if you have blocky sands and blocky overburden (blocky means fairly homogeneous over a seismically resolvable interval).

Gautam Sen

Advise/Consultant In Exploration At Independent Oil & Gas Professional

Since the synthetic data shows class1gas sands both Vp and Vs increase in the second medium. The unusual behavior in near to middle offset is an artifact. Multiples are a great source of interference in the near offset and they are rarely eliminated even with sophisticated processing. Please, check your gathers for possible multiples, including peg-legs. You can also generate synthetic multiples and then see whether the gathers are contaminated with multiples. I remembered during my stay at RIL, one company had used amplitude compensation based on a scalar derived from the data at shallower level. This was absurd for it over-compensated amplitudes for larger offsets at deeper depths. Please, check whether the gathers are being affected by a similar decision. Seismic should see what the synthetic data shows if it was acquired and processed properly. In addition, does the synthetic show polarity reversal? You may need to examine the data acquisition and processing reports.

Marianne Rauch-Davies, Ph.D.

VP of Applied Technology for Geomage

From the question, you mention that you did a synthetic model using log data. Do you have a measured shear sonic available or did you use a simple transform using your sonic? If you have a shear sonic, which type of measurement and extraction was used? My guess still is that there is something incorrect with the seismic processing. Seeing the actual seismic gather in comparison to the synthetic one and then investigating the seismic processing work flow would be very useful. That is to say, if you do not see your behavior on the synthetic model, it is most likely caused by seismic binning/processing.

Irwan Djamaludin

Principal Geophysicist (Consultant) At Etoda

Is the seismic data a 3D data? I have seen this effect very often on 3D data. You may need to see how the processor did their regularization by taking a look at the gather before and after the regularization. It is more likely an amplitude issue in the binning/regularization. It most likely caused by the near offset amplitude.

Doug Cook

Geophysicist at Newfield Exploration

The first question I would ask is, is that a land or marine dataset? If it is land data it could be caused by problems in the offset distribution, but this is very hard/impossible to diagnose without knowing the acquisition and processing history.

Nayyer Islam

Petrophysics Intern at Bp America Inc

The general trend is expected to show a decrease in RC with an increase in offset. But theoretically, an increase in RC is possible from near to Mid if the shear impedance or shear velocity is slower in the second medium as compared to first medium. We can also observe a slight increase even if the shear velocity is same in both mediums.

You can use the three term concept to understand the seismic data.

1) Near offset RC component is dependent only on P-Impedance Contrast.
2) Middle offset RC changes are caused by only Vp/Vs ratio so Vs is important here.
3) 3rd component or far offset is only P-wave velocity and angle of incidence, no effect of density here.

You may investigate the change in density and P-wave from logs. It will help a lot to understand the issue.

My conclusions is that the density is increasing from first medium to second, giving a Class I response at zero incident. Vs is decreasing from 1st medium to 2nd medium giving an increase in amplitude from near to mid and Vp is also decreasing slightly, causing decrease in the far-offset.

Class 2 AVO response

A class 2 AVO has a small trough at near offset changing to a strong trough at far offset. This is shown in Fig. 6.14.

FIG. 6.14

AVO class 2 seismic response.

If the acoustic impedance of the gas sand is the same as that of the encasing shale, as in Fig. 6.15, it is a class 2 gas sand. The slope-intercept point for class 2 gas sand lies at or near the intersection of the gas-sand trend with the slope-axis.

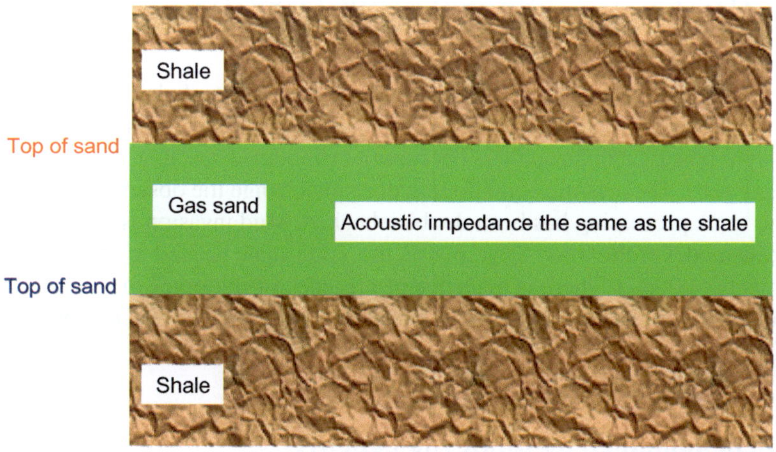

FIG. 6.15

Conceptualized class 2 AVO gas sands.

The reflection from the top of a class 2 gas sand is negligible at zero offset. But it has a negative slope, so its amplitude becomes large in magnitude with respect to the amplitude at zero-offset and negative with increasing angle. As in Fig. 6.14 a class 2P AVO has a small peak going to a large trough (Fig. 6.16).

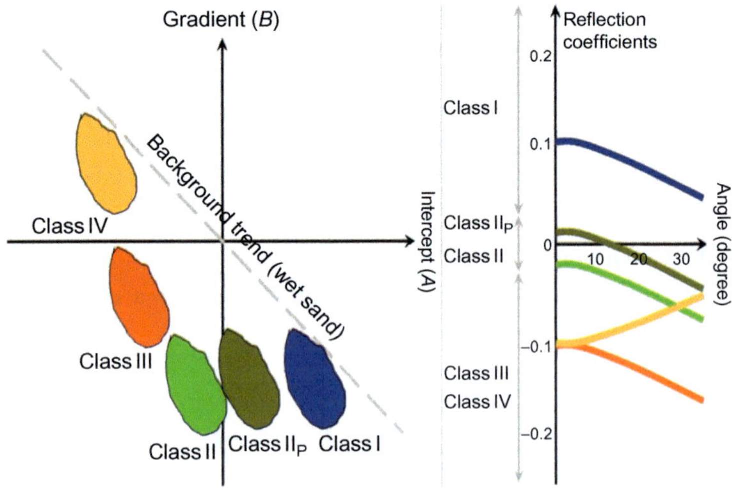

FIG. 6.16

AVO sand trend for the different class of gas sands. AVO Class 2 sand is indicated as light green, while 2p sand is indicated as dark green.

Courtesy: Jason Geoscience work bench, https://www.jason.cgg.com.

ANISOTROPY AVO MODELING AND PRESTACK GATHERS

QUESTION 79

At the top and base of a sand package (~20m) encased in shale (low AI to high AI from shale to sand), an AVO class 2p (change in polarity at higher offsets/angles) anomaly is observed in the gathers but cannot be modeled based on the log data (it is a deviated well). The seismic conditioning does not change the observed AVO response and the sonic processing has been double checked. Mud invasion is minimal (OBM). To create model of the Class 2p response, the Vp in shale needed to be increased from 2800m/s to 3000m/s. There is evidence of presence of multiples in the gathers. How does modeling in the presence of multiples affect the observed AVO response?

Hamed Amini
Geophysicist at Senergy Energy Services

ANSWERS PROVIDED BY INDUSTRY EXPERTS

Arslan Tashmukhambetov
Geophysicist at Llog Exploration Company

Hamed, Are you measuring shear in deviated section? If so, that could be an explanation of differences in observations between seismic gathers and modeling. If no, in most cases you need to double check seismic processing sequence from your vendor to be sure that they are not doing any artificial scaling to improve data appearance. The presence of multiples can also affect AVO measurements in gathers, creating polarity changes.

JP Blangy

Chief Geophysicist at Hess Corporation

Hamed, Firstly, how much deviation exists in the well? And is the bedding tilted (as in structural dip)? In getting the offset angles "right", you will need to take both possibilities into account.

Secondly, the art of estimating Epsilon and Delta with confidence can be tricky. It is likely that you only have to worry about the anisotropic parameters for the overburden/sealing shale. I recommend that you double-check the values that you are using against Thomsen's published table on weak elastic anisotropy. If you have an offset VSP in the area, that will help a great deal in establishing credible parameters.

Jorge Reveron

Senior Geophysicist at Repsol

Hamed, anisotropy could be the answer, but before using a more complicated model you have to answer the following question:

What is the original response of your synthetic gathers and how much is it different from actual seismic gathers? Are the gathers completely flat? Have you done gathers conditioning? Have noise or multiples been attenuated? Do synthetic gathers and seismic gathers have similar amplitude values? AVO class 2P is very tricky in the presence of noise because of its weak amplitudes.

Leon Thomsen

Geophysicist

Hamed, The most obvious possibility for this discrepancy is that the synthetic you have built expresses isotropic reflectivity, whereas the real seismic data express anisotropic reflectivity. The anisotropic correction to the AVO signature is large, even when the anisotropy is weak because all of the terms in the reflectivity equation are small (compared to 1). There is a new way to estimate the missing term; see: http://library.seg.org/doi/abs/10.1190/segam2013-1150.1.

Your situation is complicated by the deviated well; you will have to modify the Lin/Thomsen algorithm slightly, but the principles are the same. Note that even weak anisotropy has a strong effect on AVO.

Erick Alvarez

Senior Reservoir Geophysicist At Shell

Hamed, I have a question regarding anisotropy issue though. Have you checked that the seismic processing sequence includes anisotropic solutions to the imaging? If so, it is worth checking what parameters were used to see if this information is useful for your anisotropic AVO modeling. If not, then I struggle to see how useful anisotropic modeling would be if the anisotropy effects have not been preserved during processing.

Leon Thomsen

Geophysicist

Erick Alvarez', Seismic processing, done with or without good attention to anisotropy, concerns itself with coarse averages of elastic properties in the overburden and not the local jumps in properties at the reflector (which are encoded in the amplitudes, not the arrival times, of the various arrivals). Even if the processing flow has found optimal values of anisotropy, these are of a too low-resolution to be useful for analyzing amplitudes. This is why we have been forced to ignore the anisotropic effect in AVO (e.g. Delta-delta) for 30 years, and why the new Lin/Thomsen algorithm for estimating that parameter effect is so important.

John Chamberlain

Consultant Geophysicist

Erick, Anisotropic processing is concerned only with the kinematics, that is, the travel times. If anisotropy is not included then the travel time computations are incorrect and hence the velocities but the amplitudes will not be affected. That is, anisotropic effects will still be present and must be modeled (being careful about offset to angle transformation).

Leon, If we have a massive shale overburden, why can't we use the values of epsilon and delta from the processors? I agree they have poor vertical resolution but is it unreasonable to assume that the shale is relatively homogeneous?

Leon Thomsen

Geophysicist

Chamberlain, We all used to think that shales were uniform when we were concentrating on sandstone or carbonate reservoirs. Now that we understand that shales can also be reservoirs, we are realizing that they have a lot of internal variability. That is why companies drilling for shale oil and gas find that 80% of their production comes from 20% of their wells (or fracture stages). That is why the "factory approach" to the shale resource (drill on a grid, and fracture aggressively) will ultimately fail. As we learn how to actively explore for the sweet spots, we will drill holes only in the right places, lowering our costs, and making money instead of losing money. As we learn how to do that, an important aspect of that learning will be the application of anisotropic rock physics.

Jim Applegate

CEO, Pegasus O&G, Inc.; Explorationist, Applegate Exploration, LLC

I don't know where Leon got his numbers, but I believe his comments. Unfortunately, many companies have not figured this out, and are continuing to drill a large number of marginally economic wells. I have been 'fired' multiple times by clients for 'harping' on the need for more and better seismic and more detailed analysis to high grade acreage for the sweet spots. People do not want to hear that "it ain't a manufacturing business". It does not fit with the ideas that has been sold to the money guys.

Class 3 AVO response

The class 3 AVO is the only AVO where trough and peaks line up as shown in Fig. 6.17.

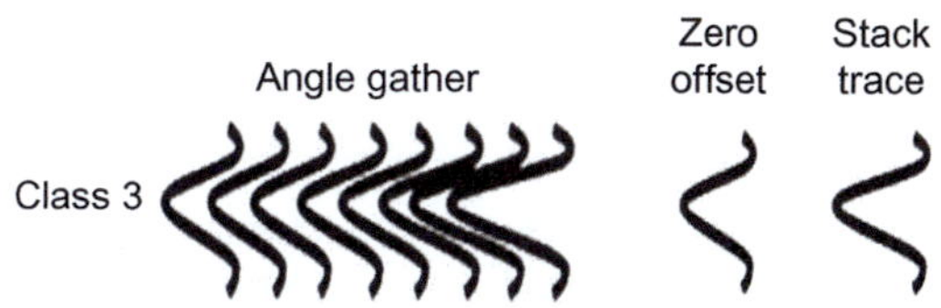

FIG. 6.17

AVO class 3 seismic response.

If the acoustic impedance of the gas sand (Fig. 6.18) is lower than the overlying shale, it becomes class 3 gas sands. The reflection from the top of class 3 gas sand has a negative intercept and slope (see Fig. 6.19). Thus, it is negative at normal incidence and becomes more negative with increasing angle.

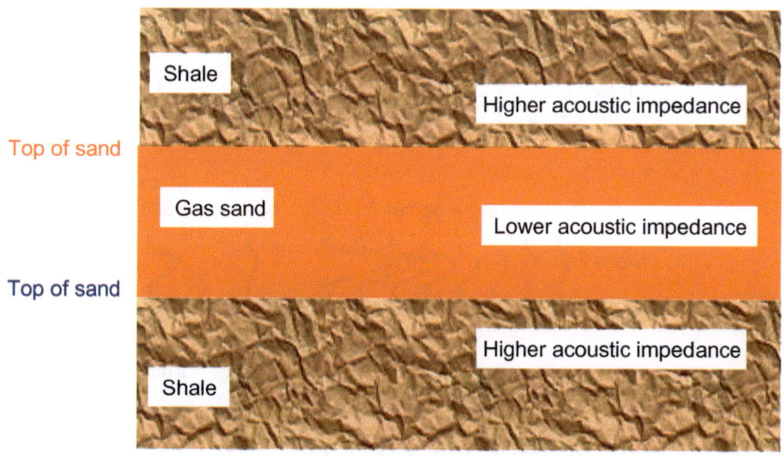

FIG. 6.18

Conceptualized class 3 AVO gas sands.

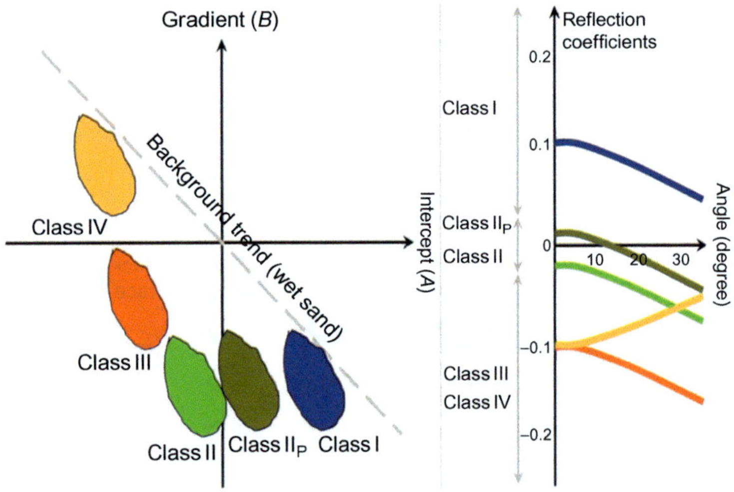

FIG. 6.19

AVO sand trend for the different class of gas sands. AVO Class 3 sand is indicated as red.

Courtesy: Jason Geoscience work bench, https://www.jason.cgg.com.

Note that class 3 AVO is the only AVO anomaly where trough and peaks line up, but a class 2 has a small trough going to a strong trough and a class 2P is a small peak going to a large trough. If we look at a class 1 AVO we get a phase reversal (Fig. 6.20).

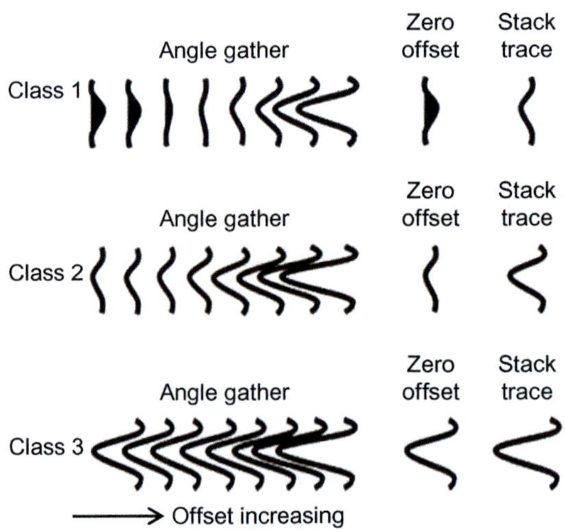

FIG. 6.20

The seismic responses for the different classes of gas sands.

QUESTION 80

What kinds of reservoirs should display a Class 3 AVO features?

ANSWERS PROVIDED BY INDUSTRY EXPERTS

Fred Hilterman

Chief Scientist at Geokinetics

If you have shale over wet sand, there is an approximation that determines if it is Class 3 or Class 4. If the overlying shale has a Vp that is 20% greater than the underlying sand Vp, then the Poisson's ratio of the shale tends to have a smaller Poisson's ratio than the sand. This leads to a Class 4 AVO signature. Also keep in mind that the AVO signature can depend on the frequency of the wavelet. If the model wavelet frequency is decreased, the seismic wavelength increases and the wavelet will sample more beds above and below the zone of interest. This can lead to a change in AVO class from a 4 to 3 (or 3 to 4).

Brian Schulte

Geophysical Specialist at Talisman Energy

When we have issues in the processing we can actually have a rotation in the cross plot and a class 3 AVO may appear to be a class 4. Establishing phase is crucial because all AVO assumptions are based upon zero-phase data. Once we have established that we have zero-phase on the stack we need to rotate the gathers accordingly. We have to look at residual NMO or trim statics and we need to ensure that our correlation window be larger than any AVO within the data. People want perfectly flat gathers but that will destroy our AVO anomalies in the data. We then need to see how the offsets were balanced. Class 3 AVO's and Class 4 AVO tend to be related but it depends upon the difference between the seal and the reservoir. When we have very unconsolidated sand with a shale as a seal we tend to have a class 3 but if we have a higher velocity rock as a seal we may get a class 4 AVO.

QUESTION 81

In partial angle stacks we have observed that in case of clean brine sand amplitude increases from near, to mid to far angle stacks. In some books it is mentioned that shear wave velocity increases as the seismic wave travels from shale to clean sand and that is why we observe the same response from a clean brine sand as a gas sand. How can we delineate gas sand in such environment? Why do water bearing clean sands show the same bright amplitude response on seismic as a gas bearing sand?

Shweta Bankhwal

Sr. Geophysicist at ONGC

ANSWERS PROVIDED BY INDUSTRY EXPERTS

Rocky Detomo

Research Lead Reservoir Monitoring at Shell International Exploration and Production

Shweta, The "bright spot" amplitude response is usually due to the fluid bearing sand being much softer acoustically than its bounding lithologies. This can be caused by a number of things, including the sand containing a less-dense fluid or gas, whose density is very low. But this can also be caused by a very porous sandstone carbonates, over-pressure, the presence of other lithologies (volcanics) or anything else that lowers the compressional velocity or density of the sandstone. As mentioned above, if you are looking at full-stack data that includes far offset data that has strongly rising AVO, you may also be fooled into thinking that this is a "bright spot". Finally, if the bounding lithologies are abnormally "hard", the bounded sandstone may look surprisingly "soft". The best thing to do is to measure or estimate rock and fluid property ranges for the sandstone, fluids, and bounding lithologies, and model a range of typical responses.

Alan J Cohen, PhD

Geophysical Consultant

Shweta, In addition to Rocky comments, what porosity does your brine sand have?

Duane Pankhurst

Principal Geophysicist at Edge Geophysics, LLC

Shweta, As Allen is suggesting, a typical AVO false positive in high porosity brines may look seismically similar to a lower porosity hydrocarbon - gas case. Modeling porosity 'sweeps' using the 100% brine case should show what porosity would achieve the same AVO character as you are observing.

Jim Applegate

CEO and Managing Member at SeismicUtensils, LLC

Shweta, Sometimes it is a case of anisotropy, particularly in the overlying rock (shale?). Blangy (1994) shows a model with anisotropic shale overlying isotropic brine-filled sand. The AVO response is that of a gas-filled isotropic shale/sand. After showing this example in a seminar, a participant sent me a real-life example from offshore UK…. bright spot, class III response - water-filled reservoir. The world is not isotropic.

Elive Menyoli

Sr. Geophysical Technical Advisor at Paradigm Geophysical Corporation

Jim brought up a good article that reminds me about that SEG/EAGE summer workshop. It was really revealing at the time. The article by Blangy may well be the answer to Shweta's question. Check it out http://library.seg.org/doi/pdf/10.1190/1.1443635

Alan J Cohen, PhD

Geophysical Consultant

Shweta, Before you hunt for anisotropy, rule out all the isotropic possibilities.

QUESTION 82

I have a very interesting set of gathers (5-45 degrees) that display a Class 4 AVO where a Class 3 AVO anomaly is expected based upon rock physics. The study is not calibrated to a well in the vicinity but, from 5 degrees to 22.5 degrees we get flat or slight Class 3 AVO anomaly and then attenuation in amplitude further out to produce a Class 4 AVO. Oil is expected in the area but Class 4 is quite rare. The interpreters insist that they are picking the correct trough. Any thoughts?

Brian OConnell

Senior Geophysicist at Repsol

ANSWERS PROVIDED BY INDUSTRY EXPERTS

Arslan Tashmukhambetov

Geophysicist at Llog Exploration Company

Brian, You need to take a look into the data processing sequence. Some of the vendors artificially boost far offset amplitudes by applying linear scaling with offset. Or your rock physics model could be wrong. Some of the tools cannot properly measure shear component for certain depths or type of sands.

Rosemary Quinn

Lead Geophysicist at EnQuest

Brian, A few questions;

- Are your gathers nice and flat, free of residual multiples and well balanced for amplitudes in general? In which case, you are in a better starting place than usual.
- Was the horizon picked on the gathers, or has it been snapped from a stack volume?
- What is the critical angle expected to be? Being beyond it often changes the gradient of your reflection curve.

And finally, maybe you do actually have a Class IV response, which accordingly to the text book models isn't that promising for oil. However, I worked in an area where you can get Class IV AVO anomalies from oil-bearing sands. We tied that response to 4 out of the 6 wells in the area. I am still not sure if we would have expected Class IV from half space modeling (we ran elastic synthetics from our well logs using logged and modeled shear velocities) or whether the Class IV was related more to bed-thickness, but it was what we had in that particular area.

Jerry Coggins

Principal Geophysicist At Coggins Geosciences, Inc

Brian, The most important factor, if you are looking for hydrocarbons, is that the AVO behavior changes along a depth contour.

Gautam Sen

Advice/Consultant in Exploration at Independent Oil & Gas Professional

Brian, First things first! Are your near offset gathers free from residual multiples as Rosemary pointed out? Is the processing sequence in conformity with amplitude preservation as Arsalan pointed out? What is the sanctity of Rock Physics in the absence of well data? Has close grid velocity analysis been done? Does it show overlying shales to have higher impedance than the sand? Is the interface a map-able reflector? Do we expect the oil to have associated gas? Lastly as Jerry points out, is the AVO behavior in conformity with the structural configuration?

Robert McGrory

Sr. Geophysical Crescent Point Energy

Brian, Class 4 basically occurs when you have soft sand encased in a relatively stiff medium. Examples include "Halfway Sands" in the West Canadian Sedimentary Basin. These are unconsolidated sands encased in tight carbonates. Other examples with younger rocks are the Mount Messenger Sands in the Taranaki Basin New Zealand. Porous low velocity sands are encased in relatively stiff silts and shales that have a significant calcite component. As Class 3 (hydrocarbon case) gradient increases towards zero the rate at which amplitudes change decreases. The geological interpretation can be varied and may included either an increase in porosity from the calibration data (lower velocity and density) or that the reservoir properties are the same and the bounding rocks (top and base seals) are becoming stiffer than the calibration data set. Or some combination of both the aforementioned and other answers provided by professionals here. I assume you have ruled out processing scaling issues and you have properly processed true amplitudes?

Jing BA

Senior Geophysicist at PetroChina

Brian, Class 4 is quite common in data from all kinds of reservoirs (limestone, dolomite and sandstone). The AVO features in the small-angle ranges should be more reliable than those extracted from large-angle gathers. First, you should check your processing sequence very carefully. Maybe you can model the AVO features with some well log data. If no log data is available, try to model the rock properties according to the drilling reports etc (actually you can get much information about the overall porosity ranges and the mineral contents of layers). This is important since it can help you confirm whether the processed gathers are appropriate.

Kester Waters

VP Global QI – Products and Solutions at Ikon Science Ltd

Brian OConnell, There are clearly a lot of processing issues that could give rise to the effect you are observing, but, let's assume your data are Ok. One thing that you should definitely do is to talk with the geologists about the depositional environment and burial history of the basin. We do a lot of frontier work here and make extensive use of information from analogues. So, do you have a good analogue basin that you could look at? Are there analogue wells in similar geological/tectonic environments from basins far away that could give insight? We have shown (as have others) that there is universality in many rock property relationships that can, when used carefully, allow extrapolation into areas without wells, providing you have (or can interpret) information on things like:

- Pressure regime - are you expecting normal or 'significant' overpressure? This in itself can seriously modify the properties of the shales and the resulting shale-reservoir interface.

- Are your shales expected to be calcite rich, which will stiffen the rock and increase velocity?
- Are your shales organic rich? This can give rise to lower than anticipated Vp/Vs (The kimmeridge clay in the North Sea can often have Vp/Vs ratio as low as the reservoirs (1.6-1.7).
- Are you in the cementation window (70 -100 degrees C and higher)? For shales this can mean onset of secondary mechanisms (e.g. smectite to illite) which itself can lead to other changes in rock framework. For sandstones, a very small amount of cement will increase velocity dramatically, and may often occur over a fairly narrow temperature window. This needs to be considered in context of burial history/kinematics.
- Are you expecting under-compaction i.e. higher than expected porosity at depth? For example, West of Shetland we see sands with preserved porosity of up to 30% at 4.5km below mudline. This is a common cause of Class IV reservoirs in deep environments (we see this in West of Shetland, North Sea, Norway, West Africa etc), the main cause being the shear impedance of the reservoir being lower than the overburden (note that the reservoir still has lower Vp/Vs than the overburden). I am not suggesting that rock physics modeling is the key to success here, but you can significantly constrain your forward models by taking information from analogue basins and reservoirs (porosity compaction trends, temperature profiles, tectonic and depositional styles etc) and then combining them with information you can derive from seismic (such as estimates of the shale velocity from seismic interval velocity, estimates of likely impedance change from relative litho-fluid inversion and so on).

Fred Hilterman

Chief Scientist at Geokinetics

Brian, If you have shale over wet sand, there is an approximation that determines if it is Class 3 or Class 4. If the overlying shale has Vp that is 20% greater than the underlying sand Vp, then the Poisson's ratio of the shale tends to have a smaller Poisson's ratio than the sand. This leads to a Class 4 AVO signature. Also keep in mind that the AVO signature can depend on the frequency of the wavelet. If the model wavelet is decreased, the seismic wavelength increases and the wavelet sample more beds above and below the zone of interest. This can lead to a change in AVO class from a 4 to 3 (or 3 to 4).

Brian OConnell

Senior Geophysicist At Repsol

Fred, Coincidentally we are getting a 20% Vp drop from encasing medium to sand. We do have a large caveat in that our velocities are taken from the seismic and then appropriate Castagnas and Gardners equations are used, but we are seeing that size of velocity drop.

Brian Schulte

Geophysical Specialist at Talisman Energy

Brain OConnell, When we have issues in the processing we can actually have a rotation in the cross plot and a class 3 AVO may appear to be a class 4. Establishing phase is crucial because all AVO assumptions are based upon zero phase data. Once we have established that we have zero-phase on the stack we need to rotate the gathers accordingly. We have to look at residual NMO or trim statics and we need to ensure that our correlation window be larger than any AVO within the data. People want perfectly flat gathers but that will destroy our AVO anomalies in the data. We then need to see how the offsets were balanced. Class 3 and Class 4 AVO tends to be related but it depends upon the difference between the seal and the reservoir. When we have very unconsolidated sand with a shale as a seal we tend to have a class 3 but if we have a higher velocity rock as a seal we may get a class 4 AVO.

David Forel

Documentation at Drillinginfo

Brian Schulte, You wrote, "People want perfectly flat gathers but that will destroy our AVO anomalies in the data." Are you saying the standard NMO equation distorts phase? Is this offset related? That is, if I use more terms in my NMO equation can I preserve phase?

Brian Schulte

Geophysical Specialist at Talisman Energy

David, If we look at the Class 2p AVO we have a small peak going to a large trough and a class 1 we have a peak going to a trough, so things like picking velocities with semblances may not work. If we try to flatten this with a small window trim what will happen? Swan's velocities actually did get better results for class 2p because it utilized the intercept and the gradient. It was trying to reduce the error within the gradient while with semblance we are trying to maximize energy.

The NMO equation is also good only to 30 degrees and the limits of the NMO equation causes the hockey sticking or VTI. We knew that anisotropy was present but anisotropy should be small so we knew that there were other factors that caused the hockey sticking. Straight ray PSTM also utilizes a two term equation and some of the hockey sticking is caused by this "curved ray" effect. This is why we created the curve ray PSTM in order to go beyond 30 degrees and get flatter gathers. Now when we do the higher order NMO our eta (estimated time of arrival) should represent rock properties. This will improve your far offsets, stabilize the phase and you should get better results from the pre-stack inversion especially for density. There are a lot of things people will argue with me on but I do believe we can get a "constrained" density by improving the gathers beyond 30 degrees and utilizing a neural net with a combination of attributes. More so, if we look at Shueys equation the near offsets are the P-impedance, the mid offsets are S-wave and the far offsets are the Vp. If we take the P-impedance and subtract it from the far offsets we get density. As we all know Vp comes from the curvature which is a small term and can be unstable. What do we have in the far offsets is noise, particularly the NMO stretch. But when we do curve ray PSTM we actually see that the gathers are flattened beyond 30 degrees because we have taken the "curve ray" or Snell's law out of the PSTM. We then do a higher order NMO and let the eta "flatten" the data further and push the NMO stretch out further. We still have a decrease in frequency because of attenuation of frequency with distance and we may have some amplitude problems. We also need to have removed the multiples in the far offsets. Multiples can really hurt us in AVO hence the current development of land SRME and also people developing inter-bed multiple programs.

Our improvement with AVO is due to the technical improvements we are making each day in processing. This allows us to do AVO better and I think it causes us to have fewer errors. AVO and processing go hand in hand. The neural net will take a series of attributes and combine them to fit a well curve. If we think about it I can run inversion many ways each time getting a different result. I know I need to have a way to substantiate which one is better so I attempt to tie it to well data. Why not utilize a smart way to do this through neural nets?

Brian OConnell

Senior Geophysicist At Repsol

Thanks for the solutions provided above. What I can say at the moment is we have properly processed true amplitudes but we have not ruled out processing scaling issues. As this is an un-calibrated study, there is a scaling problem we have to be aware of. The AVO anomalies conform to structure which is a good thing and we have zero-phase wavelets. Originally the seismic was 180degrees opposite phase. The closest possible well that might have similar sands is 1000km away, but we are looking to see if we can get possible analog information on the casing material.

Jing BA

Senior Geophysicist at PetroChina

Brian OConnell, Even though the closest well is 1000km away, you may try importing the rock properties of the target layer and the seal. They will help since you are only trying to find the type of AVO (3 or 4) rather than the more exact relations of reflection intercept and gradient versus angle. A forward modeling with the well properties will be enough to solve the problem.

Brian Schulte

Geophysical Specialist At Talisman Energy

Brian OConnell, AVO conforming to the structure is great. If I had nothing but seismic I would take the interval velocities from the seismic and take an average of the shale interval velocity above, the sand and the average of the shale below utilize Gardner's relationship for density and Castagna or Greenberg and Castagna to guess estimate the Vs. and, put it into a rock calculator and see what that tells me.

Ron Masters

Geoscience Advisor At Headwave, Inc

Brian OConnell, What you expect may not be what the earth offers. Better to investigate the seismic data than to make assumptions about AVO classes based on well control that might not be analogous. There's a simple process: Use Head-wave to compute intercept and gradient; assign unique colors to each (Intercept, Gradient) vector in the cross plot; interactively choose scales and orientation of the trend and anomaly color axes, visualize the vector colors in 3D, with transparency, optionally sculpted to conform to a reference horizon. Now you can recognize all AVO classes simultaneously; investigate fit to structure and depositional style. You do not have to worry about which loop to pick, even for dim spots or class II anomalies, where the top reservoir phase may change with fluid type along structure, because you can see the whole complex in 3D. Export trend and anomaly volumes for auto-tracking if you wish, or a volume with opacity tailored for body checking. Your grab a Headwave license and your local expert to do this. Once the attribute volumes have been calculated, you can do the analysis in 20 minutes. Do the same analysis around local well control to find out how to extrapolate the rock physics. Really, doing the rock physics first which everyone has been advocating lately, will set expectations that may not turn out to be relevant.

Gautam Sen

Advice/Consultant In Exploration At Independent Oil & Gas Professional

Ron, Do you mind suggesting some literature available in public domain for understanding the said process?

Fred Hilterman

Chief Scientist At Geokinetics

Gautam, Verm and Hilterman published the article "Lithology color-coded seismic sections: The calibration of AVO cross plotting to rock properties" (TLE, August 1995, p. 487). This should assist in describing some of the interpretation techniques that Ron Masters has suggested.

Ron Masters

Geoscience Advisor at Headwave, Inc

Gautam, here's an expanded abstract

http://www.headwave.com/article/articleview/191

And a web page

http://www.headwave.com/article/articleview/160

Many people have told me they did not get it until they saw the interactive demo. The moral is that one should not obsess about looking for a particular class. Look for all of them, and decide what makes sense in the geologic context.

Brian Schulte

Geophysical Specialist At Talisman Energy

Ron, Intercept and gradient cross plots can have many issues. Gradient is the slope of the best fit line through the amplitudes. We have many issues with that as mentioned before such as residual velocity, higher move out and also amplitudes. We can have leakage of the intercept into the gradient. This is why we should look at the correlation factor of the gradient to determine whether the gradient is good.

We also have the overburden effect which causes changes within the background. We also need to define the background which is shales and wet sands. I also feel that we should never really define an AVO anomaly on just intercept-gradient but de-risk it by using several attributes including geometrical attributes. We developed an attribute which was created by looking at wet sands and pay sands and the response of different attributes to define a zone for the pay sands. Where all of our attributes fall within this range we set a 1 and where it doesn't then we set it to 0. We then create a P10 and a P90 case by changing how tightly we constrain the attributes. We overlay it onto the seismic and see how it looks in the horizon slice and how it matches to structure. We also look at the pre-stack gathers to ensure that what we are illuminating is not due to an issue in processing.

Each one of us has a broad range of ways to define pay zones ranging from reflective attributes to inversion. Our ultimate goal is to be able to tie what we see within the seismic back to the well logs. Things like well modeling (acoustic and elastic) and calculating the type of AVO from well logs helps enhance our understanding. We are going to drill a multi-million dollar well, so we need to ensure that we de-risk and pick the correct location. We always want to pick our best locations to drill first to establish proof of concept. Asking processing companies to utilize well data to help with amplitude adjustment etc., I believe maybe a good idea if we know the processing company has the expertise. We have different tiers of processing companies and we need to treat those who do processing as internal consultants and hire them for their expertise to support us.

Gautam Sen

Advice/Consultant in Exploration at Independent Oil & Gas Professional

I would like to agree with Brian Schulte on the intercept gradient cross plot.

I have personally been involved in drilling a large number of dry wells because of false AVO and in deep waters where seismic data quality is by far the best, and I do not buy the argument of fizz gas. Unconsolidated Pleistocene high porosity brine sands show clean class 3 AVO because of decrease in Vp/Vs compared to overlying shale. DHI studies for other class sands, certainly need a far more comprehensive integrated workflow combining geology, logs and seismic including inversion, mapping, AVA, etc. to make a credible guess before suggesting drill location.

Ron Masters

Geoscience Advisor at Headwave, Inc

Brian Schulte, AVO always requires high-quality pre-stack imaging, but I would not get too hung up on the long history of complaints about fitting intercept and gradient. If you look at the distribution of the vector colors in the subsurface, you can see through stochastic noise. Near-far gain errors may trip you up if you are drawing theoretical class polygons, but if you can see all the relative behavior on all the reflections, anomalies will still show through. It may not help to impose well-based biases on processing if you are outboard or deep from well control. Better to find the anomalies, focus on the ones that make geologic sense, and then work to integrate rock physics on valuable targets.

Rob Simm

Senior Geophysical Advisor At Cairn Energy

Ron Master is right. Do not get hung up on AVO class. Look for anomalies and try and put them into a sensible rock and fluid context. All the stuff mentioned so far (amplitude conformance, the possibility that anomalies might be related to high porosity wet sand etc.) can be addressed within this context ('Route 1 AVO' I heard someone call it once). For those without the budget to purchase Head-wave software, an alternative for data scanning would be using weighted stacking on interpreted horizon amplitude maps such as advocated by Whitcombe, Connolly and others (see Pat Connolly's presentation at http://gsa.seg.org/pdf_forms/SEG%20DL%20Handout%20v1.2.pdf). At its simplest you could use a function such as 'weighted stack = ((Near*constant) -Far)'; constrain it initially from a cross-plot of near offset versus far offset in an area considered to be 'background' and spend time looking at the effects of varying the constant (note that the change between fluid and lithology dominated combinations may well be represented by a restricted range of constants). If there is one thing that has come out of this discussion for me it is that it is important to recognize the potential for a disconnect between our intuitive models and the seismic data (for various reasons including imaging issues and the effects of seismic noise). This is especially the case in areas of little or no well control. AVO scaling/calibration is a tricky issue that can only confidently be dealt with if you have some good calibration (i.e. wells on the survey in appropriate geological settings with good quality log suites including Vs).

Class 4 AVO response

Class 4 basically occurs when you have soft sand (low velocity) encased in a relatively stiff medium (carbonates, silts and shales) that has a significant calcite component. This is shown in Fig. 6.21.

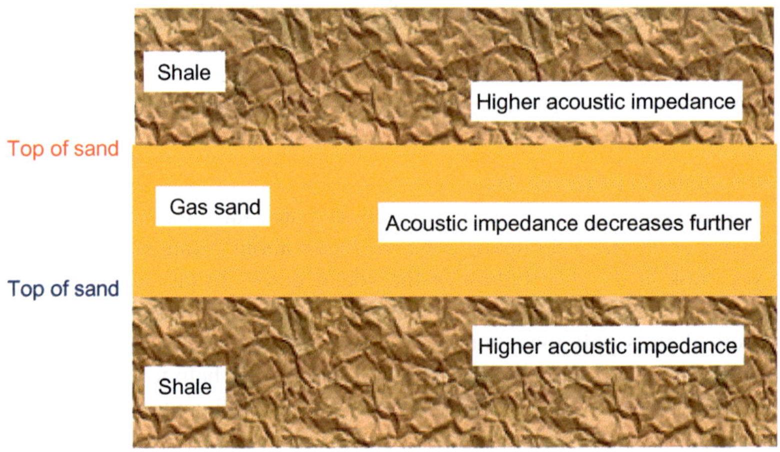

FIG. 6.21

Conceptualized class 4 AVO gas sands.

Recall that if the acoustic impedance of gas sand is lower than the overlying shale, it becomes Class 3 gas sand. Furthermore, if the acoustic impedance of the gas sand decreases further, this will move the reflection intercept-slope point up and to the left on the gas sand trend to produce a class 4 gas sand reflection. This is characterized by a negative intercept and a slope that is zero or positive (this is shown in Fig. 6.22).

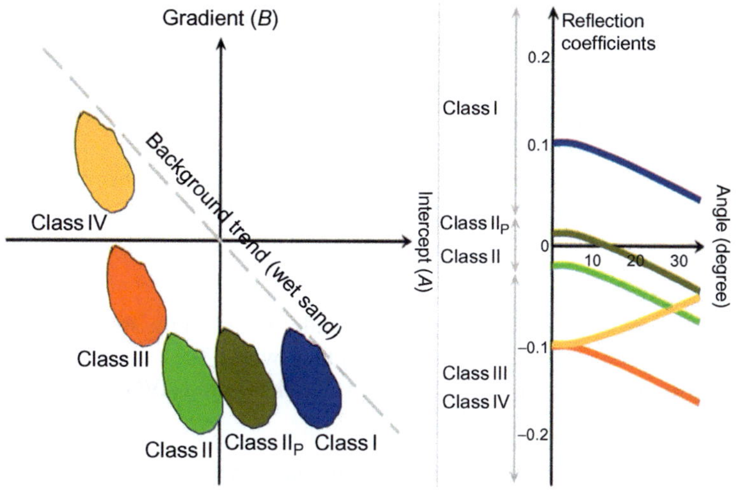

FIG. 6.22

AVO sand trend for the different class of gas sands. AVO Class 4 sand is indicated as yellow. *Courtesy: Jason Geoscience work bench, https://www.jason.cgg.com.*

The reflection from the top of class 4 gas sand is negative, but its magnitude does not increase with angle. Note that Class 3 and 4 gas sands may produce a large intercept or normal incidence reflection coefficient. Polarity is important in this case. Note also that class 3 and class 2 AVOs tend to have a negative gradient. Class 1 AVO and class 4 tend to have positive gradients. Class 3 AVO tends to have a low gradient while Class 2 and 1 have higher gradients.

QUESTION 83

I am conducting an AVO study with no well coverage. To make my job more interesting, a lot of the area of interest has imaging problems with the gathers attenuating to nearly zero after 2000m in data that is supposed to have 6000m offsets. I am seeing, lots of Class 4 AVOs, at what appears to be a layered reservoir and possibly a Class 3 AVO anomaly at the bottom of the reservoir. We expect the sand to be approximately 20% porosity. Unfortunately I have no information about what the shales bounding the sands are like.

Brian OConnell

Special Studies Geophysicist At Cepsa

ANSWERS PROVIDED BY INDUSTRY EXPERTS

Chandra Velu

Senior Geophysicist at Lundin Malaysia B.V

Brian OConnell, I am more curious to know why the bad imaging. If processing isn't an issue then is the bad imaging due to defocusing of energy due to complex geology or a scattering of energy due to gas effects? If the later then you might have a working petroleum system. The assumptions for the AVO Classes are that the shales are harder than sands. How did you come up with Class IV and III if you have imaging problems and no information about shales? If you don't have budgetary issues try PSDM first to solve the imaging issue. Bottom line, sounds like you are in frontier exploration if no wells are available, hence I would squeeze the heck out of information you can get out from seismic, provided you can afford it.

Fazrin Oktafian

Geophysicist at PT Elnusa, Tbk

Brian OConnell, Is there anomalous amplitude like a bright spot showing on preserved amplitude stack? It is important to judge where the reservoir position is located because otherwise, you might be studying the wrong AVO effect.

Gautam Sen

Advise/Consultant in Exploration at Independent Oil & Gas Professional

Brian OConnell, Are the amplitudes conforming to the structure? What is the depth of the reservoir? Only Class 3 AVO can be qualitatively inferred to as gas bearing structures provided amplitudes/bright spots conform to structure.

Brian OConnell

Senior Geophysicist at Repsol

The amplitude anomalies do conform to top and base structure. We just have some badly imaged gathers. Is there any other test that we can do?

Keith Katahara

Senior Geophysical Advisor at Hess Corporation

Brian OConnell, What is the depth to the target reservoir? Or better, what do you think the reflection angle is at 2000m offset for your target? If you have no wells, then you need to keep in mind that seismic processing may have used the wrong gain with offset, so AVO class may not be reliable (except for 2p cases where amplitude changes sign across the gather). That said, if you see an amplitude anomaly that conforms to structure, that is a very good indicator. That could be residual gas. So check for signs of breached seal. What does the seismic velocity look like? Does the seismic image look shaly or sandy to you? If you see seismically quiet zones, that may be shale, and the seismic velocity of those zones may tell you something about the shale properties. If you see an anomalously low velocity zone, that is probably shaly. It may tell you something about pore pressure too. So you do not have wells in the seismic volume, but I guess you have wells or some geological knowledge in the same basin, since you know enough to expect 20% porosity in the sands. I would take another look at the information used to come up with a porosity estimate to see if there is any information on shales to be gleaned. Swelling shales fall on different rock property trends from non-swelling shales. A 20% porosity implies significant cementation in the sands and is that quartz cement? And if so does it imply something about the temperature history? If you had swelling shales, are they through the smectite-illite transition or not?

Heath Pelletier

Lead Geophysicist at Dong Energy

Brian OConnell, I definitely agree with Keith's comment regarding getting a better feel for what your angle of incidence is at 2000m. That of course will be dependent on the velocities of your area, and it sounds like you have little information on those. That said, seismic stacking velocities converted to interval velocities along with your first break arrival analysis will give you a ballpark velocity profile. It is hard to believe that you have no well control to speak of. Even extrapolating well logs from a great distance away would be valuable. Some other recommendations;

- Where you are on the compaction curve is an obvious first step. That is typically related to depth. Would a wet (or gas) sand be lower impedance relative to a shale, therefore giving you a Class III AVO response? Also, are the impedances similar (Class II, IIp), or is your sand (wet or gas) have a higher impedance placing it into the ‘dim spot’ category, or Class I?
- If you could find some other curves, perhaps a resistivity curve which tends to be more common, you could boot strap a model together. For instance, you can create a sonic curve from resistivity using the Faust relationship, and then create a density log (using the Gardner relationship) and a shear log (using the Castagna relationship). Then you can perform fluid substitution to test the sensitivities of water, oil, and gas in that particular setting. Believe it or not this approach worked beautifully for me on a North Slope (Alaska) project where we lost all our down-hole logs except for resistivity. I managed to recreate the same Class IV anomaly seen on seismic using this bootstrapping approach.

Brian Schulte

Geophysical Specialist Quantitative Geophysics - Talisman Energy

Brian OConnell, You have near offsets and you can do acoustic impedance using relative inversion. If we look at AVO we see that the near offset data are influenced by hydrocarbons. Acoustic impedance is density times Vp and when we have a gas we have a small change in density but a big change in velocity. Exploration plays use relative inversion such as color inversion or even something as simple as integrating the data and scaling with 6 db/octave gain. You have seismic velocities so see if you can overlay the seismic velocities on the stack and as Keith mentioned see how these interval velocities match up to the seismic. Few people do this and it helps illuminate the geology.

What you are missing is key information about how the seismic ties to geology and which reflector is what. Do you have any wells in the area outside of the seismic or, even near 2D lines that intersect the seismic or pass near wells? You can gain a lot by looking for the closest wells even if it is just to understand the geology. You can try some modeling on that well and begin to understand the AVO. I am concerned about the condition of the gathers and do not know what they have done to image the data. Looking at the gathers tells a lot. Does the amplitude drop knife like? Is this effect seen on all the gathers or just in certain area? Look at the raw gathers before anything was done to see there were amplitudes beyond 2000 m offset.

Theron Edwards

Kuwait Oil Company

Brian OConnell, A near offset stack will show P reflectivity. A further offset stack (may need more than 2000m) will show mainly P reflectivity but some Vp/Vs reflectivity. By treating these two images like photographs with one being a single photo and the other being a double exposure combining that photo and another, you can try to extract Vp/Vs reflectivity. This is what a properly computed "fluid factor" stack will do. The beauty is that it requires no wells. Careful matching of frequency content is important, so that you don't just get the difference P reflectivity frequency. You will also not extract Vp/Vs reflectivity that correlates with the P reflectivity, but that is usually not what you are interested in anyway.

Brian OConnell

Senior Geophysicist at Repsol

Theron Edwards, I have tried to create a "scaled" Poisson's ratio change using these stacks and the results of intercept and gradient extractions and I am getting results. Amplitudes are conforming to structure. I am very wary of trusting these results too much.

Gautam Sen

Advise/Consultant In Exploration at Independent Oil & Gas Professional

Brian OConnell, It is good news that amplitudes do conform to the structure. Does it mean that the brine filled sands away from structure have similar impedance as underlying and overlying shales and the amplitudes are dimming? If this is true, then, we can attribute the high amplitudes to lowering of impedance due to replacement of brine with hydrocarbons over the structure. Do we see increase of amplitudes with offset on the gathers over the structure? Incase this is true then we can say that DHI supports the prospect but there is very little to tell us about its commercial potential. Please evaluate the possibility of a breached seal before making any further decisions.

JP Blangy

Chief Geophysicist at Hess Corporation

Brian OConnell, This sounds like a risky project, with no well control. How do you correlate the main reflection package to known geology and rock properties? If amplitudes on the gathers die-out fairly abruptly by 2000m that sounds more like a prestack calibration/conditioning issue. Have you checked how the gain was applied? Can you see noise trains on the far offsets or do the mid and far offsets just appear like white noise? Good to hear that amplitudes conform to structure. Do these amplitudes correspond to a negative impedance contrast? Have you checked the polarity? If all of that checks-out, do you have any idea of the temperatures and expected lithologies? Watch-out for diagenetic effects, which will conform to structure because they are temperature related. What alternative scenarios may give you the sort of amplitude anomalies that you are seeing?

William Harbert

Geophysicist

Brian OConnell, I think that the best you can do is to conduct the AVO analysis using all possible approximations using 2-term, 3-term, Aki and Richards, and Shuey. Throw everything at the problem. Your goal is to see which results are the most robust. What regions, if nay, appear in all of these? What types of combinations of parameters are the most robust (1/2(A+B); 1/2(A-B), etc.)? Try to get, from any source, a representative well log; either something from the region, or a log from a well with similar lithofacies. This could be difficult and I appreciate the excellent comments on this thread that AVO analysis without well control is Crazy-talk. Once you have your .las file you might try to use it conservatively and carefully in your field area. This is a somewhat (completely) reckless and dangerous approach and you should carefully discuss this with your supervisor before doing it. But we live in a less than perfect world and fortune (sometimes) favors the bold.

Ron Masters

Geoscience Advisor at Headwave, Inc

Brian OConnell, Turn the question around. Suppose you had excellent well control and complete confidence in the rock physics model for both clastic and carbonates, and the fluid properties. If the pre-stack imaging is unreliable, then AVO analysis would still be a waste of time. The seismic data just doesn't support a conclusion. If on the other hand, there is some evidence that the seismic imaging is valid; e.g., crisp fault definition, good NMO correction on the events you see, appropriate continuity in identifiable geomorphology, and a dominance of energy on the gathers that stacks in rather than a mess of interfering multiples and diffractions, then maybe you can draw some conclusions even without well control. Look at all the data and search for AVO anomalies. Don't worry about AVO class; you don't know what you should be looking for anyway and, any class anomaly, I to IV, can signal pay under the right conditions. Then if there are anomalies, that is, local features with different AVO behavior than neighboring regions, start asking whether they fit structure; such as what the structural/stratigraphic controls are and whether they make sense for a trap. Also what the depo-system might be and what the reservoir geometry might be. And more so, whether reservoirs might be consolidated or unconsolidated and then whether the observed AVO class makes sense.

Under no circumstances should you conduct the AVO analysis by just looking at a few isolated gathers, or by making class boundary polygons in cross plots. A few gathers won't tell you anything about geomorphology, which is really your only reliable clue here. Drawing polygons pre-supposes that rock property analysis has definitively ruled out the classes you aren't looking for, and has precisely described the required boundaries. That actually never happens, but at least in your case you know that going in. Here's the commercial announcement: an exhaustive anomaly search is still possible with Head-wave (Look for "interactive AVO" or "foolproof AVO" on Headwave.com), even without prior rock property knowledge. With graded coloring in the cross-plot, sharp anomaly boundaries in the subsurface must be changes in material properties. It's unambiguous. But if the pre-stack seismic is low quality, you are finished.

Daniel Baltar

Global Exploration Advisor at Emgs

Brian OConnell, you have some great advice on this. But I think it helps to break down the problem a little bit. The value of your AVO analysis depends on three things:

- The data (is this on-shore or off-shore data? Do you have PSDM or PSTM gathers? Do you have imaging problems? What kind of angle coverage you have?).
- The response (fit to structure, type of AVO response).
- The area (do you know your false positives/negatives, how likely each of these are and, do you have a good understanding of the rock physics? Does the AVO show up in an expected reservoir interval?).

From what you describe you have poor data, with fair quality response (fits structure), and poor knowledge of the area. So whatever you do you must be aware that the dependability of your AVO analysis will be low, i.e. you will have a large uncertainty in the end, no matter what you do. Hence the value of the AVO analysis will be low. Does this mean you should not do the AVO analysis? No, completely not. In the end you will not learn anything if you don't do it and the cost of doing it is very low when you compare it to the cost of a well. But always keep in mind the large uncertainties in your evaluation. Also, this implies that you should keep all the other parts of the prospect evaluation in mind as well, and you should never drill this prospect based on the AVO analysis alone. I cannot overstate the need for you to talk to your geologist and seismic stratigrapher in order to have the AVO analysis integrated with all the other available information (seal, reservoir, depositional environment, sediment source, timing, migration and structure). Do you have a standard seismic amplitude evaluation system in your company? If you don't, suggest to management that you get one (such as Pete Rose's SAAM, http://www.roseassoc.com/SAAM.html). It really improves evaluation consistency, which is very important in exploration. Actually you could run a simulation on the change in Probability of Success that you would get from doing AVO on your prospect.

Specifically for this case, as other people have suggested I would

- Look carefully at my seismic in order to figure out why I have poor seismic beyond 2000 m. Are you using the right processing? I would not do much until I had a clear understanding of this problem.

- Then, turn that offset into incidence angle once I have velocity field and use the seismic velocity as a proxy for shale compaction
- Take the sand compaction from whatever source you might have (analogues or far away well data), even better, keep two or three different sand compaction trends (minimum, maximum and average) and use the response away from the prospect together with the response at the prospect to improve your understanding of the lithology.

Maybe you even want to try some simultaneous inversion, given that you seem to know how to do that and the cost would be low. And above all, remember that in this case what you don't know is more than what you do know so keep the AVO as part of an integrated evaluation. Have you considered other methods for de-risking your prospect such as CSEM if you are working on off-shore data?

Binh Nguyen

Geophysicist at Bp

Brian OConnell, without well log data for AVO study, there is always a lot of room for surprises. In addition to the comments above I just want to add my check list for amplitude analysis (AVO) work.

- Data imaging conditioning (supposed done) as above commented by other colleagues.
- Is the reservoir thick or thin? Is the setting compaction dominated or not? If it is, you should see a trend in AVO over fairly narrow depth window (500-1000m).
- Which volume (near, mid or far) could you (or the interpreter) surgically map the sand both above and below the conformance contour? Does it look like a sand/reservoir in a top to base amplitude average (90-rotated volume) map?
- Is there any simple AVO chart (curve, I/G) that can be used to explain what you see?
- On this average amplitude, what are its sign and magnitude below and above the contour and are they related?
- If you are dealing with a dim spot, the AVO will be class 1 or 2p and the response (of the seismic amplitude) at the top of a brine filled sand will be positive. The response of a Hydrocarbon filled volume can be either positive or negative but will always be weak.
- If bright spot, AVO can be class 2 or 3, brine can be very week, HC top negative
- Make sure the dim or bright spot response is not coming from tuning (phase of the event gradually changing due to thickness change over distance).
- Build a set of average amplitude (top to base of 90-deg rotated volume) vs depth plots for full, near and far volumes. The change of near and far should be compatible with a normal AVO curve charts (AVO curve of brine, HC, tuning, porosity, compaction (if involved).

If all of these can be understood, the chance you have a prospect is high.

Brian OConnell

Senior Geophysicist at Repsol

In respect to the reprocessing, that is not an option at this time.

If we are to assume it is hydrocarbon that is causing the AVO anomaly, what is the best way to figure out whether it is gas or oil filled sand remembering that we have no well control?

Is it visual inspection of Bright Spots? Or the class of AVO from cross plots we are seeing (with far offset gathers ideally) or the best use of A*B, A-B, A+B, SCALED (A+B), SCALED (A-B)?

Chandra Velu

Senior Geophysicist at Lundin Malaysia B.V

Brian OConnell, I think you answered your own question. Unless you have a high GOR oil, you are not going to discern oil on seismic, unless your sealing rocks contrasts with your reservoirs. You might have a more diagnostic seismic response from gas. Please incorporate basin modeling in your project. It is important in frontier exploration and I cannot emphasize this enough.

Hemant Kumar

Technical Business Development Manager-Asiapacific

Brian OConnell, class1V AVO anomalies at shallower depth in deep-water may be associated with mineralogical changes and not true HC presence. As the anomaly sits within the structure, look at the nature the gathers at the spill point and compare them with the gathers at top of structure. In the case of reconnaissance AVO, correctly processed seismic data is the key. You may draw inferences from things that are processing artifacts and not the true anomaly signature.

Gautam Sen

Consultant at Sahara Group

Brian OConnell, It is not only class IV but even class III anomalies can be found at shallower depth and younger stratigraphy and differentiated from water bearing totally unconsolidated/uncemented sands in deep water environment. After all for AVO, all you need is a change in Vp/Vs.

OFFSET BALANCING

QUESTION 84

Assume you have a pre-stack data set where there is a slight dimming with offset within target area. You have several wells within the seismic survey. Full wave synthetics in well locations indicate that there is a need for offset balancing. The amplitudes should brighten and not dim with offset. You wish to perform joint pre-stack inversion and you split your uncorrected (without any offset balancing) dataset into five different angle stacks. For each angle stack, you estimate a wavelet.

Question: will angle dependent wavelet estimation inherently take care of the offset balancing task? What kind of effects are inherently compensated for by angle dependent wavelet estimation? What data conditioning steps can you skip, if any?

Vedad Hadziavdic

Senior Geophysicist at Wintershall

ANSWERS PROVIDED BY INDUSTRY EXPECTS

Scott Singleton

Seismic Technology Advisor at Independence Resource Management

Vedad, The high-end inversion software packages will automatically scale each wavelet and then you can separately weight or de-weight each angle stack according to some confidence criteria. For the most part, these corrections seem to adequately take care of offset amplitude problems because internally the inversion algorithm is trying to fit the intercept and gradient to reasonable rock properties so it can solve the reflectivity equation. However, I am guessing that since you have lots of well control, you are looking for an accurate inversion that matches calibration well control and blind well control. My experience is that if it is just the near angle stack or far angle stack that is not correct, merely de-weighting these will allow the remaining angle stacks to carry the solution and the result will be reasonably accurate. However, if only your intercept is correct and every subsequent angle stack has an incorrect gradient then there would not appear to be any way an inversion will give you correct rock properties since nowhere in the input data is there correct rock property response. That would mean offset balancing would be required, calibrated to a full-waveform synthetic of course.

Vedad Hadziavdic

Senior Geophysicist at Wintershall

Thank you Scott, intuition was leading me towards the same conclusion, but I could not wrap my head around it mathematically. The main point is as you say, that the Poisson's ratio is inferred from amplitude change with offset. If this change is distorted, there is no way for the inversion algorithm to get the right rock property. It would be strange if AVO class IV rock properties changed to AVO class III just because an angle dependent wavelet was extracted. I will, however, try to test this on synthetic data.

Sameh Sakr

Reservoir Geophysicist at RWE

Vedad, I have experienced data where I had control points (wells) and pre-stack data hence I had a small cube around each well and I created 5 angle stacks from each small seismic cube and tried to do seismic-to-well ties between the 5 angle stacks and their corresponding wells. After I got a wavelet for each angle stack for the same, create something incorrect wit the editing well, I overlaid them together and I could clearly see variations in amplitude as well as differences in phase. In this case I am talking about small cube of seismic around the well location which mean that there was big variation in amplitude and seismic data that is not zero phased. I had to do offset balancing, taking the amplitudes values for each angle stack at each well and then trying to scale it in a way to reach the most proper balancing. I did the AVO modeling and ran elastic inversion after applying this scale on my pre-stack data. I am still working on the inversion. However, I have tendency to feel some uncertainty regarding the output of inversion.

Rui Zhang

Assistant Professor Of Geophysics at University Of Louisiana at Lafayette

Vedad, I have similar experience as **Sameh Sakr** solving this problem by using angle varying wavelets for inversion processing. Those wavelets are generated at each angle with correlation to well log data. In my understanding, spectra variation is the bestest draw-back for inversion processing. The detail was published as: Zhang, R., M. K. Sen and S. Srinivasan, 2013, A pre-stack seismic basis pursuit inversion: Geophysics, 78, 1, R1-R11.

Ashutosh Garg

Reservoir Geophysicist Shell Technology Centre, Bangalore

Vedad, In this case you are going to get very high inverted residuals at the far stack. Offset balancing is tricky as it can introduce AVO anomalies in non reservoir rocks. You can check true AVO behavior of the seismic by compensating reservoir top amplitude with non reservoir background (Overburden/Under-burned). Sometimes anisotropic overburden shale leads to change in AVO behavior of reservoir zone. Angle dependent wavelet estimation changes AVO behavior of seismic but I am not sure if it can mimic the AVO behavior shown by your full waveform synthetic. Wavelet scaling parameter can help while choosing inversion parameters.

Scott Singleton

Seismic Technology Advisor at Independence Resource Management

Ashutosh, I completely agree that residuals are the first thing I would look at in any inversion, regardless of whether any kind of gather conditioning was applied or not. This is because the residuals will tell me whether my wavelet was appropriate for each of the angle stacks or not. I also look at the frequency spectra of my residuals because this tells me whether my wavelet is deficient in either high or low frequencies.

Earlier in this discussion there was some conversation about wavelet phase. I do not consider it a bad thing if my angle stack wavelets progressively rotate with angle. In fact, I would expect this to happen unless Q compensation had already been applied to the seismic data. It does not harm the inversion because that phase rotation is captured in the wavelet that I extract from the well tie, and thus that phase is de-convolved from the angle stack upon inversion. This is the principle that inversion is based on, and is true not only for phase but also for spectral content and amplitude. The problem comes when the wavelets being used to de-convolve each respective angle stack are not appropriate for that stack. And because all wavelets are only approximations of the true seismic wavelet, this is the reason I check my inversion residuals (as well as all other inversion QCs) very carefully.

Gautam Sen

Advice/Consultant In Exploration at Independent Oil & Gas Professional

I totally agree with Scott. If you can extract as close a wavelet to true wavelet in each of the angle stacks you do not need offset scaling. Since wavelets are neither stationary nor are the reflectivity series white, it is best to extract the wavelet in a window covering the reservoir of interest. Residuals then could be harmless in phase, frequency and amplitude distortion. It may be prudent to avoid Q compensation also unless we are very sure of the Q parameters.

Ariel Kautyian

Principal Senior Geophysicist at Total Austral

Sameh, The purpose of angle dependant wavelets is to correct/compensate any phase and/or amplitude variations not explained by geological/physical AVO behavior. Then the key issues are all extracted wavelets from each well are similar for each angle stack (wavelets near, wavelets mid etc.)? This tells you that if there is any issue, (for instance with amplitudes in the data), it is constant over the entire survey and wavelets are capturing it, so it will be corrected by the inversion. Stable and similar angle dependant wavelets with good angle dependant seismic-to-synthetic correlations are important to get a reliable inversion result. On the other hand, if observed AVO response in seismic data is consistently inverse to what is modeled from wells I should investigate the cause (signal treatment, acquisition issues etc) even if stable and similar wavelets are obtained.

Gautam Sen

Advice/Consultant In Exploration at Independent Oil & Gas Professional

Sameh, In case your wavelets extracted from the same angle stacks at different well locations for the same reservoir interval vary grossly it is best to reprocess the data. Inversion can never correct it.

Vedad Hadziavdic

Senior Geophysicist at Wintershall

Lots of nice and broad comments are presented by experts above. Offset balancing is discussed in several papers. Scott presents a nice case for offset balancing in his paper "The effects of seismic data conditioning on pre-stack simultaneous impedance inversion". He suspects that ringing on the near traces due to water-reverberations distorts near offset amplitudes at the target level. Correlation to full-wave synthetics encourages clearly application of offset balancing. Lazaratos touches also upon this in "Deterministic spectral balancing for high-fidelity AVO". He discusses a Q-compensation method which corrects relative amplitude loss between near and far-offset data due to absorption losses. That being said, I agree with those of you who call for caution when applying offset balancing. You might after all be changing the AVO response of your target and that is always tricky. An important thing is that any scaling factor estimated at the well position, in general will not be valid for the whole seismic cube. Depending on the cause of improper balancing, it might vary both laterally and vertically. If you have calibrated it for one target and applied it to the whole cube, you should not be surprised if targets at other depths end up showing strange responses.

My question was this: if your data at the target at well location show Class IV response while full-wave synthetics indicate Class III response, where only the near amplitudes matches between real and synthetic data, will angle-dependent wavelet estimation take care of the discrepancy? In other words, since these two classes assume different Vp/Vs in the reservoir as compared to the cap rock, will inversion results show an increase or decrease in Vp/Vs in the reservoir? Can you skip offset-balancing and still get a drop in Vp/Vs which correspond to class III and not IV?

My understanding can be summarized in the last two sentences of Scott's first comment. If all but the near stacks are "wrong", I do not understand how inversion algorithm can recover rock properties which are not present in the data. I do not understand how angle-dependent wavelet estimation can repair this.

Angle dependent wavelet estimation compensates for some differences between offsets. My understanding is that these differences are mainly due to NMO stretch and the fact that waves reflected from the target have traversed different paths at far than near offset. This can lead to, for example, absorption losses which will be different at far than at near. In the end, near wavelet will have both different bandwidth and possibly phase rotation than the one at the far and this has to be taken into account.

Can this however take care of the problem when the amplitude variation with offset shows a completely different character across a significant offset range?

I still don't have a clear understanding of this and I have not had time to check it in-depth. This would require generating full wave synthetics with e.g. absorption losses, creating angle stacks, inverting the synthetics and comparing it with inverted, offset balanced data. Or if someone could put forward a mathematical argument for what happens in these two cases?

Peter Mesdag

Technical Product Manager at Jason A Cgg Company

Vedad, For a single event in one well you can make the AVO/AVA response look any way you like by changing the wavelet or performing seismic balancing. You can change the response to fit any AVO class. However, one would hope that you do not take such action based on one single event in one well. The sanity of your action should be tested on several different events with varied AVO responses in several wells. If all responses need to be corrected in the same manner, this procedure can be applied. If you find that the compensations vary from well to well you will need to find the physical cause and apply the corrections consistent with that cause.

Scott Singleton

Seismic Technology Advisor at Independence Resource Management

I agree, Peter. Data processors never look at just one reflector (it is we reservoir guys who tend to do that). Thus, when deciding on processes to apply to a gather, I always look at a window of data, not individual reflectors. In fact, best practice calls for the evaluation window to avoid intervals that might contain AVO responses. Those intervals will by their very definition contain far offset amplitudes that are different from the background trend, which is what we want to normalize when conditioning gathers for inversion.

Vedad's original question, repeated above Peter's response, really seems to me to be a rock physics calibration question. If in his case he has checked the background trend and it is appropriate, or can be made to be appropriate (this is an assumption that must be met before continuing), and his reservoir contains a flat amplitude response or a dimming with offset that is not on the background trend (definition of Class IV) and his well synthetics say he should have a brightening with offset (i.e. Class III), then, personally I would do as Peter suggests. Stop everything and try to figure out why this is happening, because it is obviously very incorrect. (Actually Gautam also said the same thing above, although that was in response to Sameh's issue of a lack of stationary in his wavelet extractions, which is equally serious).

Sorry, Vedad, but the above is not a mathematical answer, just a ‘due process’ answer. To be a bit more explicit, no; pre-stack inversion with a series of angle-dependent wavelets will not correct for an incorrect reservoir response in the gathers. These wavelets only capture the proper background trend and cannot possibly deal with a specific reflector (the reservoir) that has an incorrect AVO response.

Gautam Sen

Advice/Consultant In Exploration at Independent Oil & Gas Professional

I agree with Peter. A solution could be reprocessing if log data is trustworthy.

SHALLOW GAS

Shallow gas is a relative term but is herein defined as natural gas occurring at subsurface depths of less than 4000 ft. Shallow natural gas can be generally classified as either thermogenic gas, generated at depth within the basin and migrated up faults as free gas, as gas associated with oil, or biogenic gas (Fig. 6.23).

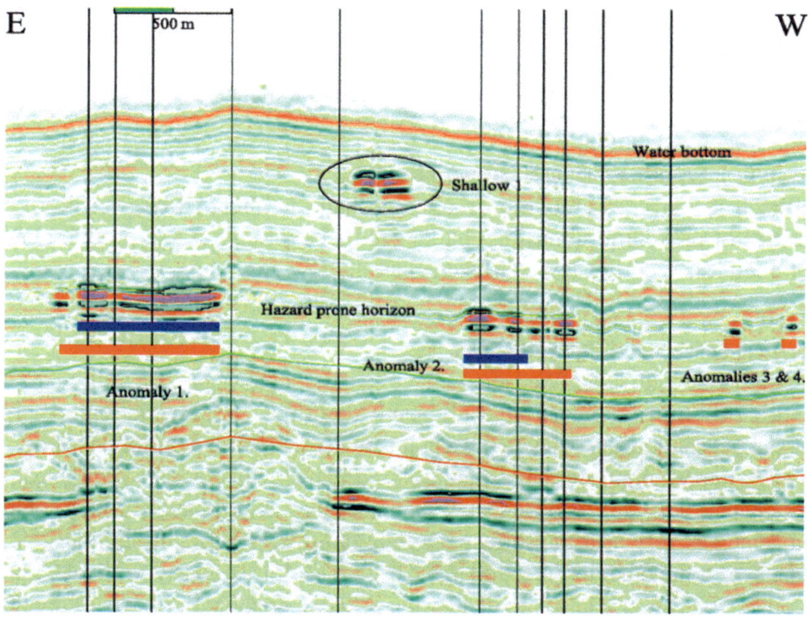

FIG. 6.23

An example study using conventional 3D seismic data to delineate shallow gas drilling hazards.

Courtesy: Petroleum Geoscience.

QUESTION 85

I am working on non-structural, high amplitude anomalies developed at shallow depth due to leakage from main reservoir. These anomalies are considered as drilling hazards due to presence of gas and also high pressure. But I am not sure the proper definition of "Shallow Gas". Does is it depend on depth? If so, then what is the range? Are there any other factors that can define "Shallow Gas anomalies in marine sediments"?

Muhammad Iqbal Hajana

Senior Geophysicist at Weatherford Petroleum Consultants As

ANSWERS PROVIDED BY INDUSTRY EXPERTS

Peter Wang, P.G.

Geophysical Technical Advisor, Paradigm

Hajana, Shallow gas or deep gas can cause blow-outs. The shallowness or deepness is not relevant. What is relevant is that you encounter an over-pressured formation and the drillers can't "mud up" fast enough to control it. That can happen shallow or deep.

Muhammad Iqbal Hajana

Senior Geophysicist at Weatherford Petroleum Consultants AS

Thank you for the valuable comments. But I want to establish a proper definition for shallow gas anomalies. It is obvious that these types of shallow gas anomalies or say gas pockets, regardless of their biogenic or thermogenic origin, are gas accumulations seen in Tertiary formations in many basins around the world but these anomalies are produced as reservoirs. So my point is when these anomalies called shallow gas?

Adeosho Ayodeji Saheed

Reservoir Geophysicist at Platinum Geoservices

Hajana, If it is gas accumulation shallower than the intended target with the characteristics of preventing (causing challenges for) exploration and production of the target, then it is qualifies to be christened as shallow gas.

Steve Adcock

Geophysical Consultant at dGB Earth Sciences

Hajana, I agree with Peter. We have frequently used shallow gas as a term in the industry to describe this risk, but it's an insufficiently descriptive term for definition purposes. The issue of "shallow gas" as you present it is really the issue of drilling hazards, which goes back to pressure differentials and geo-mechanics. Shallow is not the meaningful identifier since you want to exclude any anomalies that are producible, which could even be biogenic gas very near the surface. Drilling hazards, pressure differentials, pressure gradients, or some other term related to the hazard itself is the one you are attempting to categorize. You can, if you want, identify the pressure gradient in a basin of interest and then define some inflection point on that curve as a separator (shallower Vs deeper) for geo-mechanical behaviors of interest to you. Another approach might be to use seismic AVO response as a separator, but that leads back to the same rock, fluid, and pressure properties that are the relevant variables for all methods.

CHAPTER

GASSMANN'S FLUID SUBSTITUTION

7

CHAPTER OUTLINE

FLUID SUBSTITUTION

For most amplitude variation with offset (AVO) analysis projects, the first task is to look at the well logs. To estimate the theoretical AVO response we need accurate compressional (P-wave) velocities, S (shear) wave velocities and density curves. Shear velocities are not always measured in the normal logging suites, but these days they are becoming more generally available through full waveform logging and three-component vertical seismic profiles (VSPs). However, if shear velocities are not available, they can be estimated from a precise knowledge of the lithology

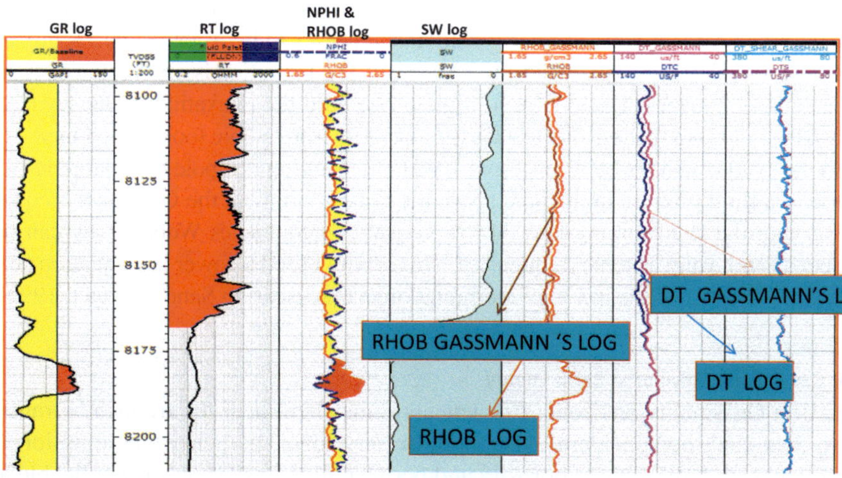

FIG. 7.1

Effect of Gassmann's fluid substitution on density (RHOB) and sonic (DT) logs.

Practical Solutions to Integrated Oil and Gas Reservoir Analysis. http://dx.doi.org/10.1016/B978-0-12-805464-2.00007-X

associated with a particular layer. Once we have density, Vp and Vs, we can then plug all of these values into the Zoeppritz equation and calculate the theoretical AVO response for a common depth point (CDP) recorded at a well location. The AVO response can also be calculated for other scenarios by adjusting the well curves to remove or add the effect of different lithologies or fluids. For example, we can replace sandstone properties with those of dolomite, or brine/water properties with those of oil, or gas, and then compute the synthetic data for these new scenarios. The technique used to do this is called Gassmann's fluid substitution.

In other words, Gassmann's fluid substitution is used to back out the effect of hydrocarbon from a well log that is hydrocarbon saturated to create a log curve that is brine saturated, or vice versa (Fig. 7.1). The brine curve is used to generate a synthetic seismic section.

QUESTION 86

One of the assumptions in Gassmann's equation is low-frequencies (less than 100Hz), but the equation is used in Xu-White model (the last step) to predict Vp, Vs logs which are in kilo Hz. I can understand the other steps or clay-sand dry frame parts of the model which are independent of frequency. However, when taking into account fluid, and trying to estimate sonic logs (at kilo Hz) from porosity, I am baffled. Please shed a light on this.

Rongfeng Zhang

Geoscientist at Geomodeling Technology Corp

ANSWERS PROVIDED BY INDUSTRY EXPERTS

Alain Brie

Give Occasional Petrophysics Training Courses at Retired Petrophysicist Expert

Zang, the Gassmann's equation is effectively the low frequency limit of the Biot formalism. This being said, at what frequency Gassmann ceases to be usable depends in particularly on fluid mobility (permeability/viscosity). The transition is often referred to as the relaxation frequency. Usually for a single phase fluid the relaxation frequency is in the 100kHz range, so that we can use Gassmann to interpret sonic tools which measure Vp around 10kHz and Vs between 1 and 10kHz (depending if dipole or monopole). In the case of a fluid mixture in the pore space the issue is not with the Gassmann's equation but with the fluid mixing law. At low (seismic) frequencies Wood's Law can be used, but it is not valid at sonic frequencies. This is why I used a power law equation (see the paper SPE-30595 "Shear Sonic Interpretation in Gas-Bearing Sands" of Oct. 1995).

Rob Simm

Senior Geophysical Advisor at Cairn Energy

Zang, Biot Gassmann puts sonic frequencies somewhere near the frequency transition so they can't be considered very high or very low. The general assumption is that for rocks with reasonable porosity and permeability Gassmann's is applicable at sonic frequencies and for the most part 'reasonable' dry rock numbers are obtained where they are expected (that is, in the clean parts of the formations). The form of the fluid mixing would be evaluated based on the data and not all practitioners would apply something like Alain's mixing law as a matter of course. Certainly, the results of applying Gassmann to tight sands are likely to carry more uncertainty. In the absence of data to constrain a dispersive correction for tight sands you might use a patchy saturation mix to offset some of the dispersive effects (that is, variable pore saturation and frequency have the same type of effect) but it is pretty much guess-work. Other aspects of Xu-White that you might consider are:

- The dry rock Poisson ratios that come out of Kuster-Toksoz tend to be lower than those found in real rocks, typically giving rise to Vs which is too high compared to measured logs.
- The rock stiffness (which controls fluid sub magnitude) is not simply related to the pore aspect ratios but also to the modulus of clay used to determine the effective mineral modulus.

Shiyu Xu

Geophysical Advisor at Exxonmobil

Zang, Both Alain and Rob have given you very comprehensive responses. I want to reinforce their opinions. First, I want to say that the frequency dependence issue is very complicated. When you are talking about frequency dependence, you have to make it clear what mechanism you are talking about; fluid flow, frictional sliding or scattering. Even for fluid flow (most people refer to this mechanism), are you talking about Biot flow or local flow? They are all different mechanisms and the answer will be different.

Secondly, what do you mean low- or high-frequency? As Alain pointed out, in order to know if your measurement is high or low frequency, you have to know the relaxation frequency. Unfortunately, the relaxation frequency is very difficult to predict and it varies significantly in real rock. I can show you that even the ultrasonic frequency (MHz) gives a low frequency response for some of the laboratory data (for example, Rathore et al., 1995). But on the other hand, seismic frequencies (tens Hz) can be high frequency for rocks such as shales. My argument is that the pore pressure cannot be equilibrated in millions of years (that is why we have geo-pressure) then how can the pore pressure be equilibrated in fraction of second? Therefore, it may not be true when you say that seismic is a low-frequency. It really depends. Therefore, if you want to be more accurate in your rock physics modeling, you have to have a way to predict the relaxation frequency. At this point, it is difficult to do it. In many cases, we just choose to apply Gassmann to the rocks. By the way, we have modified the Xu-White model during the past ten years or so. Based on the arguments above, we assume that the clay pores behave like a high-frequency component (because they are so small and tend to be associated with bound water); the sand pores are big enough for low frequency (Gassmann) to apply. We called it a Mix-Mode model. This can be easily done using inclusion-based model. But you can achieve the same thing using other approaches (for example, Brie's power exponent). This is probably true in most cases. But when you are working with tight gas sands or carbonate, you may have to think harder.

Rongfeng Zhang

Geoscientist at Geomodeling Technology Corp.

The key I so far see from the above comments from experts is that the assumption is actually based on both the frequency and the material. The same frequency could be higher for material A but not for B, as stated by Shiyu clearly. The validation seems to be based on the amount of the dispersion. Large dispersion will invalidate (directly or indirectly) the assumption/theory. The higher the frequency, the larger the dispersion, meanwhile, the more porous the medium (with fluid filled), the higher the dispersion. However, this hardly explains why tight sands are more sensitive to frequency than other rocks when using Gassmann's equation, as pointed out by both Rob and Shiyu.

Alain Brie

Give Occasional Petrophysics Training Courses at Retired Petrophysicist Expert

Rongfeng, after reading Gassmann's the next step is to look at the Biot Theory. To make it easier the equations are summarized in the Rock Physics Handbook of Mavko et Al p. 162. The problem with these equations is that the physical effect they represent is not explained simply and it is difficult to visualize what is happening. In case of an elastic wave traveling through porous rock. The wave is really a stress; say a directional pressure and variation. To visualize what happens let's consider a sample, when the compression occurs at one end of the sample the fluid moves to the other side. Then when the compression reaches the other side the fluid goes back. In other words the dry rock frame and the pore fluid move in opposite directions. This supposes that the fluid has the time and ability to move within the pore space; which is true at low frequencies provided mobility is sufficient. When the frequency increases the fluid does have the time to move, there is energy dissipation (attenuation) and the rock looks more rigid, hence wave velocity increases. The less the mobility (permeability over viscosity), the lower the frequency at which this relaxation phenomena occurs. Now you can understand why the relaxation frequency is high in porous/permeable rocks and low in tight, low permeability rocks. The magnitude of the dispersion is not a function of permeability; it only depends on the elastic properties of the rock frame and the fluid. The higher limit is that obtained from an effective medium theory that considers the pore fluid immobile, as a solid in the rock. This is for a single phase fluid in the pore space, when there are two phases it becomes more complicated as it also depends on the distribution and relative mobility of the two phases.

Rongfeng Zhang

Geoscientist at Geomodeling Technology Corp

Alain, I understand the idea now. Your clear and vivid comments really depict what an animation should do. Along with re-reading Shiyu's comments, I think I finally appreciate the idea. The only thing I would correct are the words in Alain's comments "In other words the dry rock frame and the pore fluid move in opposite directions." I think, precisely to say, they can move in opposite directions, the same directions or not move at all. I found a very good analogue to this issue that happens to relate to our daily life: the microwave oven mechanism. "In free liquid water this movement occurs at GHz frequencies (microwaves) whereas in more restricted 'bound' water it occurs at MHz frequencies (short wavelengths) and in ice at kHz frequencies (long wavelengths)." I will not copy the whole citation from the source here, except, to give the web address: http://www.lsbu.ac.uk/water/microwave.html

Within that website, there is a java applet with animation that, in some way, precisely depicts what Alain talked about. Also it explains why I suggest the correction above to Alain's comments.

QUESTION 87

How can I differentiate between a gas sand, condensate sand and oil sand using seismic and AVO response and inversion, where we have separate reservoirs or gas condensate and oil in the same reservoir?

Nasser Srour

Senior geophysicist dsq gas field (Nile Delta, Egypt) at Suez Oil Company (Suco)/RWE

ANSWER PROVIDED BY INDUSTRY EXPERTS

Arslan Tashmukhambetov

Geophysicist at Llog Exploration Company

Nasser, the easiest way to check it would be taking wire line logs from well with known hydrocarbons, making a synthetic gather in-situ and a fluid substitution gather and compare the two.

Gautam Sen

Advice/Consultant In Exploration at Independent Oil & Gas Professional

Nasser, simple crossover in logs may not happen if sands are shaly. Therefore a rock physics model for the kind of sands in question can really be the answer. Extended Elastic impedance inversion and or LMR cross plots can help in determining fluid types.

Neeharika Shukla

Geophysicist

Nasser, If we have gas condensate and oil in the same reservoir, then it can be differentiated by performing AVO inversion and I personally prefer class 2 since it can give the required difference between gas condensate and oil sands. And, if they are in different reservoirs, then the AVO anomaly curve will give the relative densities of the gas sand and oil, in different reservoirs.

Jerry Coggins

Principal Geophysicist at Coggins Geosciences, Inc

Nasser, The question is really too general to answer. Seismic won't differentiate at all for lower porosity/cemented sands. The reservoir thickness, seismic frequency content and data quality are also important. Making models as suggested is the first step. It will likely require some local calibration as the responses are not unique. This is very difficult away from well control.

Richard Loewe

Support Geoscientist with Ikon Science Ltd

Nasser, I strongly agree with Jerry. Generating multiple models for different porosity cases might be one of the most important things here. I would suggest you generate Intercept and Gradient plots to determine a Chi angle on an ortho-fluid projection line. Once you have generated the line you will see how well you can separate the two fluids or if it is impossible. Next, you may want to follow the 'workflow' proposed by Hendrickson in his paper 'Stacked' (DOI: 10.1046/j.1365-2478.1999.00150.x)

Aside from that, you might want to look out for noise, as this might shift the trends (in the Intercept and Gradient space) in the measured data, so that they appear different from your modeled ones.

John Hooper

Partner, PetroSEIS L.P

Nasser, In addition to the above comments, gas and gas condensate are generally both in the gas phase in the reservoir. To first order, gasses in the alkane series have similar compressibility (Batzle/Han, fluids project). Stochastic seismic modeling of different gas types will probably have near complete overlap regardless of rock frame stiffness. The degree of overlap in stochastic models between oil and gas depend on the compressibility of the oil. How "gas-like" is the oil sand response? You are unlikely to be able to tell dry gas from wet gas with seismic amplitude response. You may be able to distinguish gas from oil. If you have both oil and a free gas cap, the local amplitude morphology provides some clues. A pitfall in that pre-drilling interpretation is the look-alike morphology of gas/water or oil/water contacts.

Alan J Cohen

Chief Technology Officer at Sigma3 Integrated Reservoir Solutions

Nasser, Once you determine the bulk density and the adiabatic bulk modulus of the two fluid choices you can model their seismic responses and determine whether you can differentiate them or not.

Gautam Sen

Advice/Consultant In Exploration at Independent Oil & Gas Professional

Nasser, Condensate and gas probably have the same response in rock physics models and therefore cannot be distinguished.

Nepal Singh

Advisor Geophysics, at Cairn Energy

Nasser, Best way to answer the question is to model the seismic response of oil, gas condensate and gas with different porosities. Cross plots of different pre-stack inversion attributes derived from modeled seismic response will give you an idea of differentiation of different fluid types. Physical properties of gas and gas condensate may not have noticeable changes in their elastic response and hence in attributes.

Bill Wepfer

President at Eskaton Seismic, Inc

Nasser, A pragmatic, no wells approach could also be taken by effectively letting the earth do the modeling for you. This may be especially applicable if well log and other petrophysical data are lacking, but would require knowledge of the fluids in different fields in your survey area (I'm assuming that you have one or more 3D volumes that cover the various field types). Process the data through to PSTM gathers and run your AVO volumes; then see how the various attributes respond over fields with differing fluid types (assuming no 4D effects on your 3D, where the fluid of interest has been significantly depleted). Hopefully you see some differentiation between the various fluid types with a standard or advanced AVO attribute. Please note that the greatest AVO effect is seen with fizz water or fizz gas (like soda pop – mostly fluid but 10% gas). The density effect on the bulk modulus decreases the AVO gas effect on higher (and potentially more economic) gas saturations. Many times, something as simple as angle stacks can be quite diagnostic. A 30-45degree angle stack of the PSTM gathers often shows a lot of promise in distinguishing fluid types. Intercept times gradient and the standard Fluid Factor (intercept plus scaled gradient) are also favorites. Extensive modeling can be helpful, but, depending on your area of interest, the seismic processing may be your limiting factor. Consider using a few processing shops that specialize in AVO products (many offer integrated AVO services, tying in the wells for you); and it may be worth your while to use 2 or 3 processing centers to see if one of their AVO ideas works for distinguishing fluids in your area. If you see a great fluid correlation with one of the volumes, this would easily justify the additional cost, and the expenses would be a lot less than drilling a bad well.

Gautam Sen

Advice/Consultant in Exploration at Independent Oil & Gas Professional

Nasser, Bill's idea is certainly worth trying, even in areas with no or sparse well control before we go for inversion.

Rosana Polo

Senior Petrophysicist en Oxy

Nasser, You can use magnetic resonance which can give differences in viscosity and with that you can infer fluid type.

QUESTION 88

Gassmann's Fluid Replacement Modeling in a Multi-Mineral World

Andrew Wilson

Geophysical Advisor

ANSWERS PROVIDED BY INDUSTRY EXPERTS

Michel Krief

Senior Petrophysicist/Rock Physics Use in Petrophysical Analysis

Andrew, the multi-mineral analysis provides estimates of volumes of fluids and solids. The volumes can be used to compute average properties of the solid rho, k, mu as well logs. Different mixing laws may be used: VRH (Leon), HSW (Alan). If fluid saturations are known, the petrophysicist can invert the Biot-Gassmann model. The result is two continuous curves: Biot coefficient (bulk modulus) and Gassmann coefficient (shear modulus). Having B and G coefficients, the computation of the dry rock properties is straightforward and so is the fluid substitution exercise. Of course, this works best when a shear sonic log is available, otherwise shear wave velocities it needs to be estimated. This workflow was developed many years ago in the Geolog (Loglan modules) environment. It is currently being replicated in the Techlog environment (Python scripts). Not a big deal, the important thing is to accept and understand the uncertainties attached to it.

Leon Thomsen

Geophysicist

Andrew, the easy fix is to make a VRH average of all the minerals, and use that for rho-solid and k-solid in the traditional Biot/Gassmann formula, and just use your existing package. Of course, this is not exact, but that is probably the least of your multi-mineral worries. A bigger worry is that the B/G formula is not correct at all, if the mineralogy is not uniform. The proof of this is given by Brown and Korringa, Geophysics, 1975. The correct formula requires an empirical parameter k-phi, replacing k-solid in the B/G formula. This is not something that you can look up in a handbook, or easily implement in any package. It is different for every rock. Even worse, the same logic that leads B&K to this conclusion also applies if the pore space is not uniform. Of course, this is the case for every rock. So, we should use the B&K formula, instead of the B/G formula in all cases. The proof of this was given by Thomsen, SEG Abstracts, 2010.

Kester Waters

VP Global QI – Products and Solutions

Andrew, RokDoc supports multi-mineral Gassmann fluid substitution and provides you with a variety of different mixing models (for both mineral and fluid components) in order to investigate a range of different inputs/outcomes. Leon has raised some concerns with Gassmann for the multi-mineral cases and suggests the Brown and Korringa approach, 1975. RokDoc has implementations of Brown and Korringa, Ciz and Shapiro's extension (2007) to Brown and Korringa and Mavko's anisotropic Gassmann implementation (2009). There are also a variety of tools for deriving (Leon's;-) anisotropic parameters for inputs to these models (requires the anisotropy license). You could also of course use the differential effective media tools to constrain the multi-min interpretations. Some or all of these may exist (or could be coded up) in other packages such as HR, Jason, Geolog, Techlog, etc. You could also code this stuff up in the RokDoc Python interface.

John Castagna

Professor, University of Houston

Andrew, If the minerals' bulk moduli are very different from each other, then the assumptions of Gassmann's equations are violated and a simple average of bounding equations is not adequate. Read Berryman and Milton (1991) for valid fluid substitution equations in a two-mineral rock in that case. Be careful when TOC and microporous or otherwise highly compressible clays are important constituents. In that case, this effect is of great practical significance. Also, I am not aware of the general multi-mineral problem being solved.

Rob Simm

Principal Geophysical Advisor at Origo Exploration Uk Ltd

Following on from John's comment - there is clearly uncertainty in the effective mineral modulus parameter whichever software or mixing recipe is used. The practical significance of this uncertainty of course is that the mineral modulus is a variable and not just an input. That is, it impacts the stiffness of the rock and hence the amount of fluid substitution effect on Vp. It is always a good idea to keep this in mind when evaluating results of fluid replacement modeling.

Carl Reine

Sr. Geophysicist at Canadian Discovery Ltd

Andrew, A paper by Saxena, Mavko, and Mukerji (2015) looks at this exact problem. While unfortunately I can't suggest some software that implements their equations, it would be possible to do so yourself with a Geolog loglan script.

Saxena et al. calculate the exact expression for the multi-mineral fluid substitution (equation 16); however this situation requires additional parameters that are not usually known. Rather than look at effective mineral properties to simplify this problem, they calculate upper and lower bounds on the bulk modulus (equations 21-23). They show the average of these bounds is more accurate than using the bounds on the mineral properties themselves. Saxena, N., G. Mavko, and T. Mukerji, 2015, Fluid substitution in multimineralic rocks with large mineral stiffness contrast: Geophysics, 80, no. 3, L11-L33.

Andrew Snowling

Sr. Quantitative Interpretation (QI)

Andrew, RokDoc offers 4 types of Gassmann equation calculations. Dry Rock Modelling feature two anisotropic variants, and a modified heavy oil version which takes into account bitumen's shear characteristics. A key strength of using RokDoc for Gassmann is its dry rock modulus plots. These allow for a rock framework QC check between removing the insitu fluids and their replacement. This identifies how in-compressible your rock frame is and thus the effect that a change in fluid. A stiff rock will show a significantly reduced fluid affect over one which is fluid supported. It also highlights data errors since the error plot outside of the expected range (analogue points for comparison are supplied). A key strength of Ikon Science's RokDoc is rock physics and associated models, and this is incorporated in their implementation of the Gassmann equation.

FLUID SUBSTITUTION IN UNCONVENTIONAL (SHALE OIL/GAS) RESERVOIRS

Note: To learn about unconventional reservoirs, see Chapter 1 of this book.

QUESTION 89

Is there any preferred method for fluid substitution in shale oil/gas reservoirs? Is Gassmann fluid substitution appropriate? I need a lead on DT and/or density reconstruction taking into account TOC.

Pablo Patet I

Petrophysicist

ANSWERS PROVIDED BY INDUSTRY EXPERTS

Leon Thomsen

Geophysicist

Pablo, There are 3 issues with fluid dependence in shales:

1) Biot/Gassmann assumes that there is enough local permeability in the rock so that, as the primary p-wave changes temporally during the passage of the wave, it does not vary spatially, on the scale of grain size. This might not be true for shales.
2) The commonly-cited formula (from B/G) for fluid dependence of k is for isotropic rocks, not shales. Gassmann gives a generalization that is good for polar anisotropy, but not for fractured formations. Brown and Korringa (1975) give a better treatment.
3) B/G assumes that the rock matrix is uniform, on the grain scale, whereas the anisotropy that we see seismically might come, not from shales, but from layering of thin beds, each one separate on the grain scale. If true, then whatever you decide from log data needs to be up-scaled in some detail. Here is a recent reference that discusses this in greater detail: Thomsen, L., 2012. On the Fluid Dependence of the Parameters of Anisotropy: Society of Exploration Geophysics Convention Expanded Abstracts., 82.

Adrian Perarnau

Sr Petrophysicist at Pacific Rubiales Energy

Point to note for Leon Thomsen. Remember the theoretical threshold beyond which B/G is not applicable: low mobility rocks. That is, if fluid motion is not laminar during wave passage, part of the elastic energy transfers as heat to the media and is not available for mechanical interactions. Now the question remains: if we cannot use B/G, what can we use? I have no answer for that. Probably B/G is the only tool right now. You have to try it and my guess would be that it can work (sometimes). It is like using Archie for evaluating unconventional reservoirs. We can think on many reasons why Archie should not work, but somehow its answers seem close to reasonable.

Rob Simm

Senior Geophysical Advisor at Cairn Energy

Pablo, Probably the best rock physics models to start modeling this scenario are the inclusion models (e.g Xu and White, 1995, and the self consistent model of

Berryman, 1995) (see Ruiz and Cheng, 2010 Leading Edge). The problem is of course, that the fluid substitution effect will be dependent on the modeled rock compliance and assumed fluid moduli (both of which are likely to contribute significant uncertainty). Add to this the effects of azimuthal anisotropy and quantitative seismic interpretation, and this is going to be a significant challenge.

Boris Gurevich

Professor at Curtin University

Pablo, the question of whether Gassmann is applicable to shales came up in an SEG Post-Convention workshop on shales a few years ago, and was voted on. I believe the result was about 50/50. Seriously, it depends on whether fluid mobility is sufficient at a given frequency. Gassmann (or in anisotropic case, Brown-Korringa), assumes that fluid pressure is equilibrated within the representative volume within a half-period of the wave. Thus, in any rock Gassmann-BK is applicable at low frequencies but not at high frequencies. The crossover would be in multi-kHz range for rocks with millidarcy permeability, but may be much lower for shales (we must talk about matrix permeability here). If pressure is not equilibrated, something like Xu-White may be useful but needs to be anisotropic.

Julian Yao

Consultant, Geomechanics at Halliburton

Pablo, two factors make the lab validation on any fluid substitution method in shale very hard: First, the low permeability makes the pore pressure control difficult.

Second, the chemical interaction between the fluid and the solid matrix are not well understood and characterized, thus leading to great uncertainty.

This second factor is assumed not existent in B/G model. I think it's not considered in any other inclusion based rock physics models as well.

FORWARD STRATIGRAPHY MODELING

In forward modeling, we start with the geologic model of the subsurface, which we modify to reflect the information contained in both well and seismic data. The well logs provide velocity and density data as well as the vertical control we need to define formation depth. The seismic data, on the other hand, gives us lateral control. We use this control to develop a preliminary model, which includes the number of interfaces, their depth, interval geometries, interval velocities, and interval densities.

In general, forward modeling is used to determine the seismic response due to stratigraphic, structural, and hydrocarbon effects across a field. This is accomplished with an acoustic impedance log, generated from sonic and density data. From this we generate a synthetic seismogram, which we tie to the original seismic data. If the two data are similar, then we have our solution. If the observed (original seismic data) and model (trace) data are not similar, we modify our geologic model (synthetic seismogram) and produce a new geologic model output. We continue this interactive process until we produce a good match between the two data sets (synthetic seismogram

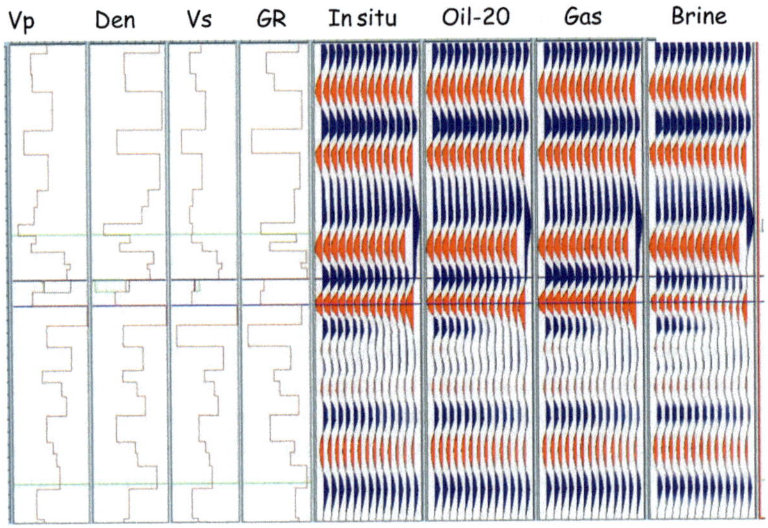

FIG. 7.2

In situ seismic trace and model synthetic seismic trace for oil, gas, and brine. Amplitudes dims with offset for hydrocarbons from near to far angle stack.

and real seismic data). Fig. 7.2 shows an in situ seismic trace and model synthetic seismic trace for oil, gas, and brine.

QUESTION 90

While doing Fluid Replacement Modeling (FRM), I use total porosity and effective water saturation. Some people are saying I should do otherwise. Can someone comment on this?

Hema Sharma
Geophysicist and Graduate Research Assistant at University of Wyoming

ANSWERS PROVIDED BY INDUSTRY EXPERTS

Colin Sayers
Scientific Advisor, Schlumberger

Hema, A key issue is whether the fluids are distributed in a homogeneous or heterogeneous manner. If the fluids are mixed in a homogeneous manner, so that the different fluids are mixed uniformly on a length scale smaller than the characteristic diffusion length, the differences in pressure induced by an elastic wave in the different fluids will have time to equilibrate during the period of the wave. If, however, the fluids are mixed in a heterogeneous manner on a length scale larger than the characteristic diffusion length, the patchy nature of the saturation leads to an effective velocity that is higher than if the saturation is homogeneous. See, for example, http://www.intlgeophysics.geoscienceworld.org/content/63/3/918.short

Albert K.

Geophysicist

Hema, it depends on what inputs you are using for FRM analysis. For clean sand, PHIE=PHIT (effective porosity equals total porosity, so it does not matter whether you take PHIE or PHIT & SWE or SWT). But for shaly sand, since you are not going to replace fluid within shale, use PHIE & SWE. But without rock physics modeling how will you define elastic properties (like bulk modulus) of clay? So best way to do FRM is to first do rock physics modeling and then go for FRM (Fig. 7.3).

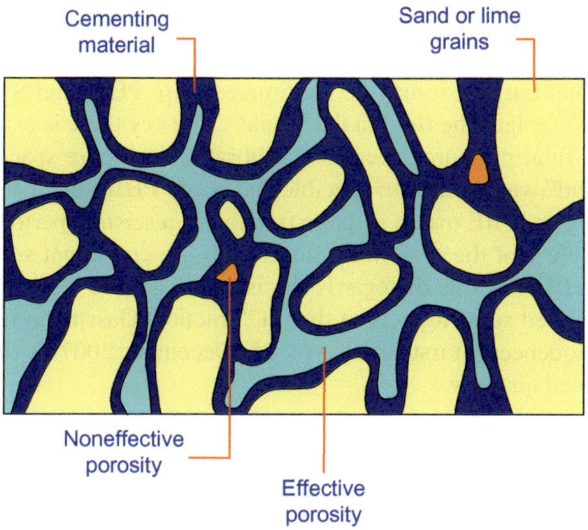

FIG. 7.3

A conceptualized porous sandstone.

Hema Sharma

Geophysicist and Graduate Research Assistant at University Of Wyoming

Colin: Thanks for your suggestion and the link. I am working on a gas reservoir in consolidated sands. The fluid mixture is quite homogeneous in nature. So I think that I can work with the Gassmann's fluid substitution equation. Can you recommend some more things which I should be concerned about?

Ainul: I am working with gas sands having few laminated shale intervals. It's a stratigraphic reservoir with good gas sand thickness. There seem to be no faults or discontinuities within the reservoir at least within the limits of seismic resolution. I am using HRS for fluid replacement modeling. And I have the elastic logs and reservoir parameters for calculating matrix and fluid properties before doing fluid substitution.

Keith Katahara

Senior Geophysical Advisor at Hess Corporation

Hema, It is certainly true that you should use either PHIT with SWT or PHIE with SWE. To use PHIT with SWE is to mix inconsistent variables. The key with either is to properly specify the corresponding solid phase properties. PHIT goes with a clay volume and you have to specify clay mineral properties. PHIE goes with a shale volume and then you have to specify the shale properties. In general, it is not possible to derive "shale" properties in a rigorous way without a bunch of core measurements, or well penetrations of the same pay zone with and without hydrocarbons. Certainly the properties of nearby shale may have nothing to do with the "shale" in the shaly sand. So I think that when Albert says to do rock physics modeling first, he is really saying to play around with the "shale" properties to get an answer that you think is reasonable. I agree with that as long as you don't fool yourself into thinking that you are doing something rigorous.

I want to point out that it may be legitimate to use PHIT and SWT, even if it is true that we can't replace the fluid in the "shale". The key there is exactly what Colin explained - the fluid pressures need to equilibrate in the pore space. So you could have "clay-bound" water that is irreducible and not in PHIE, but that will equilibrate with the pressure in PHIE over a suitable fraction of a seismic period. Of course, we need to make sure that the new water saturations are consistent so that we replace only fluid in PHIE. All this disregards the issue of getting a robust solution. Rob Simm, has discussed some aspects of that in "Practical Gassmann fluid substitution in sand/shale sequences", First Break, vol. 25, December 2007. A final point: PHIE is a poorly defined quantity.

Hema Sharma

Geophysicist and Graduate Research Assistant at University Of Wyoming

Keith, The gas reservoir I am talking about has few drilled wells at different places. We also have core data in one of them. I have a Vshale log which matches the core data very well. So about the matrix property I think there should not be any problem. XRD data is also present in most of the wells so I would also get an estimate of mineral composition. Should I move forward with this approach by taking PHIE and SWE as the input logs? As you know, elastic properties of clay are not fixed and vary from field to field. So If I am doing fluid substitution in shaly sand by using default value of elastic properties of clay, then my answer would have lot of uncertainty. The main reason for this is "Elastic properties of clay for that field are unknown to me". I am just using a default value of the elastic properties of clay which may not be right for my study area. So to find out these parameter (elastic properties of clay and aspect ratio of clay related pore), I prefer to do rock physics modeling in a well where I have both P sonic and S sonic. With the help of this model, I will generate synthetic P and S sonic, and try to calibrate with the measured P sonic and S sonic. With the help of this rock physics model, I have fixed elastic properties of clay for that field. Now I also have proof as I generated a rock physics model for this field by using these elastic properties of clay and my measured model data are well matched.

So I have confidence in the elastic properties of clay which I am going to use for fluid and fluid substitution for this field. If it is a clean sand or limestone, around 5 to 8% clay then elastic properties of clay will not affect too much on result. But once it's high, 30 to 40%, then a better option is go for rock physics model and fix elastic properties of clay.

Mukesh Gupta

Senior Petrophysicist

Hema, essentially you want to replace the movable fluid in the reservoir. You should not be playing with clay bound water. Keep it as it is and just replace free water with HC using effective reservoir properties.

Rob Simm

Senior Geophysical Advisor at Cairn Energy

Hema, A point on the fitting of rock physics models. There is usually more than one degree of freedom in the model fitting. For example, both clay mineral moduli and pore geometry factors are key variables so different combinations of these might give similar fits, but different stiffness (and stiffness effectively controls the magnitude of fluid substitution on the velocity log). However one approaches this problem it seems to me to come down to what you think is a reasonable stiffness for a shaly sand of a given porosity. The normalized modulus plot (Phi vs Kd/K0) is a good way to appreciate/evaluate the effects of your Gassmann input choices. My view is you can approach the Gassmann fluid substitution problem either using Phie/Swe or PhiT/Swt but do not mix them and be careful not to substitute irreducible water. A practical step is to check the velocity curve after fluid substitution. If there is a greater fluid substitution effect in the shaly sands when compared to the clean sands it is unlikely to be correct. Also check the Poisson ratio log especially when substituting gas.

Nepal Singh

Advisor Geophysics, at Cairn Energy

Hema, regarding the PHIT (total porosity)/PHIE (effective porosity), PHIE is the reservoir property used for the in-place HC volumetric calculation since we are concerned with the production issue. PHIT is the bulk property of the rock (may not be concerned with the development/production issue of the reservoir having interconnected pores) affecting the elastic property of the rock. If we go to the Biot-Gassmann equation, one basic assumption of the low frequency approximation, interconnectivity of pores, becomes important so that during the wave propagation there should be pore pressure equilibrium during the time period of the wave. In that concern, PHIE should be the porosity that we should be concerned with during FRM. However the recoverable/irrecoverable issue is still there which is related to the residual saturation. So I agree with Rob that saturation of irreducible water should be kept in mind and considered when performing FRM and a normalized modulus plot (Phi vs Kd/K0) is a good way to appreciate/evaluate the effects of your Gassmann input choices as QC step. Hence in my opinion PHIE/SW can be the right combination.

POROSITY CONCEPTS IN FLUID SUBSTITUTION

Hamed Amini

Reservoir Geophysicist at Senergy Energy Services

I'd like to share some thoughts about "Porosity concepts in fluid substitution". Some pictures relevant to the discussion below can be found here:

https://www.flickr.com/photos/hamedamini/sets/72157645696780941/

It is accepted, according to the seismic petrophysics literature, that different porosity choices (total or effective) exist for fluid substitutions. However, a thorough assessment of these models is not covered, and both approaches attempt to validate Gassmann's model with only a passing reference to the underlying porosity basis. Following Wang (2001), Gassmann's model requires that the pore fluid pressure induced by the passing wave be in full equilibrium within the time frame of half of a seismic wave period. To fulfill this criterion, all the pores need to be interconnected. Wang (2001) and Simm (2007) reported that the effective porosity model fulfils the underlying assumptions regarding the full pore-space interconnectivity, and the clay-bound water should be considered as a part of the dry frame. However, using numerical and analytical analyses, Grechka (2009) showed that under some conditions (as long as the aspect ratios of the pores are greater than approximately 0.2); the presence of disconnected porosity is unlikely to invalidate Gassmann's predictions. The use of total porosity has also been reported in a number of fluid substitution studies and laboratory measurements (e.g. Grochau and Gurevich, 2009). It should be noted that maintaining the effective porosity conditions for laboratory measurements is difficult; hence most laboratory measurements tend to be based on total porosity (Taggart, 2002).

Yan et al. (2013) suggested that the porosity in Gassmann's equation include the volume fraction of pore fluids that is connected and relaxed and the other fraction of pore fluids should be included in the rock matrix. They used irreducible water saturation to quantify the pore fluids that cannot be driven out by hydraulic force. Thus, in addition to total and effective porosity, a third choice also exists for porosity in Gassmann's model. In fact, similar to clay-bound water, capillary-bound fluids will not be replaced during fluid substitution. Therefore, one can limit the fluid substitution exercise to the portion of the pore space that is occupied by movable fluids only, that is, effective porosity less irreducible water and residual hydrocarbon. In this setting, both the clay-bound water and capillary-bound fluids are considered to be a part of the rock frame. The uncertainties associated with the estimation of the irreducible water saturation and residual oil volume fractions is one of the issues in the application of the movable fluid model. Amini and MacBeth (2015) utilized the movable fluids model in a fluid substitution exercise. They reported that compared to the effective porosity model, this model derives a more reasonable range for the dry-rock bulk moduli. Also by replacing gas with brine, they reported a lower fluid effect (i.e. less change in impedance) compared to the model based on effective porosity, and argued that the results based on the movable fluid model are more realistic.

QUESTION 91

What is the link between petro-elastic models and fluid flow models in generating synthetic seismic model? What are the best practices, references, concerns and shale content? How do I deal with fluid flow characteristics?

Rafael Medeiros De Souza

PhD Candidate (Centre for Petroleum Geoscience and Co2 Sequestration) at University of Western Australia

ANSWERS PROVIDED BY INDUSTRY EXPERTS

Hamed Amini

Reservoir Geophysicist at Senergy Energy Services

Rafael, I refer to this task as simulator to seismic modeling (sim2seis) in my explanation. Normally a petro-elastic model (PEM) is calibrated in the petrophysical domain where the effect of the different components of PEM on the 4D response is examined. However an essential criterion of a suitable PEM for sim2seis is its applicability to the simulation model. Petrophysical evaluations offer different porosity basis with the corresponding rock constituents that can be adapted for PEM calculations. Petrophysicists look at rocks through mineralogy, grain size, sorting and fluid content, etc., whereas, the engineers' grasp of porosity and other rock constituents is defined from a fluid flow perspective. Therefore it is very important to understand the definition of the parameters in the engineering domain, and establish the underlying relationship with the petrophysical domain. This allows for the same formulation that is developed in the petrophysical domain to be applied to the parameters in the simulation model; indeed, the consistency of the petrophysical analysis and the properties in the simulation model need to be addressed through such analysis.

Some of the challenges that you need to address in such studies include:

- Relating the rock typing in the simulation model to your PEM parameterization
- Assigning the elastic parameters to the inactive cells
- If you are working on CO_2, normally you are dealing with a compositional simulation model, where the Batzle and Wang equations for acoustic properties of the fluids are not directly applicable as they use black oil parameters.
- The vertical scale difference between the log data and the simulation model cells is another concern. There are a couple of publications on "scale-dependent PEMs" (e.g. Menezes and Gosselin, 2006; Alfred et al., 2008). However, they are based on Backus averaging, which I think is not applicable to sim2seis applications. If possible, my recommendation would be to work at a fine scale simulation model to avoid the artefacts
- Overburden properties etc.

Rafael Medeiros De Souza

PhD Candidate (Centre for Petroleum Geoscience and Co2 Sequestration) at University of Western Australia

Thank you, Hamed. You are correct. But let's dig a bit deeper into one of your comments on assigning the elastic parameters of the inactive cells. What did you usually do? Simulation just shuts down some cells that, in practice, are not affecting the flow and I am not sure what to do. Obviously your impedance distribution will have that hole there (that is what it is) and then your amplitudes will be useless. I don't want to talk about the procedure itself but I am interested in knowing which property would you consider? Would you extrapolate? Another good question is: what made the simulator shut that cell down? Was the porosity too small there? My particular concern about the vertical resolution is about how realistic or how far could I go to compare my FWD seismic with my real one. My current approach is to use the simulator outputs to model seismic and I am not comparing the derived impedances/amplitudes with any log derived attribute. In regards to the Backus average, Karl Stephen has been applying it for awhile (http://goo.gl/jrTijL). I agree in relation with your advice to work to the fine scale. However, how can you afford to run those simulations in a history matching procedure?

Hamed Amini

Reservoir Geophysicist at Senergy Energy Services

Rafael, Inactive cells are decided by the thresholds/cutoffs set in the simulator (say Eclipse *.data file) for pore volume (NTG*PHIE) or permeability. The issue regarding the elastic parameters of the inactive cells depends on the geological settings. What lithology(s) are making the inactive cells? During the PEM calibration stage, you should not ignore the lower quality/non-reservoir materials. In the North-Sea sandstone reservoirs encased by shale, using shale depth-trends for inactive cells is a reasonable approach. But in more laterally heterogeneous settings you may want to consider using the seismic inversion results. The comparison between synthetic and observed seismic depends on how the reservoir model is tied to the 3D seismic data. Normally reservoir properties that are distributed in the reservoir model using geostatistical methods fail to generate a seismic response close to the observed seismic. For a one-to-one quantitative comparison in the seismic domain, you can use closing-the-loop approaches to fully condition your model to the baseline seismic first, before moving to 4D. However, in the absence of a fully conditioned reservoir model to compare to the seismic data, making qualitative map-based comparisons is a good starting point. Performing the comparisons in the impedance domain is less costly than seismic-history-matching workflows.

CHAPTER

SEISMIC INVERSION

8

CHAPTER OUTLINE

Seismic stratigraphic modeling is used to extract stratigraphy data from seismic data. This method, which is used for small-scale tasks such as prospect evaluation, has two approaches: forward stratigraphy modeling and inverse modeling.

In forward modeling, we start with a geological model of the subsurface and simulate its seismic response. We then compare the synthetic output to our real seismic data.

In inverse modeling (seismic inversion), we start with our seismic data, estimate reflection coefficients and then transform geophysical data into a geologic model of the subsurface, which we then compare to our real geologic data such as wells. The model geologic output often consists of acoustic impedance or synthetic sonic logs (Fig. 8.1).

DATA NEEDED TO DO SEISMIC INVERSION

QUESTION 92

What are the data needed in order to do seismic inversion? And what are the best seismic inversion techniques?

Laouini Ghassen

Seismic Interpreter

Practical Solutions to Integrated Oil and Gas Reservoir Analysis. http://dx.doi.org/10.1016/B978-0-12-805464-2.00008-1

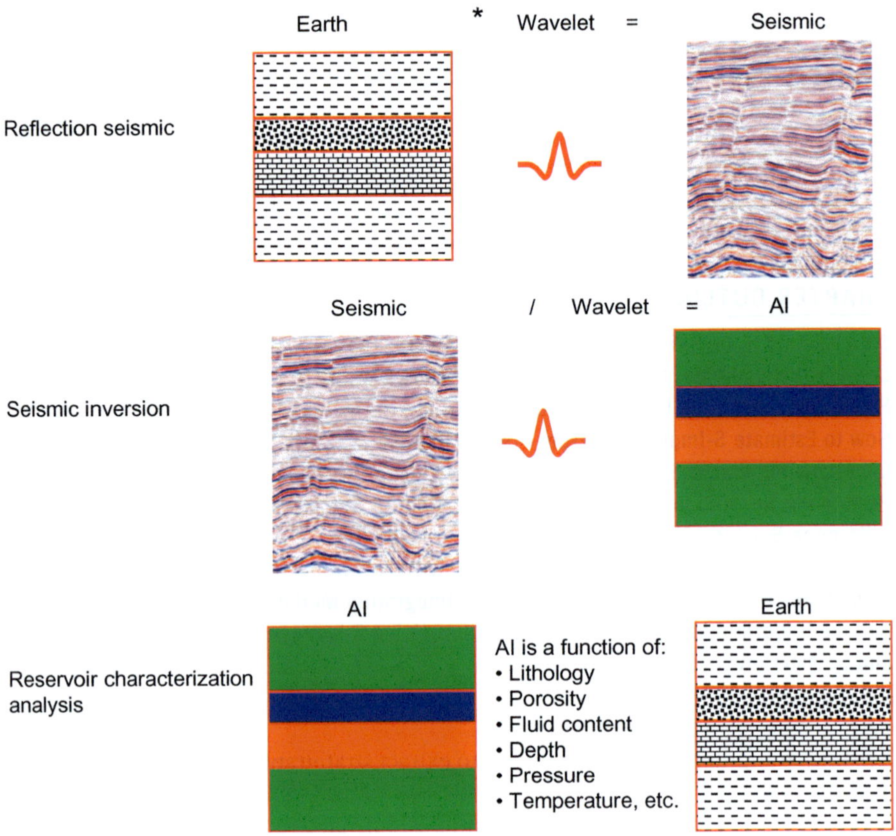

FIG. 8.1

Conceptualized seismic inversion.

ANSWERS PROVIDED BY INDUSTRY EXPERTS

Erik Johnson

Fugro Geoscience Division

Laouini, there are a large number of inversion techniques available for many different purposes. The selection of technique depends on what you are trying to accomplish. For conventional reservoirs there is cross plotting P-Impedance vs. S-Impedance logs, with shape coding to lithology and color coding to hydrocarbons (maybe Resistivity, or Sw). Examination of this plot will tell you how well you can isolate the zone of interest, and whether you need pre- or post stack inversion to isolate the lithology type of interest. Typically, only younger clastic reservoirs (Gulf of Mexico and West Africa) respond well in post stack, so in most cases you will want to run pre-stack to get the most lithology information out of the seismic data. For this you need

- 3 or more partial angle stacks (5-6 preferred).
- At least one well's data with a complete log suite including density and dipole sonic.
- You will also need interpreted horizons with which to build a low frequency model unless you are doing more of a "reconnaissance" type inversion to find potential candidates in large volumes of data.

Gautam Sen

Advice/Consultant in Exploration at Independent Oil & Gas Professional

Laouini, will you be carrying out inversion on pre-stack time migrated data or depth migrated data? Depth migrated gathers which are reasonably flat are the best candidates in case you are looking at doing it pre-stack. You will need a reasonable velocity depth model and you also need to have a good well-to-seismic tie, which is key to the inversion. You will be lucky if you are working in areas where anisotropy can be neglected.

LOW-FREQUENCY MODEL USE FOR SEISMIC INVERSION

Seismic inversion does not produce a true sonic log. The synthetic sonic log is really an acoustic impedance log as it includes a density component from the original seismic trace, while a sonic log from a well consists only of velocity data. Because the earth and the seismic recording instrument act as high-frequency filters, seismic inversion data does not contain the high frequencies found in a sonic log taken from well-logging/measurement.

Since the sonic log from well-logging records the very low- and high-frequency components of velocity that the seismic method does not record, we must model the missing low-frequency component and add it to the model synthetic sonic trace before we can match it with the true sonic log curves (from well-logging) (Fig. 8.2).

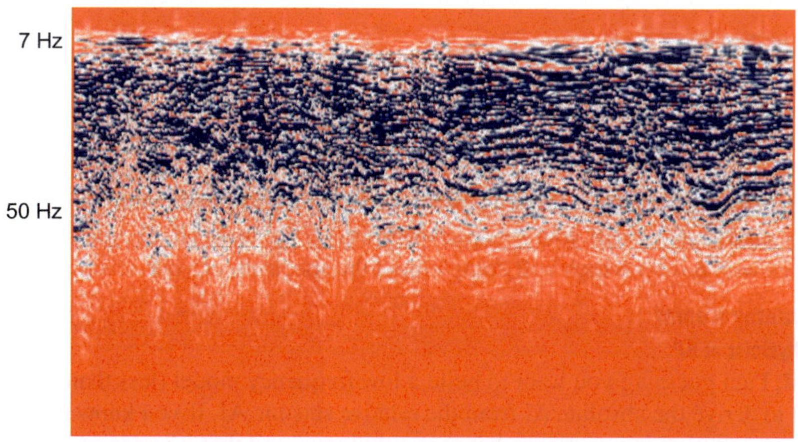

FIG. 8.2

Seismic data frequency spectra and the bandwidth that can be seen in the data.

HOW TO DETERMINE THE LOW FREQUENCY MODEL

QUESTION 93

Which methods are most accurate in estimating the low frequency component from well data, finding a linear trend, smoothing or applying low pass filter of acoustic impedance logs knowing that there are strong variations in the acoustic impedance between very thick shale layers and the reservoir sandstones followed by low acoustic impedance from alternating sand and shaly sand layers. Also are they another way to get LFM?

Yousf Abushalah

Lecturer at El Mergib University

ANSWERS PROVIDED BY INDUSTRY EXPERTS

Marcin Przywara

Senior Geoscientist in PGS Reservoir Ltd

Yousf, If you have sufficient well control over an area, just low-pass filtering of AI logs is what you need and what you should do. But if you don't have well control, you can make an attempt to create some layer-cake model of your subsurface perhaps with some regional approximations of the velocities layer by layer. Then, once again, you can apply a low pass filter. But I don't think this LFM will be of any use, and strongly advise you to limit your inversion to relative inversion in this case.

Yousf Abushalah

Lecturer at El Mergib University

Marcin, Yes, I will most likely invert seismic data to RAI. But, I thought it would be good practice to ensure that the inversion was accurate enough (directly compared with AI logs) as I experienced an excursion on the inverted impedance.

Marcin Przywara

Senior Geoscientist in PGS Reservoir Ltd

Yousf, if you have some well control over the area you plan to invert, you can easily make some tests at well locations. Even if you are not going for absolute AI, it may still be worth checking on how it would match your absolute impedance curves and for that reason, you should simply low pass filter your AI logs. Mind you, well control is needed just to fill your seismic spectrum on the low end. So you should filter AI logs with a filter which will complement the spectrum of your seismic.

Dhananjay Kumar

Geophysicist at BP

Yousf, I am guessing you want to build a low frequency model (less than say 10 Hz) of AI to derive the absolute AI from the relative seismic AI. If you know that there is a strong variation in impedance between lithologies and that it varies spatially, that's important information. First, you need to do well-ties to select the best wells based on well-tie and log quality. When deriving LFM try all wells to get a model and also try just 1 well (or 2-3 wells) to build the Low Frequency QC model and compare the result with the multiple models. Along with good wells, you also need a good set of horizons picked on the same seismic volume you are inverting for impedance. If you have a velocity model after seismic migration, try including this velocity model in building the LFM. Simply use well logs (VP vs AI cross-plot) to derive a relationship to go from Vp to AI. On another note, depending on the objective, you can also use relative AI to derive your rock properties (like, porosity).

Loic Michel

Regional Manager at CGG

Yousf, In addition, the amount of structural information you input, will have a large impact on the interpolation of P-Impedance. Also, interpolating before or after low pass filtering will impact the final model. Then, depending on the inversion software you are using, the model to be used may have a significant impact (assuming you use a model based inversion).

QUESTION 94

How do we choose the low frequency band (Low Frequency Model) of seismic inversion model if we assume that our surface seismic has a bandwidth of (10-50 Hz)? And what are the QCs that we should do on this model?

Mustapha Chaibi

Geophysicist Engineer at Sonatrach/Exploration Top Contributor

ANSWERS PROVIDED BY INDUSTRY EXPERTS

Alexey Sokolov

Reservoir Engineering Manager at CGG-Vostok

Mustapha, choosing the low frequency band for seismic inversion is an uncertain business; there is no universal recipe for this. The general idea of LFM is to provide a trend to extrapolate away from wells, so you have to understand what you are modeling first. I would start with a high-cut filter at 10 Hz for your seismic band. I would slightly vary this value and see if resulting LFM variability matches my understanding of the reservoir. As for QC, your LFM should follow the same trends that are evident from well logs. Make sure that your LFM limits parameters within the same bounds as the original data does (not always possible). Make sure in addition, that Vp/Vs in your LFM is greater than 1.41 (sanity check), and increases down the time section. I assume you have already extracted the proper wavelet.

Mohamed Badawy

North & East Africa Business Manager

Mustapha, there are rules for low frequency modeling. It's either you already have it recorded using broadseis technology, or you iteratively model it. Generally speaking the modeling will depend on the velocity logs in the well so it is better to use a few wells and have a blind well test for verification. The key point is that you should be able to build a model that matches to the geology and not get misled by the modeling process itself.

Hamed Amini

QI Geophysicist at Senergy Energy Services

Mustapha, we try to minimize the user input into inversion process and try to avoid over-constraining the inversion. The number of wells and horizons to be used in LFM depends on the complexity of the geology. You should not ignore what is given by the input seismic data in terms of structure and seismic signature variations. That will help you to choose the key horizons. Obviously that also needs some geological insight, where the input from geologist is useful. Inversion is not going to do magic and extract the information if the key elements are missing in the LFM. Once you are done with inversion, subtract the LFM from your inversion results and map the "relative rock physics" templates on the seismic volumes to double check how your lithology/fluid cubes will be affected. This allows you to check the genuine contribution of the seismic data into your inversion results.

Gautam Sen

Advice/consultant in exploration at independent oil & gas professional

Mustapha, high density velocity analysis is the input for LFM, beside well logs. In case there is a decent coverage of well data, (multiple wells), time and spatial variation of the extracted wavelets is another QC check for usually inversion is carried out using an average wavelet .Inverted data must not only follow logs but also needs to more than geologically mimic the seismic section. Also inversion could be carried out for a targeted horizon only, to ensure that the wavelet used for inversion is a good representation for areas with widely time and spatially varying wavelets.

Dhananjay Kumar

Geophysicist at BP

Mustapha, another item to QC: use only one good well to build LFM. Sometimes, if you use multiple wells, there are issues in interpolating among wells, so try one well only and extrapolate by following a few horizons. Obviously you are going to output QC at multiple wells.

QUESTION 95

I am doing a band limited inversion (no wells in the area) over a reservoir that has a strong impedance contrast with the overlaying shale (hard kick). The strong seismic reflection at top of the reservoir causes a strong excursion of impedance values, which is not real. What can I do to eliminate or attenuate the excursion?

Jorge Reveron

Senior Geophysicist at Repsol

ANSWERS PROVIDED BY INDUSTRY EXPERTS

Rui Zhang

Assistant Professor of Geophysics at University of Louisiana at Lafayette

Jorge, you may check the near stack or angle gathers excluding offsets larger than 15 degrees.

Scott Singleton

ResSCAN Technical Manager at ION Geophysical

Jorge, In line with Rui's response, are you doing a pre-stack or post-stack inversion? Regardless of which one you are doing, I would suggest you go back and look at your AVO response. Does that match what you would expect from your conceptual model of what is happening at the shale/sand interface? Is it even a reasonable AVO response, meaning are the near and far offsets well-behaved? However, if you only have post-stack data, then you have no recourse but to go with what you have. After data gets stacked you're stuck with the volume in your hands.

Michael Doty

Geoscience Consultant

Jorge, consider the following points;

1. Band limited inversion, especially in the absence of low frequencies, can result in "impedance" values that look much different than expected. Since you have no wells I suggest you create a simple 1D impedance model over your zone of interest and then filter the impedance to a similar bandwidth on your inverted data for comparison.
2. Not sure how confident you are with the phase or whether it's land or marine data, but non-zero phase data can lead to strange looking inverted impedances.

Erik Johnson

Fugro Geoscience Division

Jorge, how thick is the expected reservoir, and how much does the thickness vary laterally (as expected from geology)? The answers to this will go a long way to answer whether or not you will get anything reliable out of an inversion. Band-limited data will naturally have a large overshoot into the layer below the strong contrast. A possible analog I have seen in the North Sea is eolian sands below a high impedance anhydrite layer. In order to achieve a successful characterization there, an interactive-update approach to building a low frequency model for the inversion was required. Of course, to do this accurately, you need a well with a dipole sonic, which you don't have yet. One "guesstimate" approach (to elaborate on Michael's suggestion above) would be to take the stacking velocities, convert them to interval velocities, and guess at a compaction trend for the density profile. Make a ~1Hz low frequency model out of that for use in the inversion. This may improve the image enough for you to re-pick the horizons in the area and update the low frequency model. Does your geotechnical team acquire slim-hole data for shallow water hazards (assuming this is offshore)? If yes, some of these drill-ships are now capable of drilling and logging 2500m below the seabed. Depending on how deep your target is, this data may also help develop a low frequency model that will help with the inversion for placement of your first well. In any case, you are probably looking at an inversion that will not give you a single answer. It can confirm that your hypothesis is possible, but it may leave many other possibilities as open well.

Gautam Sen

Advice/Consultant in Exploration at Independent Oil & Gas Professional

Jorge, it is not clear whether the overlying shale has higher impedance or the reservoir has higher impedance. This can be checked from the seabed reflection in offshore or the reflection from the weathered zone in onshore data. The Velan plots can also help besides generating low frequencies as suggested earlier. Assuming it is a pre-stack inversion, it will be interesting to see if the excursions you mentioned are present in all the inversions or are confined to near stacks only.

Jorge Reveron

Senior Geophysicist at Repsol

Thanks for your comments, they are very useful. The inversion is post-stack. However, the gathers behave as we expect for AVO class I at the top of the reservoir. The reservoir is hard sand underlying soft shale, so there is a positive impedance contrast. The seismic is almost zero-phase, as computed from an offset well with different lithology.

Erik Johnson

Fugro Geoscience Division

Jorge, how thick is the sand in relation to the tuning thickness? If it is at or below tuning thickness, then the inversion may help you to separate the lateral thickness changes (which cause the tuning) and the lateral property (porosity) changes. You can do this with post stack inversion, but broadband will be more useful than band-limited for reducing the tuning effect and attenuating the excursions. You can use a basic background low frequency model based on expected petrophysical properties of the sand and shale. With this, you can probably make reasonable quality relative porosity maps (relative to the low frequency model - the porosity will be accurate to the extent the LFM is accurate). If you use the relative porosity maps from inversion and fluid indicators from AVO analysis, it can help you pinpoint thicker, more porous, pay-indicated zones for your first well.

HOW TO ESTIMATE S-IMPEDANCE FROM P-WAVE DATA

QUESTION 96

How can we estimate S-wave where we have only P-wave? And would you please give us more explanation of angle stacks and low frequency modeling procedures?

Noomen Dkhaili
Reservoir Geophysicist/Reservoir Department/Etap

ANSWER PROVIDED BY INDUSTRY EXPERT

Erik Johnson
Fugro Geoscience Division

To estimate S-impedance from P-wave data, you would make 4-6 partial angle stacks from the processed gathers. Since the first trace in your gather will probably be 5-10 degrees, you don't 'really' have 100% P-wave data either. All the gather traces are at some angle between 0 and 90 degrees. When you make, say, 6 partial stacks, you improve the signal to noise by combining 5-10 adjacent traces into one angle trace for the inversion. To make a full-bandwidth inversion result, you will need to supply the low frequency models for P-impedance and S-Impedance. These are usually built with a geological modeling program using the interpreted seismic horizons with P- and S- sonic and density well log(s). The well log data is interpolated through each layer of the model based on the seismic horizons. If you are using pre-stack depth data, or if you are using offset instead of angle gathers, you will also need a velocity model. In simultaneous inversion, the software will fit an AVO curve to the results from the 6 partial stacks and extrapolate to find the normal incidence P-impedance and the S-impedance. Then as a check, it will internally forward model its results and check them against the input data, updating its estimates if the error is outside preset limits. The more sophisticated methods allow you to set rock physics relationships and give you greater control when the seismic data isn't of the best quality. So to summarize, you do not need to record S-wave data to estimate the S-impedance. But for best results you will need a well log with a shear sonic in order to estimate the angle wavelets and build the low frequency S-impedance model.

SIMULTANEOUS ELASTIC INVERSION AND EXTENDED ELASTIC INVERSION

QUESTION 97

Which techniques is best to use and why: fluid and lithology indicators from AVO analysis or Elastic inversion?

ANSWERS PROVIDED BY INDUSTRY EXPERTS

Giorgio Cavanna
Geoscientist at ENI E&P

There are many, often equivalent, ways to combine AVO attributes or inverted elastic properties to discriminate fluids and lithologies. Choosing one or another may depend on pragmatism, company traditions, simplicity, robustness, or, effectively, by a higher potential. I am referring to numerous AVO fluid factors, LMR, Poisson Impedance, AVO Impedance, EEI, etc.

Sagnik Dasgupta

Senior Rock physicist at Schlumberger

Absolute properties are better than the relative properties for quantitative prediction. AI, SI or PR or other LMR are better than EEI or fluid factors as the definition of those properties are more robust than the relative estimates.

Giorgio Cavanna

Geoscientist at ENI E&P

Absolute properties are better in principle but many times the low frequency model can be highly uncertain leading to incorrect absolute properties. In such cases, relative amplitudes and impedances are more reliable. EEI, function of Vp, Vs and density are an absolute properties as well.

Sagnik Dasgupta

Senior Rock physicist at Schlumberger

I know that low frequency is big problem in calculating absolute properties. But for quantitative estimation it is necessary because porosity and fluid saturations are not relative properties. Regarding EEI, I have reservation. Mixing two different properties (AI & SI) may not as valuable as those two properties separately, even if we can estimate EEI so called in an absolute sense.

Robert McGrory

Sr. Geophysical Crescent Point Energy

Giorgio, in an exploration problem, where hard data (well logs, core etc.) are not readily available or where well sampled relative attributes such as fluid factor, A, B, are missing, Poisson's reflectivity, Rp and Rs can be easily and generally, robustly estimated (notwithstanding processing issues). However, these attributes provide only qualitative reservoir parameters. EEI, AI, SI, LMR etc. all provide truly quantitative estimates of reservoir properties. However, their estimations can be fraught with uncertainty. To overcome these issues, I take a statistical approach by estimating as many different attributes as possible, relative or absolute, and compare them in a matrix. This can be very useful in de-risking a particular seismic anomaly. This method is predicated on having done some extensive modeling up front where attributes (relative and absolute) are estimated prior to and compared with, estimated or inverted results from actual data. My lowest risk anomalies are generally the ones that show consistent behavior between expected modeled response and estimated or inverted results across many different attributes be they relative or absolute.

However, if the project/problem is a development type problem then a different approach could be used. Modeling and well logs, cores etc., can be mined for information and robust statistical models can be developed. If seismic data quality is excellent then advance pre-stack inversion techniques can be applied and rock properties estimated robustly. This can be taken further by post-stack and/or pre-stack stochastic inversion/modeling where the uncertainty around a specific set attributes can be assessed away from well control. Again, that is not a substitute for understanding the limits of certainty in the actual data by using cross-plots of the attributes derived from well logs and cores. For example there is uncertainty in lithology determination using cross-plots of say, Vp/Vs vs AI or AI vs SI or cross-plotting of the Lame parameters. There is often, overlap between clusters of data for sand and shale. That is a level of uncertainty that is over and above the level of uncertainty in the inverting of the data. The most important problem we tackle in earth science is categorizing and estimating uncertainty.

Gautam Sen

Consultant at Sahara Group

Robert McGrory's statistical approach to exploration problems makes practical sense and can be adopted as an effective workflow.

More so, uncertainty as mentioned, on account of other than accurate LFM in absolute properties estimation and cluster overlap, is a matter of concern. With all its limitations Vp/Vs can still be relied upon but not density. I agree also with Roberts workflows even for development problems.

Jorge Reveron

Senior Geophysicist at Repsol

What about the use of EEI to estimate elastic properties (K, mu, lambda etc.) or petrophysical properties (Vsh, phi) by finding the chi angle, which maximize the correlation between EEI and the properties? Certainly there is uncertainty because this is an estimation from a correlation, but in some cases with sparse information it's the only way. Model based inversion with information from only a few wells could be biased, especially SI and density estimation from simultaneous inversion, because their response could approach the well log response. At the end, it's a very difficult issue and well control is the best weapon that we have to reduce the uncertainty in our results.

QUESTION 98

Let's say that based on well-log analysis, Vp/Vs ratio is the best discriminator for lithology/fluid in my reservoir of interest. I therefore want to invert my seismic data to some form of Vp/Vs ratio (i.e., directly to Vp/Vs or by an equivalent EEI volume). I have the option to run a simultaneous inversion of my angle stacks (near, mid, far) to AI and Vp/Vs ratio (as can be done in Hampson-Russel, Jason, Rokdoc etc), or go through the Extended Elastic Impedance (EEI) workflow for a chi angle of ~45 (which gives the highest correlation to my Vp/Vs log). What are the benefits and limitations of simultaneous AVO inversion versus extended elastic impedance (EEI) for Vp/Vs? Which should give me the best result in your opinion and why? Which is likely to be more noisy? For EEI I need a good gradient stack, but correct alignment of my angle stacks is still important for simultaneous inversion.

Matthew Saul

Geophysicist at Chevron

ANSWERS PROVIDED BY INDUSTRY EXPERTS

Efthymios Efthymiou

Geophysicist at Chevron

Matthew, there are 2 issues here: first, which AVO model to use (EEI linear approximation based on 2 terms Shuey, or 3 term Aki Richards, Fatti etc). Second, how to apply the chosen model to the data? Which model to use, I think, is of minor consequence for real life data. The differences are more evident with high quality long offset data. The inversion method is where all the differences lie. With the Jason philosophy/workflow, you have a lot of control on the process (how much weighing to give to different stacks for the gradient calculation and also you constrain the calculations on your rock physics model). With a simple 3 stack EEI calculation, any misalignment errors will affect your gradient a lot, and you have no QC ways to assess the magnitude of the error.

Hema Sharma

Senior Development Geophysicist

Matthew, when you go for 3-term EEI, noisy far stack data due to wave interference is always a problem. Two-term EEI is a greater approximation and is valid only for near to mid offsets. On the other hand Three-term EEI is affected by noise. So in my opinion you should go for 2-term EEI in comparison with simultaneous inversion. In 2-term EEI as well there is a constant term Vp/Vs which can be sensitive to noise. Also, when doing simultaneous inversion, deriving a low frequency model is a challenge. You can take a small area at first and run both the processes by using a blind well and taking a look at the differences.

Brian Taylor

Operations Geophysicist at GDF SUEZ

Matthew, How successful you are going to be will depend on:

- How well conditioned your seismic data is (you need to understand how reliable your datasets are by examining the gathers).
- How you choose to scale it. Leon gives one reason why the 'standard' technique of fixing it in the inversion package is unreliable in 'On the use of isotropic parameters to understand anisotropic shale behavior' (http://library.seg.org/doi/abs/10.1190/segam2013-1080.1).

In the presence of laterally varying anisotropy the blind test may not be conclusive. In other cases as Hema states, it is probably the best assessment you have. The pragmatic benefit of EEI is that you can test a lot of options (varying Chi, comparison with logs and maps).

Leon Thomsen

Geophysicist

Matthew, this thread reflects old thinking by failing to recognize the major role played by anisotropy in AVO. A recent paper (Lin and Thomsen, SEG Convention 2013) confirms the previous theoretical result that anisotropy, even if small compared to one, makes major contributions to AVO since all of the terms in AVO reflectivity are likewise, small compared to one. Hence, the anisotropy term in the AVO gradient cannot be neglected and in general, all isotropic AVO analysis (for the last 30 years!) is suspect. Most of the anomalous AVO behavior discussed in Chapter 6 becomes understandably straight-forward, by including the neglected anisotropy term in the AVO gradient. If you want to estimate lithology, you should do it via its correlation with anisotropy, rather than with Vp0/Vs0, since the former correlation is stronger, and more direct. More substantive discussion, with a data example, is given in Lin and Thomsen (2013).

Theron Edwards

Kuwait Oil Company

Matthew, I appreciate Leon Thomsen's comments on how suspect our AVO calculations are. I would add that they can be suspect for more reasons than just anisotropy such as, converted waves, multiples and any other non-primary reflection energy. If we lower our expectations and aim for a more qualitative, rather than quantitative approach I think trying EEI with a variety of chi angles and trying to see which makes most geologic sense would be more flexible than a deterministic single answer Vp/Vs ratio. At the end of the day, we see Vp on near angle stacks and we see the influence of Vp/Vs on far angle stacks, so perhaps if we first isolate that part (separation of images in a double exposure) before we try to quantify what it means would be useful. This approach requires no 2 or 3 term approximations and if done properly, will give the optimum chi angle. To visualize this, imagine you have a density and a neutron log that have no scales and have had unknown depth dependant scaling applied. Your job is to find crossover. You would scale (depth dependant) the two curves so that track each other as much as possible and thereby isolating the parts that don't track. As long as these zones are small compared to your scaling window, you will find the interesting parts. What is lost in this approach is the ability to quantify the answer.

Igor Escobar

Geophysics Team Leader at Maersk Oil

Matthew, one of the interesting aspects of a simultaneous elastic inversion, as opposed to EEI, is that one tries to simultaneously explain all the angle stacks at the same time, bringing a pretty decent amount of data control to the prediction (provided the data quality is there). Indeed this is not the case of EEI where each sub-stack is "inverted" independently and later remixed with a chi angle. There are, despite the statements on Lin's & Thomsen's paper, some advantages in keeping the wavelet angle dependent and using that as a way to accommodate for NMO stretching, to some extend anisotropy, and all the bit and pieces we did not managed to correct during processing. I found interesting the rather bold statement on the effect of anisotropy on AVO behavior, and yes indeed, we all know to a good extent it matters and drives some of the AVO behavior. However, could we really comfortably say that all the AVO work done until now is to be suspect? I reckon we could easily agree to that in places like deep-water turbidites in West Africa with 1000's of meters of Miocene shales sitting on top of Oligocene reservoirs showing rather large anisotropy. But this extreme values are not observed everywhere especially if we consider (after the mentioned paper) that sin (angle) ^2*tan (angle)^2 is way smaller than sin (angle)^2. I found the claim that all 3 terms are of the same magnitude to be unclear. I guess it might hold right at 45 degree.

Leon Thomsen

Geophysicist

Matthew, I agree, with Igor Escorbar that wavelets vary with offset, and are further stretched by NMO removal. Lin and Thomsen (2013) ignored this effect, in order to concentrate on the major effect of anisotropy. I do think that all isotropic AVO analyses are suspect; although they are not necessarily wrong. The reason they are suspect is very simple: they ignore a term which is plausibly as large as the retained terms. You don't know if the neglected term is actually negligible until you do an analysis. Note that Lin and Thomsen (2013) did not claim that the curvature term is comparable to the gradient term (it is not, for the reason that Igor Escobar stated above). Rather, the claim is that all 3 terms which comprise the gradient coefficient are plausibly of the same magnitude.

QUESTION 99

What inversion method will be best when using AVO analysis to simultaneously invert for Vp, Vs and density simultaneously?

ANSWER PROVIDED BY INDUSTRY EXPERT

Brian Schulte

Geophysical Specialist - Talisman Energy

How we get Vp, Vs and density in simultaneous inversion is through inverting the intercept, gradient and curvature. The intercept is gathered from the near offsets (0-15 degrees), the gradient is from the mid angles (15-30 degrees) and the curvature is from the far angles (30-45 degrees). It is essentially the intercept minus the curvature which gives us density but the curvature is a small term and it comes from the gathers where we see most of our problems such as multiples, NMO stretch, changes in the wavelet due to NMO stretch, higher order move out etc. Cambois argued that when we look at intercept, gradient and curvature, these attributes are statistically correlated together and are affected by wavelet changes and also residual velocity and higher order move-out. Cambois pointed out that when we have residual move-out, we have leakage of the intercept into the gradient. He argued what we should do instead of using gathers is to utilized angle stacks and then estimate the wavelet per angle stack. By stacking the data, we remove some of the noise issues and can estimate a better wavelet.

QUESTION 100

What approach should I use for inversion of limited offset stacks? Is it possible to perform elastic or pseudo-elastic inversion? I don't have the gathers, only the full stack and near, medium and far limited offset stacks.

Rafael Pabon Manzano
MS in Geophysics candidate at University of Tulsa

ANSWERS PROVIDED BY INDUSTRY EXPERTS

Marcin Przywara
Senior Geoscientist in PGS Reservoir Ltd

Rafael, having near, mid and far stack, technically speaking you can do virtually all the same workflows as with gathers. All the AVO analysis/inversion relies on is the possibility of extracting AVO equation parameters (intercept and gradient, sometimes also curvature), with 3 term linear AVO equations. Although curvature estimation is rarely (if ever) stable, when you having three data points you are able to do a linear fit, so you can derive AVO parameters. One word of warning though, you are talking about offset stacks. As you probably know, AVO equations are the function of incidence angle, not offset itself (contrary to what the AVO acronym stands for), but efforts to replace the term AVO with the more accurate AVA have failed to make an impact in industry. What you should do in order to evaluate quantitatively meaningful AVO analysis is to perform ray tracing, using a velocity model, and subsequently create common angle stacks. For this you would need gathers, though. As it is not the case from what you said, you will have assume that the average incidence angle is at your target level on every of your partial stacks and accept the fact that your AVO investigation will be flawed from the very beginning.

PERFORMING SEISMIC INVERSION ON HIGH-RESOLUTION 2D SEISMIC DATA

QUESTION 101

I am looking into the possibility of performing seismic inversion on newly acquired 2D high resolution seismic data from within a field. Two gas wells were drilled but are now P and A. There are still several other untested shallow gas reservoirs at depths ranging 300m to 900m sub-seabed. What are the challenges or limitations we can expect when performing seismic inversion or other QI workflows within the shallow section (within 1500ms TWT) using 2D high resolution seismic data?

Zhi Xin Teh
Geophysical Reporting Manager

ANSWERS PROVIDED BY INDUSTRY EXPERTS

George Yao
Manager of Depth Imaging

Zhi Xin, since two wells were drilled, you can design a synthetic model to test tuning effects on the gas zone thickness. From this test, you can decide whether you have enough resolution on the "newly acquired 2D high resolution seismic data within the field" to perform "the seismic inversion or other QI workflows within the shallow section (within 1500ms TWT)".

Erik Johnson

Fugro Geoscience Division

Zhi Xin, be sure to look at your pre-stack gathers to see what fold and angles you have in the data. Also, check for flatness of the gathers in the shallow section. Unless the data was specifically processed for the shallow section, the velocity picks may not be accurate enough to support quantitative work in that depth. I have done seismic inversion looking for shallow gas pockets as part of a shallow hazards evaluation, and it can be done successfully if your data is properly prepared and the geology allows for it (AVO class, tuning thickness/frequency content, sufficient angles in the data).

Zhi Xin Teh

Geophysical Reporting Manager

Erik, the 2D high resolution seismic data that we collect is specifically processed to image the seabed and the shallow section. Target depth is within 1000ms TWT. Data beyond 1200ms TWT suffer frequency and amplitude attenuation due to the small 150 cu-in gun that we use. And we are using a relatively short streamer (1200m). The number of fold is 48 acquired using 96 recording channel. We could adjust the acquisition geometry to get 96 fold. Frequency range in the seismic data is 20-200 Hz (dominant frequency is approximately 100 Hz). I guess the inversion would rely heavily on the well data for the low frequency. Did you perform the inversion for the shallow section using the conventional 3D data in deep water environment?

Erik Johnson

Fugro Geoscience Division

Zhi Xin Yes, the project I referenced was in the Gulf of Mexico and was conventional 3D. I had to re-flatten the gathers in the shallow section before making my angle stacks. I was lucky to have sonic and density in the shallow section in one of the wells to get the wavelet. But I was looking for shallow gas pockets to avoid, where you are looking for shallow gas to exploit.

Maher Maklad

President at Signal Estimation Technology Inc

Zhi Xin, several factors contribute to a good inversion:

- Input seismic data should be properly processed. Common processes such as f-x decon, spectral whitening and even frequency domain implementations of surface consistent deconvolution may add artefacts to the data or enhance coherent noise.
- The wavelet should be accurately estimated.
- The inversion algorithm should be accurate.
- Well logs should be used to validate the inversion.
- We have had good success in detecting shallow and deep gas (both for exploration and drilling hazards) using spectral attenuation in conjunction with inversion.

Peter Rowbotham

Lead Geophysicist at AGR Petroleum Services

Zhi Xin, following on from Maher's comments on using well logs to validate the inversion, the well logs will also be used for wavelet estimation and potentially for building a low frequency earth model depending on your inversion engine of choice. My concern would be on the quality of well logs over your shallow interval, because of the borehole conditions and on whether acquisition of good shallow logs was a specific requirement at time of logging.

Zhi Xin Teh

Geophysical Reporting Manager

I agree with you, Peter, that the sonic log quality within the shallow interval may not be good enough due to the unconsolidated sediments. We may have difficulty building a rock physics model for this shallow section as well. The two wells drilled however, were indeed targeted at the shallow section. We may have to merge the low frequency model at a higher threshold due to the lack of low frequency content in the high resolution seismic data. What are the consequences of merging the low frequency model at a higher threshold; say at 20Hz?

Loic Michel

Regional Manager at CGG

Zhi Xin, one issue with the Low frequency model is well data interpolation/extrapolation. When seismic has no low frequencies you need to interpolate well data or use a kriging technique. Depending on the geology, validating your "interpolation" can be very hard if you do not have any geological information to help constrain it. Some may say that you will over constrain the inversion. You may consider applying a looping technique where you run a band-pass inversion, interpret the results and create a first model then, re-run inversion and update your model as you get better understanding of the geology and spatial variations. Have you tried AVO first? What is the water depth?

Zhi Xin Teh

Geophysical Reporting Manager

Loic, I remember I used the Multi Attribute Well Interpolator (MAWI) before and the results seemed to be more or less conformable to the geology. Not sure if CGG is still using this module or EMERGE? AVO will be the first analysis to be performed in my workflow proposal. The concern is the 2D high resolution seismic data may not have enough traces for the analysis. I hope that the gather is sensitive enough to show AVO anomalies in such a short offset. The water depth is less than 100 m.

George Yao

Manager of Depth Imaging

Zhi Xin, since water depth is less than 100m and your target depth is within 1000ms TWT, I think the AVA (in the ANGLE domain) should work even with AVO (in the OFFSET domain) anomalies and such a short offset. The AVA should work on the untested shallow gas reservoirs you describe. I hope that you have enough NMO resolution on the newly acquired 2D high data, within the field to transform your gathers to AVA.

Loic Michel

Regional Manager at CGG

Zhi Xin, if you are confident that well interpolation is close to the geology you are on track. As George replied, AVO response should be able to record fluid effect (assuming favorable rock physic and tuning thickness). You can find papers about AVO using shallow seismic data, Paternoster, 2008 - EAGE is one of them.

DIFFERENCE BETWEEN AVO ANALYSIS AND SEISMIC INVERSION

QUESTION 102

Could you kindly describe AVO and inversion in simple and easy words.

Muhammad Muti Ur Rahman

Student at Bahria University

ANSWER PROVIDED BY INDUSTRY EXPERT

Celestine Eze

Geologist

AVO anomalies occur when amplitudes vary as a function of offset (or angle). The particular way the amplitude changes take place may be an indication of hydrocarbons. Normally we expect the reflection amplitudes for the interfaces between the cap rock and the gas (or oil) filled reservoir to increase with offset (or angle). The same will occur for the gas/oil, gas/water or oil/water contact, but with the opposite polarity to the cap rock/gas interface. On the other hand, seismic inversion is one of the means of extracting rock properties (such as acoustic impedance, shear impedance, porosity volume etc) from seismic data with the integration of geologic horizons and well information.

COLORED SEISMIC INVERSION

QUESTION 103

What is colored seismic inversion?

One of the seismic inversion algorithms is called "colored" inversion. It is performed in the frequency domain and the point is to build an operator that directly transforms a seismic trace into a reflectivity trace. The question is, however, what does "colored" mean in this context and why was the specified technique named in this way?

Colored inversion is designed to approximately match the average spectrum of inverted seismic data with the average spectrum of observed impedance (Lancaster and Whitcombe, 2000). The earth's reflectivity can be considered fractal, and the resulting amplitude spectrum favors high frequencies (spectral blueing). If there was

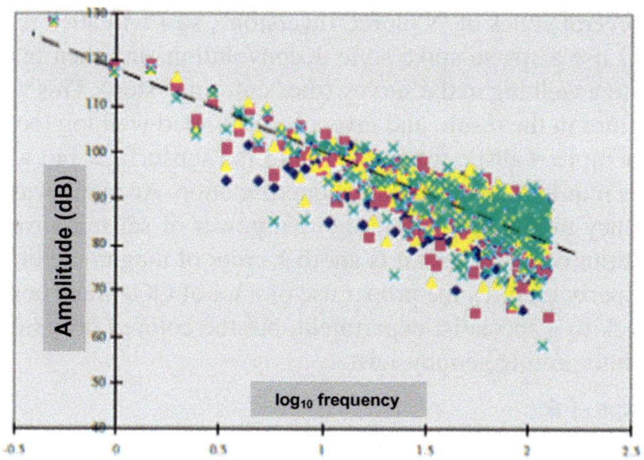

FIG. 8.3

Acoustic impedance from four sites in the North Sea.

From Lancaster, S., Whitcombe, D., 2000. Fast-track 'colored' inversion. SEG Technical Program Expanded Abstracts 2000. pp. 1572–1575.

no preferred frequency, then you would have a "white spectrum," but as there are some frequencies with more energy, it is called "colored" (Fig. 8.3).

ADVANTAGES OF COLORED INVERSION OVER TRACE INTEGRATION METHOD

QUESTION 104

What is the advantage of colored inversion over the trace integration method? Are you aware of any experience/case studies/published paper comparing these two methods?

Rajesh Soni

Reservoir Geophysicist In Geophysics For Geomodeling Team At Total Sa

ANSWERS PROVIDED BY INDUSTRY EXPERTS

Erik Johnson

Fugro Geoscience Division

Rajesh, trace integration makes the assumption that every peak and trough in the seismic trace corresponds to a real reflection point in the earth, which is a really poor assumption. It also assumes your data is 0 or 180 degree phase. Trace integration essentially gives you a 90 degree phase rotation, which may be helpful to (for example) a geologist doing interpretation, as now you will not be picking peak-to-trough units, but rather peaks and troughs become geologic units, and you are picking the zero crossings as the layer boundaries.

There are several styles of "Colored Inversion", so I will stick to generalities. A typical one will use a sparse-spike style deconvolution, and then match the impedance spectrum to a well log in the survey (the "coloring" step). This will suppress the side-lobe signature in the result, and give a band-limited well log look to the results. This will give a more realistic interpretation of the subsurface, but will still lack the low frequencies required for quantitative interpretation. Accurate wavelet estimation and low frequency models are where the real power of seismic inversion comes in to play for quantitative work. But it is another order of magnitude of work to do this and to do it properly requires the proper use of a lot of QC. Many companies restrict this kind of work to a specialist department, but the colored inversion can typically be done by an interpreting geophysicist.

Ajay Badachhape, PG

Senior Reservoir Geophysicist at Global Geophysical Services

Rajesh, actually, I would say that most colored inversions take an average of the spectra of the impedance logs (Vp * Rhob) and compute an inverse filter (operator) that is applied to the seismic. The seismic is thus shaped to match the average impedance log spectra. One thing that many do not know or seem to notice is that trace integration necessarily reduces your bandwidth and therefore, the resolution. To alleviate this, it might be advantageous to simply rotate your seismic to -90 degrees (you need to have a good estimate of the phase of your data, whether by careful wavelet extraction or other methods). As Erik points out, neither of these really provides good estimates of the low frequencies which are necessary for a good full-bandwidth model of the geology in the subsurface. These are quick and dirty, techniques and pretty much any geoscientist can do them.

Mike Currie

President and Principal Consultant at Predictive Geoscience Solutions LLC

Rajesh, either method is a stop gap measure at best for precisely the reason that both previous respondents have said; no low frequency information in the model. And, as Ajay, pointed out, resolution suffers. My personal preference is very much the opposite direction - bandwidth extension (either post-stack or pre-stack), followed by the appropriate model-based inversion. Building a low frequency model (assuming you have well control) and careful wavelet extraction are not major issues given the plethora of good software out there. So, while colored inversion and/or trace integration may give you a quick answer, they are virtually guaranteed to be very limited in usefulness.

REDUCE UNCERTAINTIES IN RESERVOIR PREDICTIONS USING SEQUENCE STRATIGRAPHY AND SEISMIC INVERSION

QUESTION 105

How can one reduce uncertainties in reservoir predictions using sequence stratigraphy and seismic inversion?

Michael Nwosu

Geologist

ANSWERS PROVIDED BY INDUSTRY EXPERTS

Dr. Ali Jaffri

Geological Consultant at Applied Stratigraphix LLC

Michael, sequence Stratigraphy can be a predictive tool if you use it correctly (it is also super easy to mess up). I am assuming you only have seismic data. In which case either do the following steps yourself, or have someone help you:

- Perform basic structural analysis on your seismic line. You should always do this before sequence analysis because reflections that are offset by faults appear to create reflection terminations which you will use to mark key surfaces.
- Mark reflection terminations such as on-lap, down-lap, top-lap, off-lap and erosional truncation. These will help you pinpoint your sequence boundaries and maximum flooding surfaces.
- Based on the external geometry and internal reflection configuration (amplitude, continuity, and configuration) divide similar looking packages into seismic facies.
- Interpret seismic facies and key surfaces to figure out:
  (a) Whether you have carbonates or siliciclastics
  (b) Whether you are on the shelf or basin, or in the case of carbonates your platform morphology (ramp vs isolated platforms etc)
  (c) Gross Depositional Environments (GDE)
- Make horizon slices and strata slices, and then choose an attribute that works well for that GDE.

QUESTION 106

I am running CSSI using statistical wavelets around wells (average). My angle stacks are quite balanced therefore I expect to see good resolution (band width 8-70 Hz). The results however are narrower in bandwidth (8-50 Hz) and I can see big residuals for all stacks. I applied high pass filter to the data and I was able to see geological features at 60 Hz. My question is why the CSSI algorithm did not accept high frequencies although they are useful (not noise)? And also, how does CSSI determine the noise threshold in the minimization function since it cannot be controlled by the user?

Samer Hafez

Student at King Saud University

ANSWERS PROVIDED BY INDUSTRY EXPERTS

Haider Alabdulaal

Samer, CSSI should honor those high frequencies if they are a contribution to signal. I would suggest revisiting your wavelet first and increasing your syn-matching in that frequency range. Then, you may run a stable inversion. Also, try to optimize the inversion parameters. CSSI usually iterates around your modeled to determine the inversion; hence the accepted threshold. Most commercial software such as Jason or H-R have their own functions to determine the threshold.

You can also adjust the smoothness of your model to produce low noise vs. high residual section.

Ruth Kurian

Project Geoscientist at CGG

Samer, you should revisit your wavelet. In Jason, there are QC's applied when you are estimating your wavelet which check the frequency spectrum in the wavelet vs the seismic. Also, if the amplitude was wrong in any frequency range (underestimated in high frequencies) you would still have residuals.

Theron Edwards

Kuwait Oil Company

Samer, I agree that a low frequency wavelet is the most likely culprit; however, inversion enhances low frequencies because it is similar to integration. The spectrum of a log should not be flat.

Samer Hafez

Student at King Saud University

Thank you for your comments and suggestions. I have checked the wavelet spectra and they do match the seismic and this is expected as they are statistical. However, I am still losing the high frequencies.

Gautam Sen

Consultant at Sahara Group

Samer, coherent noise could be the culprit. Does it be helpful if you carry out a tau-p transform and use velocity filters and then redo the inversion to check if high frequency residuals still persist?

Robert McGrory

Independent Geophysical Consultant

Samer, something else to consider is the data itself. Residuals at higher frequencies may also be caused by some un-removed coherent noise (multiples, peg-legs, inter-bed etc.). They would live in the higher frequencies and not be accounted for in your initial model but would be seen in the residual. The strategy to use would be to take a well and estimate a pre-stack synthetic gather that matches your acquisition geometry and compare the gathers visually. Do both the synthetic and the real data look the same? If not, why? Try running a full elastic synthetic gather using the full Zoeppritz equations with multiples and converted waves left in and compare. Wavelets can be the cause of residuals especially if the wrong spectrum is used and not extracted. However, my experience suggests that problems with data are more often than not the culprit. Another cause not yet discussed is an improperly built initial model where wells were not properly tied to the data. There could be all sorts of reasons for that. For instance improperly applied check-shot data, datum of the seismic and check-shot switch are not the same. These are just some other avenues for you to consider if the wavelets are not the culprit.

SECTION

3

Pore Pressure Prediction

CHAPTER

PORE PRESSURE AND PORE PRESSURE GRADIENT

9

CHAPTER OUTLINE

Practical Solutions to Integrated Oil and Gas Reservoir Analysis. http://dx.doi.org/10.1016/B978-0-12-805464-2.00009-3

Pore pressure is one of the most important parameters in designing a drilling plan and in geomechanical and geological analyses. Pore pressure is the pressure of the fluid in the pore spaces of porous formations. Pore pressure varies from hydrostatic pressure to severe overpressure (48%–95% of the overburden stress). If the pore pressure is lower or higher than the hydrostatic pressure (normal pore pressure), it is called abnormal pore pressure. When pore pressure exceeds the hydrostatic pressure, it is called overpressure. The fundamental theory for pore pressure prediction is based on Terzaghi's and Biot's effective stress law (Biot, 1941; Terzaghi et al., 1996). This theory indicates that pore pressure in the formation is a function of total stress (or overburden stress) and effective stress. The overburden stress, effective vertical stress and pore pressure can be expressed in the following form:

$$p = \frac{(\sigma_v - \sigma_e)}{\alpha} \tag{9.1}$$

where p is the pore pressure; σ_v is the overburden stress; σ_e is the vertical effective stress; and α is the Biot effective stress coefficient. It is conventionally assumed $\alpha = 1$ in a geopressure situation. Pore pressure can be calculated from Eq. (9.1) when one knows overburden and effective stresses. Overburden stress can be easily obtained from bulk density logs, while effective stress can be correlated to well log data, such as resistivity, sonic travel time/velocity, bulk density and drilling parameters (e.g., D exponent). Fig. 9.1 demonstrates the hydrostatic pressure, formation pore pressure, overburden stress and vertical effective stress with the true vertical depth (TVD) in a typical oil and gas exploration well.

The pore pressure profile with depth in this field is similar to that in many geologically young sedimentary basins where overpressure is encountered at depth. At relatively shallow depths (<2000 m), pore pressure is hydrostatic, indicating that a continuous, interconnected column of pore fluid extends from the surface to depth. Deeper than 2000 m the overpressure starts, and pore pressure increases with depth very rapidly, implying that the deeper formations are hydraulically isolated from shallower ones and the surface. By 3800 m, pore pressure approaches a value close

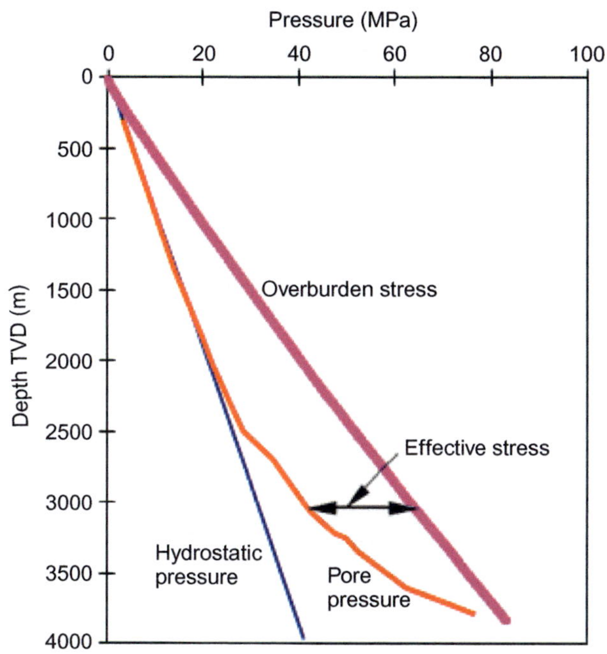

FIG. 9.1

Hydrostatic pressure, pore pressure, overburden stress, and effective stress in a borehole.

to the overburden stress, a condition referred to as hard overpressure. The effective stress is conventionally defined as the subtraction of pore pressure from overburden stress (refer to Eq. 9.1), as shown in Fig. 9.1. The increase in overpressure causes reduction in the effective stress. The pore pressure gradient is of more practical concern in drilling engineering, because the gradients are a convenient way to determine mud weight (or mud density), as shown in Fig. 9.2. The pore pressure gradient at a given depth is the pore pressure divided by the TVD. The mud weight should be appropriately selected based on pore pressure gradient, wellbore stability and fracture gradient prior to setting and cementing a casing. The drilling fluid (mud) is applied in the form of mud pressure to support the wellbore walls, to prevent influx of formation fluids into the wellbore and to prevent wellbore collapse during drilling. To avoid fluid influx, pressure kicks, and wellbore instability in an open hole section, a higher mud pressure than the pore pressure is needed. However, when mud weight is higher than the fracture gradient of the drilling section, it may fracture the formation, causing mud losses or even lost circulation. To prevent the wellbore from hydraulic fracturing when a high mud weight is needed because of overpressuring, casing needs to be set to protect the overlying formations from fracturing, as illustrated in Fig. 9.2.

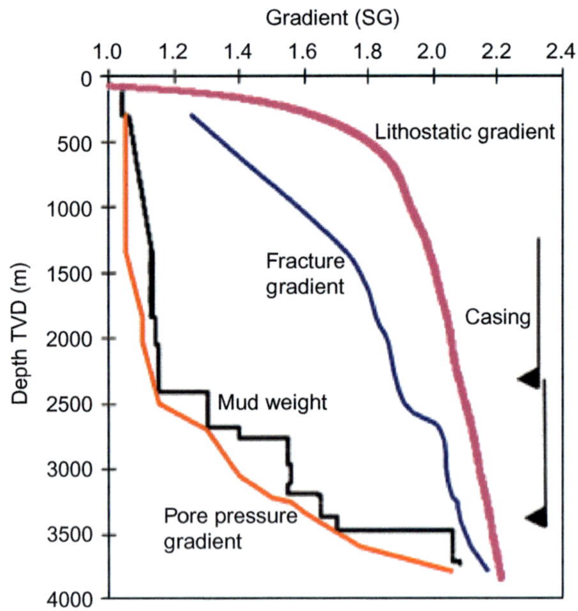

FIG. 9.2

Shows pore pressure gradient, fracture gradient, overburden stress gradient (lithostatic gradient), mud weight, and casing shoes with depth. In this figure the pore pressure and overburden gradients are converted from the pore pressure and overburden stress plotted in Fig. 9.1.

Pressure gradients and mud weight are expressed in the metric units of g/cm^3 (also called specific gravity or SG) in Fig. 9.2. Pressure gradients and mud weight are often reported in English or American units in the oil and gas industry. The pressure gradient conversions between American and metric units are listed in Table 9.1.

Table 9.1 Conversions of pressure gradients in American and metric units

Conversions	Conversions
1 g/cm^3 = 9.81 MPa/km	1 ppg = 0.051948 psi/ft
1 g/cm^3 = 0.00981 MPa/m	1 ppg = 0.12 g/cm^3
1 g/cm^3 = 1 SG	1 ppg = 0.12 SG
1 MPa/km = 0.102 SG = 0.102 g/cm^3	1 ppg = 1.177 MPa/km
1 MPa/km = 1 kPa/m	1 ppg = 1.177 kPa/m
1 g/cm^3 = 8.345 ppg	1 psi/ft = 19.25 ppg
1 g/cm^3 = 0.4335 psi/ft	1 psi/ft = 2.31 g/cm^3
1 SG = 8.345 ppg	1 psi/ft = 22.66 MPa/km
1 SG = 0.4335 psi/ft	1 psi/ft = 2.31 SG
1 SG = 62.428 pcf (lb/ft^3)	1 ppg = 7.4805 pcf

Pore pressure prediction from well logs: Methods, modifications, and new approaches by Jincai Zhang.

Pore pressure analyses include three aspects: predrill pore pressure prediction, pore pressure prediction while drilling, and postwell pore pressure analysis. The predrill pore pressure can be predicted by using the seismic interval velocity data in the planned well location in addition to geological, well logging and drilling data in the offset wells. The pore pressure prediction while drilling mainly uses the logging while drilling (LWD) measurement while drilling (MWD), drilling parameters, and mud logging data for analyses. The postwell analysis analyzes pore pressures in the drilled wells using all available data to build a pore pressure model which can be used for predrill pore pressure predictions in the future wells.

MAJOR CAUSE OF OVERPRESSURE

Overpressures can be generated by many mechanisms, such as compaction disequilibrium (undercompaction), hydrocarbon generation and gas cracking, aquathermal expansion, tectonic compression (lateral stress), mineral transformations (e.g., illitization), and osmosis, hydraulic head, and hydrocarbon buoyancy (Swarbrick and Osborne, 1998; Gutierrez et al., 2006). In nearly all cases where compaction disequilibrium has been determined to be the primary cause of overpressuring, the rocks are geologically young. Examples of areas where compaction disequilibrium is cited as the primary reason for abnormal pressure include: the U.S. Gulf Coast, Alaska Cook Inlet, Beaufort Sea, Mackenzie Delta, North Sea, Adriatic Sea, Niger

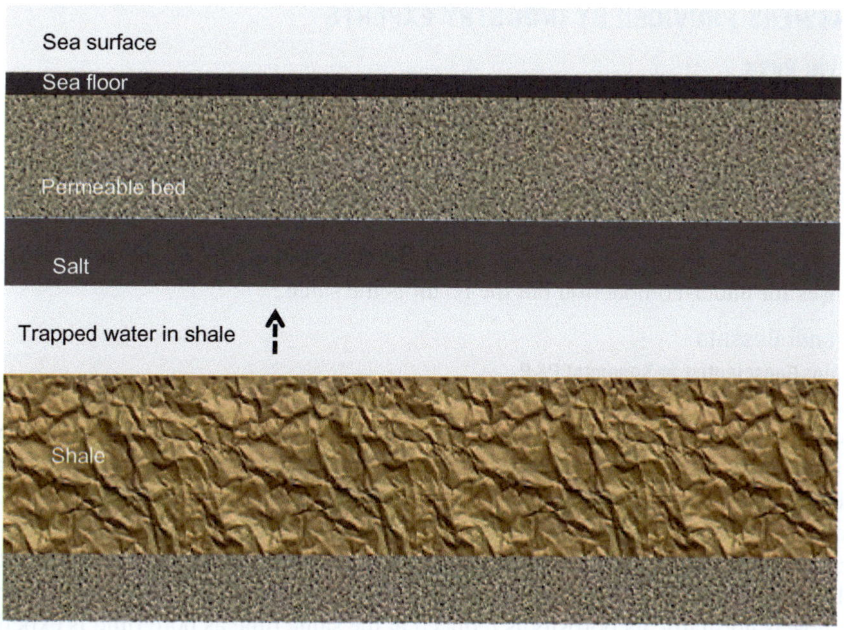

FIG. 9.3

Conceptualized cause of overpressure.

Delta, Mahakam Delta, the Nile Delta, Malay Basin, Eastern Venezuelan Basin (Trinidad) and the Potwar Plateau of Pakistan (Powley, 1990; Burrus, 1998; Heppard et al., 1998; Law and Spencer, 1998; Nelson and Bird, 2005; Morley et al., 2011). In these areas, the abnormally pressured rocks are mainly located in Tertiary and late Mesozoic sedimentary formations, the depositional settings are dominantly deltaic and the lithology is dominantly shale.

Under normal conditions, the porosity of shale decreases with depth while the density increases. But if it happens that the shale succession is interrupted by an impermeable layer (salt, for example), then the water in the pores in the shale cannot be squeezed back into the sea (as in a marine/offshore environment). In other words, when low permeability sediments are buried rapidly, water cannot escape quickly enough for compaction to occur. The water is therefore trapped in the pores and resists their closing. The water is now contributing to the support of the overburden rather than the grain structure of the rock; it is overpressured, holding the grains apart and therefore keeping the density of the overpressured zone to lower than normal.

QUESTION 107

What is the meaning of under-compaction of shale and how can we distinguish it on seismic sections?

Raouf Nasr

Geophysicist

ANSWERS PROVIDED BY INDUSTRY EXPERTS

Silviu Rusu

Wellsite Geologist-Consultant at RTimeGs

Raouf, shale compacts as it is buried deeper and deeper by sediments. Normal compacted shale is assumed to have a pore pressure equal to the hydrostatic pressure of a water column of equal height to the burial depth of the shale. So, under-compacted shale has bigger porosity than a normal compacted one. As a result the bulk density of the shale will decrease so waves will travel more slowly etc. There are many causes for under-compaction but the result is the same.

Ngondi Bessala

Senior Geoscientist at Sonangol P&P

Raouf, compaction is the reduction in volume of sediment due to compression and the first stage of which is marked by reduction in pore volume. Generally, compaction is due to burial and is the result of the mechanical action of the formations and the weight of overlying sediments accumulated during subsidence. In order to understand these processes more readily, it seems useful at this point to restate some of the pertinent aspects of clay mineralogy. The simplest clay mineral formula is written Al2 [Si4O10] (OH)2. Clay minerals therefore hold interstitial water. So when shale compacts normally, its porosity is reduced at the same time as pore fluid is expelled. During burial, increasing overburden is the prime cause of fluid expulsion. If the rate of sedimentation is slow, normal compaction occurs, that is to say that equilibrium between increasing overburden and the ability to expel fluids is maintained.

To the contrary, when the sedimentation rate is greater than that which allows clays opportunity for dewatering, abnormal pressure occurs.

To simplify, under rapid subsidence and reduced drainage capacity, clay minerals will not have time to expel a great part of their water and interstitial fluid. This excess fluid ends up supporting part of the overburden and there is under compaction.

In other words this is translated by Terzaghi's formula

$$S = P + \sigma$$

Where:

S = overburden pressure (or total stress)
P = formation fluid pressure
σ = pressure supported by the matrix (or effective stress)

In compacted conditions, rock fluids are not supporting the overburden and there is normal pressure. But when the interstitial fluids support part of overburden pressure there is abnormal high pressure.

Geoffrey Thyne

Principal Technologist at Esal

Raouf, to throw more light to the comments above, the water in clay is structural not interstitial. The swelling clays (smectite) do undergo a chemical transformation into illite with burial. This transformation does release a small amount of water into the pore spaces over the transition window (80 to 120C). The original idea was proposed by Burst (1969) for Gulf of Mexico (GOM) clays and has not proven out as the cause of overpressure. It is more likely that compactional disequilibria as mentioned above are the main cause. Another possible contributing factor is gas generation during kerogen maturation. This is why looking for under-compacted shale can be used to prospect for a good gas play.

QUESTION 108

How can I recognize the causes of geo-pressure and how can I determine which pore pressure prediction method to use?

Francisco Velez

wellsite geologist at PDVSA Petroleos de Venezuela S.A

ANSWERS PROVIDED BY INDUSTRY EXPERTS

Julian Coker

WSG/PP

Francisco, the simple rule of thumb is if it is a tertiary basin with clastics (in particular CLST) you can use either vertical or horizontal methods. If not, then you can only use horizontal methods. Which, is a roundabout way of saying always use a horizontal method like Eaton, and if you are lucky you might be able to use, say, Millers/Bowers/Equiv Depth. If you can't use vertical methods, use multiple trend lines in Eaton, and then I would assume that the PP is due not only to undercompaction. Depending on the basin, HCO generation, clay diagenesis, etc., you need to know the Temperature/Total organic content/clay type.

Oops, I was very non-pc by saying 'break' the trend line, I should have said use multiple trend lines.

Steve Hobart

Sr. Pore Pressure Advisor

Francisco, depending on the experts you consult, you will find that geopressure generation mechanisms are classed in different ways. For myself, I generally consider three classes of mechanism:

(1) External squeezing of the pore space, either by the sole action of increased overburden stress (the most commonly recognized mechanism) or by the composite action of overburden stress in conjunction with lateral tectonic compression
(2) Permeability-related mechanisms which include centroid effects and hydrocarbon buoyancy effects.
(3) Mechanisms associated with the microenvironment of the pore space, often, but not always, geochemical in nature: oil generation, gas generation, aquathermal pressuring, diagenesis of clay and other minerals, hydrate dissolution, osmosis and others which have generally not been of much interest to the petroleum community.

The prediction method, as you already suspect, should be appropriate to the mechanism(s) involved.

For external squeezing, there are numerous techniques available based on Terzaghi's definition of effective stress. These techniques generally require an implicit relationship between porosity and effective stress. Porosity tends to be reduced as effective stress increases. This compaction process is primarily due to consolidation of the grains into a more compact packing. The relationship between porosity and effective stress can be complicated by microfracturing and breakage of the grains, pressure solution, dissolution and precipitation processes as well as the compressibility of the fluids and solids. However, the effect of these processes at the macro scale is generally assumed by practitioners to be captured by mineralogically-dependent "normal compaction trend lines" or by empirical "calibration parameters". Proxies for porosity which are commonly used include resistivity, velocity, sonic and drillability (D-exponent, Dxc). Methods published by Eaton and Bowers as well as the Equivalent Depth method, are among the most popular of the approaches which assume that vertical squeezing is the geopressure generating mechanism. When porosity is not primarily a function of effective stress, such as in many environments containing carbonates, the state of the art is such that these empirical methods are not expected to be as robust. Further complicating the use of these methods are the non-porosity influences on the indicators, e.g. temperature, salinity, hole conditions, varying mineralogy, depth of investigation, etc.

For "permeability-related" mechanisms, knowledge of the fluid density (or densities, when more than one fluid of type is involved, the fluid column height(s) and at least one pore pressure measurement within the hydraulically connected system will suffice to predict pore pressure distributions. Although simple to execute mathematically, the more challenging aspects of this approach include identifying fluid contact depths, establishing the extent or presence of hydraulic connectivity and dealing with formations which have been subjected to fluid extraction, usually hydrocarbon production. In this latter situation, since the pressure equilibrium which is presumed to have existed prior to production has been disturbed, a reservoir modeling approach may be required.

For the mechanisms associated with the pore's microenvironment, also sometimes called "secondary pressuring mechanisms", an explicit predictive approach based on the geochemistry is generally not attempted due to numerous unknowns. Since these mechanisms reduce the grain-to-grain stress, they are also often lumped together as "unloading" mechanisms. It may also be the case that several of these mechanisms are in play simultaneously with the relative contribution of each mechanism not being well constrained. For the present, many of the practitioners who address the influence of these secondary pressuring mechanisms adopt an empirical approach based on modifications to the Eaton methods (e.g. changing the normal compaction trend or increasing the exponent), or using Bowers' suggested modification of his "virgin compaction relationship". Other methods also exist. These various secondary mechanisms may have different effects on the relative amounts of solids and fluids present in a reference volume of rock, and consequently may affect porosity in different ways. Within geologically related formations, the assumption is made that the empirical, "calibrated", adjustments that work in an offset well will also work at other nearby locations.

The bottom line is that the local geology will determine which mechanisms are in play and which methodologies will work best.

Jorge Reveron

Senior Geophysicist at Repsol

Francisco, you can consult the paper SWARBRICK, R., 2012, Review of pore pressure prediction challenges in high – temperature areas: The Leading Edge, no. 11 p. 1288-1294, http://dx.doi.org/10.1190/tle31111288.1. It has a complete description of abnormal pore pressure mechanisms. You can also check the UCV undergraduate thesis: GENERACIÓN DE UN VOLUMEN DE GEOPRESIONES DEL CAMPO FRANQUERA, ESTADO TRUJILLO, author: Jesús Andrés Borges, 2013.

Mark Tingay

Senior Geomechanics specialist at Chevron

Francisco, you can usually work out the origin(s) of overpressure in your basin by first understanding the geology and from this, which mechanisms are most likely.

This can then be augmented by some petrophysical analysis, particularly vertical effective stress-sonic velocity cross plots and density-sonic cross plots. The main overpressure mechanisms all have different signatures that you can sometimes tease out on these cross-plots. But the key is still the geology - young basin or old basin? How much recent burial? Is there any recent tectonic compression? What is the present-day stress regime? Where is the overpressure located (geographically and stratigraphically)? Is overpressure associated with source rocks? What temperatures are overpressures found at?

The Swarbrick paper above is a good one for understanding different mechanisms. Webster et al., 2011 in Geofluids is another excellent reference. You can also look at my 2013 AAPG Bulletin paper on overpressures in the Malay Basin as an 'example recipe" for determining overpressure origin in a region.

Regarding an actual method; this depends on overpressure origin and available data. The main thing is to think about what you are doing and put it in geological perspective. There is really no method out there that always works well or that is theoretically correct to use but some will give useful results.

QUESTION 109

Why is it that pore pressure is estimated in shale but measured in sand"? Is there any geological/geophysical reason for that?

Benedict Amobi-madu

Research Geoscientist at SPDC of Nigeria

ANSWER PROVIDED BY INDUSTRY EXPERT

Julien Martineau-Fabre

Sr. RT Geomechanics Eng.

Benedict, I will try to be clearer regarding my points (estimate PP in shale vs. measure PP in sand). Most of our techniques to estimate PP rely on the shale losing its water in one hand, and losing some "porosity" on the other hand. To develop, the theory behind this is double pronged: the shale loose water while being compacted. Theoretically, at great depth shales should have an infinite resistivity, as they should have expelled all the water from when they are deposited. Any change in the "normal compaction trend" would mean the shales do not expel all their water content, hence are not "de-watered" properly. You estimate your PP based on potential of un-expelled water within the formation (actual values vs. expected value, using Eaton's method). From the seismic (and the acoustic) side, the theory is: at great depth, shales should not exhibit any sign of porosity. Any change in the "normal compaction trend" would mean that the shales exhibit some "residual" porosity (in the sense it should not, as the porosity decay should increase with depth). As above, you estimate the PP based on the presence of "porosity". Also, PP is not measured in shales, as no customer would deploy a pressure measurement tool to acquire PP, as you need (1) porosity and (2) permeability. As such, PP is measured from potential reservoir zones, as they should exhibit porosity and permeability.

QUESTION 110

Why is it that almost all pore pressure calculations are based on the predictability of shale and clay rocks, not sandstone, siltstone, marl, limestone and other rock types?

Ekwereubong Ini Jonah

Petroleum Geoscientist

ANSWERS PROVIDED BY INDUSTRY EXPERTS

Steve Hobart

Sr. Pore Pressure Advisor

Jonah, different mineralogies compact differently under the same effective stress. At initial deposition they will have different initial porosities. During subsequent burial, the change in porosity in response to changes in effective stress will also be different. The response of porosity-sensitive indicators (e.g. Resistivity, Sonic, Velocity, Dxc) to changes in effective stress will therefore be different in different rocks. Compaction parameters (e.g. the Eaton or Equivalent Depth normal compaction trendlines, Bowers' A and B parameters, Miller's Vma and lambda) will also be different in different mineralogies. In principle, if you have a compactible lithology and can mathematically model its porosity response to changes in effective stress, you can use it for estimating effective stress according to Terzaghi's principle (assuming you know the overburden stress, of course).

Quartzitic sands are made of very durable grains and are resistant to deformation and compaction. Over a wide range of effective stresses, the porosity in a sand of consistent composition is not going to change as much as a claystone/shale undergoing the same stress changes. Thus, porosity-sensitive indicators in sands are not as sensitive to changes in effective stress as are they are in shales. Formations containing calcium carbonate are affected by the solubility of the mineral. Due to the fact that it readily dissolves and (re-)precipitates due to changes in water properties, the response of porosity-sensitive indicators in carbonates can be inconsistent due to this factor alone. Thus, Resistivity, DT, etc., in carbonates may better reflect the geochemical history of a formation than its stress history. The fact that porosity development and preservation in carbonates can also be affected by characteristics of the initial material further clouds any potential relationship between porosity and effective stress in carbonates (note the many different kinds of porosity in carbonates).

Shales/claystones are convenient because:

- They are very common
- They are relatively inert geochemically-speaking (smectite-illite-kaolinite diagenesis notwithstanding)
- They have a wide range of porosity over the range of effective stresses encountered
- They are relatively easy to identify on petrophysical logs.

Not all "shales" are alike. Each of these formations can be expected to have different porosity responses to changes in effective stress: shaly sands, sandy shales, silty shales, volcanic tuffs. Good shale-point picking will ensure that the compaction parameters you are using will be calibrated to a mineralogically consistent, compactible lithology.

Richard Swarbrick

Managing director at Swarbrick Geopressure consultancy Limited

Steve that is a fantastic summary of the state of play on the issue and it is useful to the entire pore pressure community, as a reminder of what works well for pressure prediction (i.e. shales/clays) and what does generally not (i.e. sandstones; carbonates). Although there are some exceptional cases I have seen where both sandstone and chalk with early oil emplacement do appear to show a meaningful relationship between porosity and (vertical) effective stress but that is phenomenally rare.

And the reminder that not all shales are the same is very helpful too.

Ayman Khorshid

Subsurface Geologist team & Senior Operation Geologist

Jonah, clay minerals group are the one that have tendency for pressure accumulation. There are different methodologies used to predict sigma for other sedimentary rock by establishing a normal pressure trend for every rock type separate from the others.

QUESTION 111

What are the limitations/disadvantages of using regional geologic information to model pore pressure for an exploratory well, where there are no offset well data?

Divine Somiari

Experienced Drilling Engineer

ANSWERS PROVIDED BY INDUSTRY EXPERTS

RM Iman Argakoesoemah

Senior Petroleum Exploration Geologist

Divine, the accuracy is much less than if there is a well control. You could optimize the use of seismic lines located nearby your proposed exploration well. If you could correlate any seismic events regionally and tie it to a well even though it is far, far away (at basin scale), it would be better as long as you feel confident to follow the events. It is important to bring and establish the extent of formation objectives into your area of interest. Any seismic events you interpret should be tied to the existing regional stratigraphic column to ensure the presence or completeness of formations in your area. Pay attention on any specific dominated rocks in the formation that could be "correlatable" with the seismic responses. For example: a thick shale (soft) over a thick carbonate (hard), etc. You may want to run seismic modeling to predict lateral and vertical extent of the formations, and then use software to estimate the pore pressure. Hence, you need to ensure the presence of any possible stratigraphic contacts caused by tectonics or depositions, i.e large unconformity, significant SB, MFS, etc.

Selim Simon Shaker

Consulting Geoscientist at Geopressure Analysis Services

Divine, there is no limitation of using regional geology, always there is advantage.

In addition to RM Iman Comment, I might add the importance of regional geology, even in case of missing offset, from seismic only:

(1) Deposition thickness/subsidence/environment (Sv). Subsidence increase thickness, temperature and Sv.
(2) Basin Accommodation (Sh max/Sh min orientations). Sh max tends to coincide with depositional/compressional axis
(3) Tectonics (diapirism, overthrust) and, is PS (principal stress) > OB or not ??). Active Diapirism increases the principal stress and sometimes it is larger than the OB. Overthrust turns the PS and Sh min 90 degrees,
(4) Sequence Stratigraphy is very important for guessing reservoir Seal or Breach. Read Shaker (2004) "Trapping vs. Breaching due to salt-sediment interaction" GCAGS 2004.
(5) Faults type is a good predictor of reservoir communication and P transgression vs. regression. Buried and antithytic faults have more sealing capacity than active and synthetic faults.
(6) No.4 and 5 give you a good estimate for the disparity between the PPP shale vs. PP sand.
(7) Running velocity profiles to predict the Top of Geopressure. Use fast track RMS. It is easy and cheep to process.
(8) Contingent on the proposed location, more information can be obtained and used to enhance the PPP and predict the challenging spots on the well trajectory

Anjanava Das Purkayastha

Geomechanist-Pore Pressure Prediction, Shell

Divine, there is no disadvantage in using regional geology in PP-FG prediction. It is always advantageous. I think, your question was more related to the level of uncertainty rather than limitations. Yes, in case of no offset wells - pure wildcat exploration scenario you will have high level of uncertainty in all the modeling stages of PP-FG prediction. In this case, you will only have seismic data to go by. Mr. Selim Shaker and RM Iman have already pointed out the ways on which you can start working. Basin Modelling will be the best approach to start with. Also, sequence stratigraphic analysis will be of utmost importance. Also, you can look for regression models between velocity and formation pressure from analogous basins and use that as another scenario. In fact in such a wild cat scenario, understanding of geology will be more helpful than anything else. The credibility of Regional geology also lies in the hands of the interpreter.

QUESTION 112

Which techniques will be appropriate for a given situation in pore pressure prediction using Geophysical methods? A standard approach for pressure prediction is to use conventional stacking velocity analysis and convert the stacking velocities to Dix-Equation-corrected interval velocities. Beyond this simple approach to velocity analysis lies a range of more sophisticated techniques including horizon keyed velocity analysis, refraction and reflection tomography, post-stack inversion and pre-stack inversion. These techniques can increase the accuracy of the velocity analysis but require additional analysis and processing, often at higher cost.

Ekwereubong Ini Jonah

Petroleum Geoscientist

ANSWERS PROVIDED BY INDUSTRY EXPERTS

Julio Laredo

Senior Geomechanics Consultant At Halliburton

Jonah, complex techniques can be expensive and need enough data from the field which are not always available. Even with the best interval velocity, it's not a rule that abnormal changes at depth are related to abnormal pore pressure. Before start with Vi inversion for pore pressure prediction, a geomechanical evaluation is important and would be the first step.

Shundong He

Development Geologist at Chevron

Jonah, there are so many different kinds of seismic velocity, and so many different ways to build and calibrate the velocity models. Selecting a fit-for-Pressure-analysis velocity depends on:

- The understanding of lithology, tectonic history, and geomechanics
- The calibration of the velocity model
- Normal velocity trend model
- Assumptions of vertical and horizontal stress ratio
- Also the offset wells information.

If possible, the best velocity for pressure interpretation, in my opinion, will be the velocity model for time and true depth conversion.

Selim Simon Shaker

Consulting Geoscientist at Geopressure Analysis Services

Jonah, as Julio and Ashundong mentioned, there are some factors beyond pore pressure that can affect Vi (Interval velocity) changes. However the inexpensive corrected Vi is the backbone of pore pressure prediction before drilling and these factors should be considered:

- The normal moveout and RMS velocities can only used for a quick look and regional assessment. However, interval velocity (Vi) is the velocity recommended for pore pressure prediction for 1D and 2D interpretations.

- Quality control displays such as NMO-corrected gathers should be used to check the accuracy of velocity picks. Seismic data must processed for de-multiple, DMO (dip move-out) and pre-stack migration. Converting the time – velocity pair to depth - velocity pair probably is the first objective and they need to be calibrated against check-shot, velocity survey and/or sonic logs from offset wells.
- Outlier velocity values representing calcareous beds, salt inter-beds; gravel, thick sands and pay zones should be eliminated from the paired data sets (depth-Vi).
- Geologist/geophysicist/processor (velocity pickers), in a coordinated fashion, using one of the software packages must pick the right velocity that fits the horizon curvature on the stacking velocity semblance display.

Chapter # 18 in the AAPG memoir 76 by David W. Bell "Velocity estimation for PPP" is a simple and a great to read. Also there is a good article by Chopra and Hoffman in TLE, 2006 pp. 1502-1515 "velocity determination for PPP."

J. Grant Brydle, P. Geoph.

Interpretation Geophysicist

Jonah, the higher cost and time required for tomographic velocity analysis is well worth it to derive accurate results and prevent a blow out like BP's Macondo well. If you consider the extra costs of a well disaster including fines, liability lawsuits and cleanup costs, the extra data processing costs are very minor. If your goal is not the highest accuracy possible, then why even bother doing the pore pressure prediction at all.

I've commented previously on the risks of using Dix-derived interval velocities. They are often very erroneous, as Dix's formula has several inherent assumptions which frequently do not fit your project's geological environment.

QUESTION 113

I have the results of my pore pressure predictions. How do I validate what I have calculated? Does it have something to do with reservoir pressure or est. pore pressure from pressure log?

Bobby Siswanto

Geophysicist at ConocoPhillips Indonesia

ANSWERS PROVIDED BY INDUSTRY EXPERTS

Julien Martineau-Fabre

Sr. RT Geomechanics Eng.

Bobby, assuming you have computed pore pressure from porosity sensitive curves (Seismic, DXC, RES, ACO), it is important you note that you estimate pore pressure (PP) in shale, while you measure PP in sand. As such, your validation can use well indicators (gas, PUSO, cavings and PVT) and extra measurements. With more details, PP can be better constrained in Shale based on gas events (NB with ECD/ESD), cavings and PUSOW. No direct measurement will tell you the actual PP regime. Within a potential reservoir, PP can also be estimated from well events (influx vs. ESD or ECD), as well as gas events (again, check ESD/ECD). But a direct measurement will give you the exact PP. Moreover, we would tend to assume the PP from Shale=PP from Sand at the formation boundary. This is assuming there is equilibrium between the two formations. But quite often, this assumption can be wrong, especially in case of dipping structure, with buoyancy effect or fluid transfer. It's not necessarily because you don't reach PP Shale=PP Sand that your model/prediction is wrong. Look into regional geology to see what could cause it!

Selim Simon Shaker

Consulting Geoscientist at Geopressure Analysis Services

Bobby, In addition to Julien's excellent comment, Shut in Pressure (SIP) is a good calibration tool between the ESD and the pressure differential at the shale - sand interface.

Ngurah Beni Setiawan

Geomechanics (petro Technical Expert) at Schlumberger

Bobby, if you are using seismic interval velocity to predict the pore pressure then I would also recommend you to extract interval velocity at (at least) 5 different locations nearby your primary location. You can apply the same parameters to those interval velocities to ensure consistency of your predicted pore pressure. Then, as explained by Julien and Selim, you can calibrate your predicted pore pressure using all drilling experiences related to pore pressure and any available pressure related measurement. Please always keep in mind that your predicted pore pressure is a SHALE-based pore pressure (as Julien mentioned above).

Bobby Siswanto

Geophysicist at ConocoPhillips Indonesia

Beni, right now, I am making the velocity model. Yes, exactly the same that you've explained. I apply the same parameter to extract interval velocity at 5 different (well) locations nearby my primary location. But I have a question. I know it's not quite good, but can I use only one interval velocity for several deeper layers? It is because it is only this well which contains p-wave data and also reaches deeper depth.

Ngurah Beni Setiawan

Geomechanics (petro Technical Expert) at Schlumberger

Bobby, sometimes (and even most of the times) we have to generate something from whatever we have. Even though it is probably not the best cookies in town, at least you have a cookie.

The point that I want to make is that you have to understand the quality of your data. And if you do, then you know how much "uncertainty" it will bring to your model and you know the limitations of your model. If you know the uncertainty of your model, then it can be used with caution. Lastly, any model that you generate from your (limited) data should be consistent with any regional information or experience. Then the geologist is probably the best person to talk to.

Yogesh Kulkarni

Pore Pressure Analyst/ Field Engineer, Geoservices, Schlumberger Company

Bobby, it would also be helpful to study the gas behavior (gas signatures if available) from offset wells while validating PP model. It will help to separate out effect of secondary generation of PP (column height, etc.) from porosity related PP. And it should also help to reduce error while calibrating PP model as shale PP may not be always equal to pressure recorded in an adjacent sand interval.

QUESTION 114

How can I do 'Impedance Based Pore Pressure Prediction', including workflow, methodology, and type of data needed?

Benedict Amobi-Madu

Research Geophysicist at SPDC, Nigeria

ANSWERS PROVIDED BY INDUSTRY EXPERTS

Erik Johnson

Fugro Geoscience Division

Benedict, fluid pressure is a function of effective stress, which in turn is correlated to velocity. You can get velocities from seismic, but they would need to be calibrated. If you have at least one well with logs in the survey area, then you have some data to use to calibrate the seismic velocities.

Per Wylie, Fluid Pressure (FP) = Overburden Pressure (OBP) - Effective Stress (ES).

Then you can set up two equations to calculate the OBP and ES:

OBP = d1(Z-WD)^d2 + (WD*1.64), where WD = water depth and Z = depth below mudline, and

(Sonic-53us/ft)/100 = A0Z^a1ES^a2, where Z=Depth/1000ft and ES =Effective Stress/1000psi

Using multi-linear regression in the log domain, you can calculate A0, a1, and a2. d1 and d2 can be calculated from the density logs, also with a regression fit.

Solving this and back-substituting will produce a low frequency volume of Fluid Pressure. To calculate required mud weight multiply FP*6.3/Z.

Now you have established some key relationships if you want to proceed to a higher frequency result using impedance from seismic inversion. Pressure is related to velocity, not impedance, so you have to develop a relationship between velocity and density to back out the density from the impedance. Again, a well log will help with this relationship. If you can develop a geologic model/explanation as to what happens with the sonic and density contrasts as you enter and leave the reservoir zone, you will be able to make more confident pressure predictions for the well and reservoir zone you expect to encounter. If after investigating your velocity and density profiles you find that density is smooth and well-behaved, then you can apply the methods described in the SEG paper (Reference: Banik, N., Koesoemadinata, A., Wagner, C., Inyang, C., and Bui, H. (2013) Predrill pore-pressure prediction directly from seismically derived acoustic impedance. SEG Technical Program Expanded Abstracts 2013: pp. 2905-2909.). But for me, I wouldn't want to make that assumption ahead of time unless I was in the Gulf of Mexico (GOM) or other well-behaved basin.

Gerardo Clemente Martínez

Interprete de atributos sísmicos en PEMEX

Benedict, one approach to predicting pore pressure is based in the behavior of the seismic velocities in presence of over pressured shale. There, the seismic velocities will be slow, when compared with a normal compaction trend. Obviously, the seismic velocities play a very important role in this approach. To have a correct velocity model for pore pressure prediction, we used to make high density velocity picks based on semblance and performed in super gathers. But now we prefer to use tomography. The approach you are looking to work is based on seismic inversion. You may be able to derive the velocity from the impedance because the impedance is the relation between density and velocity. Remember always you will need well information to calibrate and tie the model.

QUESTION 115

Huge canyons or erosional cuts are common features in the offshore tertiary of Bay of the Bengal, India. Thorough understanding of how they will affect the pore pressure generation is of critical importance for wells targeting reservoirs beneath these geological features. I need your view and experiences on this.

Nitin Bhardwaj

Senior Operations Geologist (Geopressure and Geomechanics)

ANSWERS PROVIDED BY INDUSTRY EXPERTS

Phil Holbrook

Chief Scientist and Educator, Force-Balanced Petrophysics

Nitin, The erosion of canyons automatically decreases the overburden. This decreases the average confining load, S (ave) which is a term in the effective stress theorem. This will almost certainly put the remaining near surface sediment on the un-loading limb of a (stress/strain) curve. The un-loading limb intersects the loading limb as described in my 1995 US "unloading" PATENTS. The references in these patents will fill you in on the rest of this mechanical story.

Selim Simon Shaker

Consulting Geoscientist at Geopressure Analysis Services

Ninita, you describe a wildcat well penetrating a buried Channel/Canyon/erosional cut. There are several scenarios which can impact your pressure profile contingent on the geological and paleo-environment of the erosional cut:

A. Pressure Transgression: Mudflow fill, Bottlenecks, Abandoned distributary and meander (i.e. Ox-Bow).

B. Pressure Regression: Gravels and coarse sand fills, Up-dip channel communication.

C. Channel cut exists above the top of the geopressure (TOG) zone: possible loss of circulation if MW is higher than the pressure of the encased overbank shale/mud.

D. Channel cut exists below TOG (overpressured section): Possible strong kick associated with abandoned distributary channels (filled with course sediments) where higher pressures were entrapped.

Possible loss of circulation if erosion cut is in communication with up-dip open braided channels and faults. This is intricate subject and there needs to be a regional sequence stratigraphy/paleoenvironment mapping and study before the borehole trajectory penetrates the buried channel. Velocity analysis might not be able to predict the pressure in some of these cuts due to the ratio of wave length to the size of cuts.

Nitin Bhardwaj

Senior Operations Geologist (Geopressure and Geomechanics)

Phil, can you share with us some of your experience in such cases? What we have observed or experienced is that after the canyon/erosion cut event, the underlying sediments does not compact further even after the canyon cut is filled with new sediments.

Phil Holbrook

Chief Scientist and Educator, Force-Balanced Petrophysics

Nitin, Rock or sediment compaction is an almost purely plastic i.e. irreversible process. Further compaction does not re-start until the porosity on the unloading limb equals that of the loading limb for that particular lithology.

RELATIONSHIP BETWEEN SHALE ANISTROPY AND PORE PRESSURE

QUESTION 116

Is there any direct relationship between shale anisotropy and pore pressure prediction? Normally we apply Anisotropy correction during seismic processing.

Iftikhar Satti

Research Scholar at Petronas University, Malaysia

ANSWERS PROVIDED BY INDUSTRY EXPERTS

Colin Sayers

Scientific Advisor, Schlumberger

Satti, Two sources of anisotropy are mechanical compaction and the smectite-to-illite transition. As shales become compacted, anisotropic clay platelets become more aligned, leading to anisotropy. Since compaction is governed by effective stress, the anisotropy in overpressured areas due to this mechanism tends to be lower than in normally pressured areas, since the effective stress will be less.

A more important source of anisotropy is the smectite-to-illite transition. As smectite transforms to illite, the anisotropic illite platelets tend to form with orientation perpendicular to the maximum stress, resulting in anisotropy. The smectite-to-illite transition also leads to an increase in pore pressure, as the overburden load is partially transmitted to the pore fluid, so in this case an increase in anisotropy accompanies the increase in overpressure.

More so, if you do not account for anisotropy and calibrate a velocity or resistivity to effective stress transform using sonic or resistivity logs measured, for example in a vertical well, and then apply that transform to predict pore pressure based on sonic or resistivity logs measured in a deviated well, you will be in error. Accounting for anisotropy is not just important, it is essential.

Iftikhar Satti

Research Scholar at PETRONAS University, Malaysia

Colin, In my work I also noticed that Velocity-EStress transformations using vertical wells do not give me good pore pressure predictions when applied to deviated wells. That's why I want to know the relationship between anisotropy and pore pressure.

Phil Holbrook

Chief Scientist and Educator, Force-Balanced Petrophysics

Iftikar and Colin, Resistivity and Sonic traces do not bear loads. Minerals and fluids do. Pore fluid pressure affects everything in subsurface mechanics. Anisotropy affects the trace reading but that totally misses the fundamental mechanical point.

Go back to the effective stress theorem, (S ave) = (Sigma ave) + pore fluid pressure describes this mathematically. If you have a mechanical question, search for a mechanical solution. You're looking in the wrong domain.

Colin Sayers

Scientific Advisor, Schlumberger

Phil, We are using the effective stress principle. The reason we are interested in velocity, resistivity, etc., is that we want to estimate effective stress so that it can be subtracted from total stress to determine pore pressure. This we do by calibrating a velocity to effective stress transform using offset well data. You can also use resistivity, etc. If the transform is calibrated using data from a vertical well, for example, and applied to predict pore pressure based on sonic or resistivity logs measured in a deviated well then there will be an error in the predicted pore pressure, unless anisotropy is accounted for, hence our interest in anisotropy.

Phil Holbrook

Chief Scientist and Educator, Force-Balanced Petrophysics

Colin, you are not correct. Pore fluid pressure is physically limited by Fracture propagation pressure. Both are scalars. Fractures perpendicular to the least principal stress are opened by fluid pressure and the natural fluid pressure until the fractures close. See (Walsch, 1981) shown as Fig. 7.1 in Pore Pressure through Earth Mechanical Systems. See leakoff tests and Extended Leakoff Tests (XLOTs) of this Pore Fluid Pressure interest group. This brings the governing physics into the pore fluid pressure discussion.

Colin Sayers

Scientific Advisor, Schlumberger

Phil, Fracture propagation pressure is determined by the minimum principal stress. Except for cases in which the minimum stress is vertical, the magnitude of the minimum principal stress tends to increase with increase in pore pressure, and depends on the anisotropic properties of the rock.

Mitchell Alan

Manager New Wells, BP

Satti, from purely operational point of view shale anisotropy should be taken into account when calibrating pre-drill shale velocity estimations from seismic to realtime or post drill using sonic log values. Errors of more than 10% are common due to anisotropy - basically a seismic P wave travels vertically and a sonic measurement is horizontal (more or less).

FRACTURE PRESSURE AND FRACTURE GRADIENT

Fracture pressure is the pressure required to fracture the formation and cause mud loss from the wellbore into the induced fracture. Fracture gradient can be obtained by dividing the TVD into the fracture pressure. Fracture gradient sets the maximum mud weight; therefore, it is an important parameter for mud weight design in both the drilling planning stage and while drilling. If mud weight is higher than the formation fracture gradient, then the wellbore will have tensile failure (be fractured), causing losses of drilling mud or even lost circulation. Fracture pressure can be measured directly from a downhole leak-off test (LOT). There are several approaches to calculating fracture gradient. The following two methods are commonly used in the industry: the minimum stress method and the tensile failure method.

Minimum Stress for Lower Bound of Fracture Gradient

The minimum stress method does not include any accommodation for the tensile strength of the rock; rather, it represents the pressure required to open and extend a preexisting fracture in the formation. Therefore, the minimum stress represents the lower bound of fracture pressure. The minimum stress is the minimum principal in situ stress and typically is equal to the fracture closure pressure, which can be observed on the decline curve of a LOT following the breakdown pressure (Zhang et al., 2008). The minimum stress method, as shown in the following, is similar to the methods proposed by Hubbert and Willis (1957), Eaton (1968), and Daines (1982).

$$\sigma_{min} = \frac{\nu}{1-\nu}(\sigma_v - p) + p \qquad (9.2)$$

where σ_{min} is the minimum in situ stress or the lower bound of fracture pressure; ν is the Poisson's ratio and can be obtained from the compressional and shear velocities (V_p and V_s), and

$$= \frac{\frac{1}{2}\left(\frac{V_p}{V_s}\right)^2 - 1}{\left(\frac{V_p}{V_s}\right)^2 - 1}$$

When open fractures exist in the formation, the fracture gradient may be even lower than the minimum stress. Matthews and Kelly (1967) introduced a variable of effective stress coefficient for fracture gradient prediction:

$$P_{FG} = K_0\left(OBG - P_{pg}\right) + P_{pg} \tag{9.3}$$

where P_{FG} is the formation fracture gradient; P_{pg} is the formation pore pressure gradient; OBG is the overburden stress gradient; K_0 is the effective stress coefficient, $K_0 = \sigma'_{\min} / \sigma'_v$; σ'_{min} is the minimum effective in situ stress; σ'_v is the maximum effective in situ stress or effective overburden stress. In this method, the values of K_0 were established on the basis of fracture threshold values derived empirically in the field. The K_0 can be obtained from LOT and regional experiences.

Formation Breakdown Pressure for Upper Bound of Fracture Gradient

For intact formations mud loss can occur only after a tensile failure appears in the wellbore. In this case, fracture pressure/gradient can be calculated from Kirsch's solution when the minimum tangential stress is equal to the tensile strength (Haimson and Fairhurst, 1970; Zhang and Roegiers, 2010). This fracture pressure is the fracture breakdown pressure in LOT (Zhang et al., 2008). For a vertical well, the fracture breakdown pressure or the upper bound of fracture pressure can be expressed as follows:

$$P_{FPmax} = 3\sigma_{min} - \sigma_H - p - \sigma_T + T_0 \tag{9.4}$$

where P_{FPmax} is the upper bound of fracture pressure; σ_H is the maximum horizontal stress; σ_{min} is the minimum horizontal stress or minimum in situ stress; σ_T is the thermal stress induced by the difference between the mud temperature and the formation temperature; and T_0 is the tensile strength of the rock. Neglecting tensile strength and temperature effect and assuming σ_H is approximately equal to σ_{min}, Eq. (9.4) can be simplified to the following form:

$$P_{FPmax} = 2\sigma_{min} - p \tag{9.5}$$

Substituting Eq. (9.2) into Eq. (9.5), the upper bound of fracture pressure can be expressed as:

$$P_{FPmax} = \frac{2v}{1-v}\left(\sigma_v - p\right) + p \tag{9.6}$$

Comparing this upper bound of fracture pressure (Eq. 9.6) to the lower bound of fracture pressure (Eq. 9.2), the only difference is the constant in front of effective vertical stress $(\sigma_v - p)$. This upper bound of fracture gradient can be considered as the maximum fracture gradient, or the bound of lost circulation (Zhang et al., 2008). Therefore, the average of the lower bound and upper bound of fracture pressures can be used as the most likely fracture pressure, i.e.,

$$P_{FP} = \frac{3v}{2(1-v)}(\sigma_v - p) + p \quad (9.7)$$

where P_{FP} is the most likely fracture pressure.

Pore Pressure in a Hydraulically Connected Formation

Before introducing methods of well-log-based pore pressure prediction, it should be noted that the methods are based on the shale (mudrock) properties, and the pore pressures obtained from these methods are the pressures in shales. For the pressures in sandstones, limestones or other permeable formations, the pore pressure can be obtained by either assuming that the shale pressure is equal to the sandstone pressure, or using the centroid method (Dickinson, 1953; Traugott, 1997; Bowers, 2001) and using a lateral flow model (Yardley and Swarbrick, 2000; Meng et al., 2011) to do the calculation. In either the centroid method or lateral flow model, the principle is similar and based on the following equation (Eq. 9.8) for a hydraulically connected and fully saturated formation.

Where a laterally extensive inclined aquifer or a hydrocarbon-bearing formation exists, deep overpressured regions are connected to shallower regions by a permeable pathway. Fluid flow along such an inclined formation can cause pore pressures at the crest of a structure to increase. The pressures in a hydraulically connected formation can be calculated based on the difference in the heights of fluid columns, i.e.,

$$p_2 = p_1 + \rho_f g(Z_2 - Z_1) \quad (9.8)$$

where p_1 is the formation fluid pressure at depth of Z_1; p_2 is the formation fluid pressure at depth of Z_2; ρ_f is the in situ fluid density; g is the acceleration of gravity.

Therefore, for a permeable aquifer or hydrocarbon-bearing formation, if we know the formation pressure at a certain depth, then the pressures at other depths can be obtained from Eq. (9.8). The calculation and principle is relatively simple. However, to perform this calculation, we need to know the connectivity and extension area of the formation. In other words, we need to distinguish each individual fluid compartment and seal, which can be determined from regional geology, well-logging data and drilling data (Powley, 1990). Fig. 9.3 shows an example of calculating formation pressure in an oil-bearing sandstone using the fluid flow model (Eq. 9.8). When the formation pressure and fluid density in well 1 are known, we can calculate the pressures in other wells using Eq. (9.8). When the formation is hydraulically connected and saturated with the same fluid, the formation pressures in the four wells should follow a single fluid gradient. Fig. 9.4 demonstrates that Eq. (9.8) gives an excellent prediction. Therefore, when the geologic structure, fluid pressure and density in a well are known, the fluid pressures in other wells located in a hydraulically connected formation can be fairly predicted. It should be noted that the permeability magnitude and its variation may affect the hydraulic connectivity of a formation.

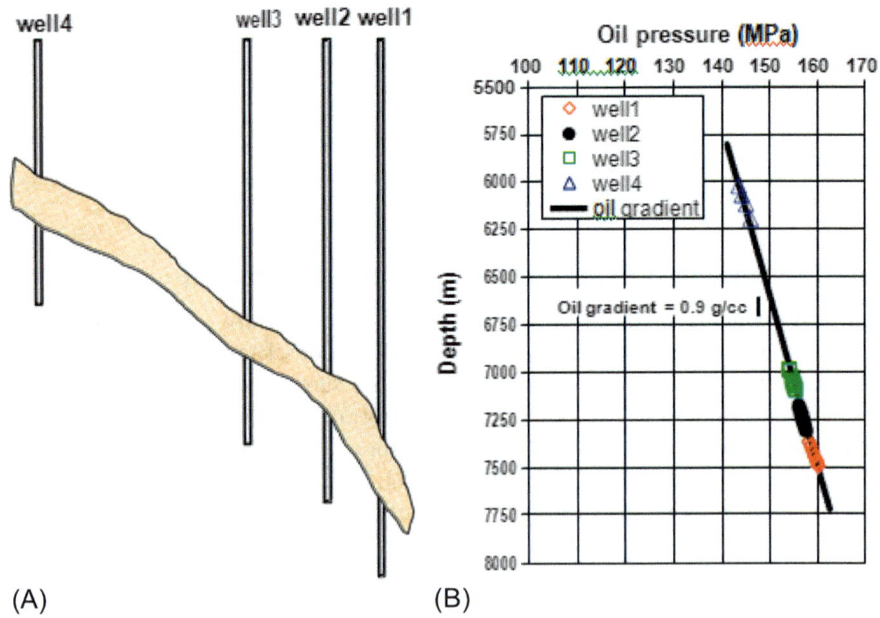

FIG. 9.4

Schematic cross section (A) showing four wells in a hydraulically connected oil-bearing sandstone compartment and the fluid pressures in different wells (B). Measured fluid pressures *(dots)* in these wells match the calculated pore pressures *(line)* with an oil gradient of 0.9 g/cm^3.

For extremely low permeable formations (e.g., shale gas formation), the applicability of Eq. (9.8) may be limited. Pore pressure gradient is different in a formation when it is saturated with different fluids. In each fluid column, the pore pressure can be calculated using Eq. (9.8) with the density of the fluid saturated in this column. Fig. 9.5 shows a hydraulically connected formation filled with gas, oil, and water/brine. If we know the fluid pressure at a depth, fluid densities, and depths of water-oil contact (WOC) and oil-gas contact (OGC), then we can use Eq. (9.8) to calculate the pressures at other depths. Therefore, in Location A (the crest) the gas pressure can be expressed as follows:

$$p_A = p_B + \rho_g g (Z_A - Z_B)$$

where ρ_g is the in situ gas density; p_B is the pore pressure in Location B; p_A is the pore pressure in Location A.

In Location B (oil-gas contact), the oil pressure can be obtained from the following equation:

$$p_B = p_C + \rho_o g (Z_B - Z_C)$$

where ρ_o is the in situ oil density; p_C is the pore pressure in Location C; p_B is the pore pressure in Location B. If the formation is only saturated with water, then the pressure in Location A is:

$$p_{wA} = p_C + \rho_w g (Z_A - Z_C)$$

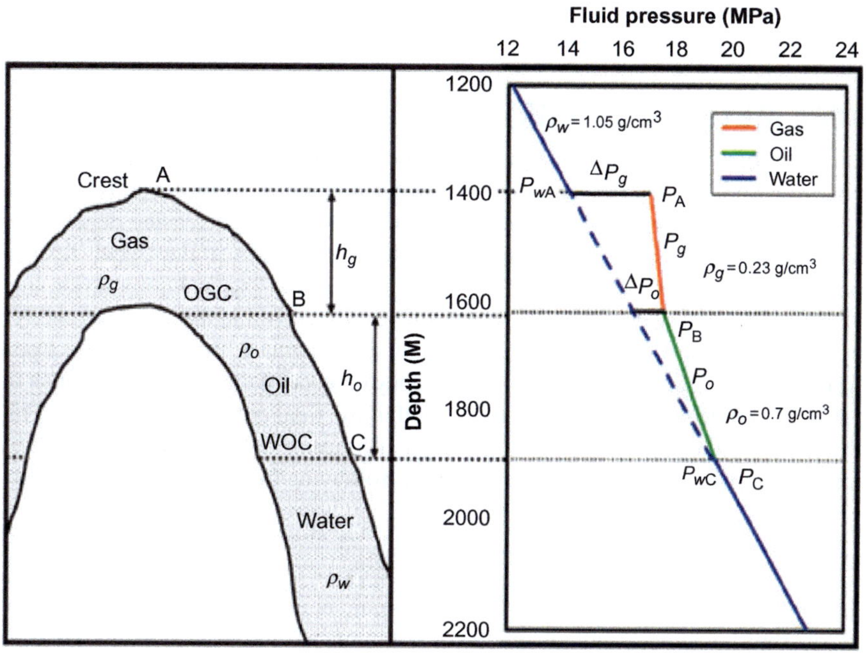

FIG. 9.5

A schematic reservoir saturated with gas, oil and water (left) and pore pressure elevated by oil and gas columns and density contrast between water and oil, gas in a reservoir. This density contrast causes pore pressure increase in Location A compared to that caused only by water/brine gradient. The right figure shows pore pressure elevated by hydrocarbon columns in a hydraulically connected formation.

Comparing the pressure (p_A) in the crest to the water/brine pressure (p_{wA}) at the same depth, the pore pressure increment induced by the oil and gas column in the crest (location A in Fig. 9.5) can be expressed in the following form (i.e., $\Delta p_{og}=p_A-p_{wA}$):

$$\Delta p_{og} = (\rho_w - \rho_o) g h_o + (\rho_w - \rho_g) g h_g \tag{9.9}$$

where Δp_{og} is the pore pressure increment induced by the oil and gas columns; ρ_w is the in situ water/brine density; h_g is the height of gas column; h_o *is* the height of the oil column. It should be noted that gas density is highly dependent on pressure. Therefore, the in situ gas density should be used for the calculations. Normally, the gas column height is not very large; hence, in the aforementioned equations, gas density is assumed to be a constant value. This pore pressure elevation (Δp_{og} in Eq. 9.9) is caused by the hydrocarbon buoyancy effect due to density contrasts between hydrocarbon and brine. The pressure elevation due to the difference in densities gradually decreases from the maximum value at the top of the reservoir to zero at the water and hydrocarbon contact.

SOME METHODS OF PORE PRESSURE PREDICTION

Hottmann and Johnson (1965) were probably the first to make pore pressure predictions from shale properties derived from well-log data (acoustic travel time/velocity and resistivity). They indicated that porosity decreases with increasing depth by analyzing acoustic travel time in Miocene and Oligocene shales in the Upper Texas and Southern Louisiana Gulf Coast. This trend represents the "normal compaction trend" as a function of burial depth, and fluid pressure exhibited within this normal trend is hydrostatic. If intervals of abnormal compaction are penetrated, the resulting data points diverge from the normal compaction trend. They contended that porosity or transit time in shale is abnormally high relative to its depth if the fluid pressure is abnormally high. Analyzing the data presented by Hottmann and Johnson (1965), Gardner et al. (1974) proposed an equation that can be written in the following form to predict pore pressure:

$$p_f = \sigma_v - \frac{(\alpha_v - \beta)(A_1 - B_1 \ \ln \Delta t)^3}{Z^2} \tag{9.10}$$

where pf is the formation fluid pressure (psi); σ_v is expressed in psi; α_v is the normal overburden stress gradient (psi/ft); β is the normal fluid pressure gradient (psi/ft); Z is the depth (ft); Δt is the sonic transit time (μs/ft); A and B are the constants, $A_1 = 82{,}776$ and $B_1 = 15{,}695$.

Later on, many empirical equations for pore pressure prediction were presented based on resistivity, sonic transit time (interval velocity) and other well-logging data. The following sections only introduce some commonly used methods of pore pressure prediction based on the shale properties.

Pore Pressure Prediction From Resistivity

In young sedimentary basins where undercompaction is the major cause of overpressure, e.g., the Gulf of Mexico and North Sea, the well-log-based resistivity method can predict pore pressure with fair accuracy. Eaton (1972, 1975) presented the following equation to predict pore pressure gradient in shales using resistivity logs:

$$P_{pg} = OBG - (OBG - P_{ng})\left(\frac{R}{R_n}\right)^n \tag{9.11}$$

where P_{pg} is the formation pore pressure gradient; OBG is the overburden stress gradient; P_{ng} is the hydrostatic pore pressure gradient (normally 0.45 psi/ft or 1.03 MPa/km, dependent on water salinity); R is the shale resistivity obtained from well logging; R_n is the shale resistivity at the normal (hydrostatic) pressure; n is the exponent varied from 0.6 to 1.5, and normally $n=1.2$.

Eaton's resistivity method is applicable in pore pressure prediction, particularly for young sedimentary basins, if the normal shale resistivity is properly determined (e.g., Lang et al., 2011). One approach is to assume that the normal shale resistivity is a constant. Another approach includes to accurately determine the normal compaction trendline, which will be addressed later in Eaton's resistivity method with depth-dependent normal compaction trendline.

QUESTION 117

How can I calculate pore pressure based on resistivity using Drillworks prediction software? When I use resistivity, which Normal Compaction Trend (NCT) should be used?

Ahmed El Faramawy

Lead Petrophysicist at Pico International Petroleum (PIP)

ANSWERS PROVIDED BY INDUSTRY EXPERTS

Steve Hobart

Sr. Pore Pressure Advisor

Ahmed, to follow the Drillworks workflow use the main menu selection "Analyze > Pore pressure" (or its equivalent toolbar icon, and select "Eaton Resistivity" in the first window of the wizard that opens. In the second window select which well and wellbore you wish to perform the pore pressure calculation in. The third wizard window is for providing the input parameters: the porosity sensitive dataset (generally filtered deep resistivity shale points), the normal compaction trend, the Eaton exponent, the normal pore pressure gradient or equivalent mud weight for the normal pore pressure, and the overburden gradient dataset. The last window is for naming the new dataset (a default name is provided, e.g. "PP res e1.2" for an Eaton exponent of 1.2) and for providing the display parameters (color and line style).

After creating the Resistivity-based pore pressure dataset, add it to a track containing calibration information, such as mud weights, kicks, formation pressure measurements, minimum static pressure while drilling data points. Taking that information into account, along with events such as connection gas, differential sticking, etc. manipulate/edit the normal compaction trend graphically and/or adjust other parameters using the track menu item "Edit special" to improve the ability of your Eaton model pore pressures to account for the calibration information.

There are many issues to be addressed prior to using the Eaton resistivity method within the Pore Pressure wizard. A partial list would include:

- how to calculate the overburden gradient
- how to generate the best possible density profile for calculating the overburden gradient
- how to select shale intervals
- how many segments are appropriate for the normal compaction trend

- how to select an appropriate filter for the resistivity shale points
- whether normalization of resistivity values for temperature and or salinity is warranted
- selecting the weight to be given to each of the various forms of calibration information, e.g. "Are formation pressure measurements in permeable zones really equal to the pore pressures in the overlying and underlying shales?", "Is the connection gas due to swabbing or diffusion?", "Are some of my pore pressure indicators reflecting a pore pressure generation mechanism other than shale undercompaction and thus not valid for calibrating the Eaton resistivity model?"

Drillworks provides tools to help you address some of these questions. The software is intended to be a tool box and not a black box. If thoughtful attention is not given to issues such as those listed above, the analyst runs the risk creating an inappropriate model.

Besides the Eaton model, Drill works Predict also provides the Equivalent Depth Method, which is also based on the assumption that the pore pressure generation mechanism is shale under-compaction. Most of the issues listed above will also apply to this method. I would not consider an Equivalent Depth method based on shale resistivity to be valid unless normalization for temperature and salinity had been performed. There is a temperature gradient and a salinity gradient with increasing depth in clastic sediments. These are normally "hidden" in the Eaton normal compaction trend and not given much thought. Since the Equivalent Depth Method references a normal compaction trend at a depth other than the depth of interest, failure to normalize shale resistivity for temperature and salinity will result in faulty pore pressure calculations.

Allan Reyes

Senior Geomechanics Engineer at Schlumberger

Ahmed, Drillworks Predict has a workflow to do PP using resistivity or sonic as well as other logs. First use gamma ray log to draw shale lines. Then create shale points on resistivity log and filter these points creating a new log associated with your original resistivity. In practical terms you smooth your log based in shale occurrence; finally you can draw a TNC over that resistivity smooth log and estimate pore pressure. Remember that you first need an overburden estimation prior to estimating pore pressure.

Sarah Sausan

Associate Geoscience Consultant, Halliburton Landmark

Ahmed, if you use resistivity logs as the porosity logs for your PP prediction in Predict, then the NCT should be drawn manually. Look at your resistivity log, and examine an interval over which the pore pressure seems normal. Then make a line group that follows the slope of the interval. Extend the line from the mudline to TD. If there is any overpressured section then that section should read less in resistivity than the line you created, i.e., "trend deflection". I know it sounds less automatic, but that is the tricky part if we use resistivity. I personally prefer to perform prediction from sonic log data whereas resistivity log is just for guidance.

There is one more automatic way to create NCT on resistivity in Predict, which is by using Semilog NCT on resistivity. But this is risky because the Semilog trend is not as widely applicable as other methods (if I am not mistaken Semilog is only appropriate in deepwater area).

Pore Pressure Prediction From Interval Velocity and Transit Time (Sonic Log)

Pore pressure prediction using Eaton's method

Eaton (1975) presented the following empirical equation for pore pressure gradient prediction from sonic compressional transit time:

$$P_{pg} = OBG - (OBG - P_{ng})\left(\frac{\Delta t_n}{\Delta t}\right)^3 \quad (9.12)$$

where Δt_n is the sonic transit time or slowness in shales at the normal pressure; Δt is the sonic transit time in shales obtained from well logging, and it can also be derived from seismic interval velocity. This method is applicable in some petroleum basins, but it does not consider unloading effects. This limits its application in geologically complicated areas such as formations with uplifts. To apply this method, one needs to determine the normal transit time (Δt_n).

QUESTION 118

I would like to draw a synthetic graph from well data with: OBG pressure, hydrostatic pressure and pore pressure. For that I have a sonic log and the Eaton equation. It gives me really weird results (including negative values for Ppg). I have some really basic questions here: Are OBG and Png constant or variable in the equation? Also where can I find Dtm and Dtml for my specific case?

Caroline Guion

Development Geologist at Lundin Petroleum

ANSWERS PROVIDED BY INDUSTRY EXPERTS

Julien Martineau-Fabre

Sr. RT Geomechanics Eng.

Caroline, prior to making your PP from Acoustic data using Eaton's method, a few things you need to check are:

- What is your OBG based on? OBG should not be a straight line/constant with depth. It does evolve with depth, with a tendency to increase drastically in shallow sediment, while increases reduce with depth. Although you may find "more gentle increase" with depth, depending on lithology.

- Which gradient are you using for NHG? Depending on area, you may have great changes.
- If not directly measured (in offset wells), then you could extract data from your seismic data - assuming you have some!
- Which Eaton's exponent are you using? Acoustic data is decreasing with great depth, so you should have a "minus" sign for your EE. Eaton used -3 for the GoM, which works well but may have to be adjusted to your region/area.
- Eventually, how does your NCT for acoustic behave? NCT should decrease with depth, i.e. top of your NCT for high acoustic value and bottom of your NCT with lower value.

Cory Moore

Senior Real Time Pore Pressure Analyst, Ikon Science

Caroline, The problem you describe with the Eaton method sounds like a lithology discrimination issue. i.e., the shale baseline on your gamma ray (assuming you have a GR) is set too low, and the model is calling all lithology shale. If you are doing that, yes you will get sub-hydrostat/negative values.

Julien Martineau-Fabre

Sr. RT Geomechanics Eng.

I don't understand Cory's point, as GR has nothing to do with the Eaton's equation. Even though you would "select" your entire well as shale", then you would compute PP. I think the problem would be more in your NCT placement and EE sign.

Cory Moore

Senior Real Time Pore Pressure Analyst, Ikon Science

Caroline, you are modeling shale undercompaction, so yes; lithology has everything to do with it. If you do not fully understand the process (shale undercompaction) which I refer to, an excellent paper to read is

Swarbrick, R.E. and M.J. Osborne, 1998, Mechanisms that generate abnormal-pressures: an overview, in Law, B.E., G.F. Ulmishek, and V.I. Slavin eds.,Abnormal pressures in hydrocarbon environments: AAPG Memoir 70, p. 13–34.

Julien Martineau-Fabre

Sr. RT Geomechanics Eng.

Caroline, I did not say that shale selection is not important. As it is indeed important since Eaton refers to the compaction disequilibrium in shale - and this only (no secondary process). However, what I meant is that the GR is not involved in Eaton's equation.

QUESTION 119

I am working on pore pressure prediction using Eaton's method with sonic log and density log data. The problem is that the sonic log readings are very scattered. Readings change from 55 micsec/ft to 170 micsec/ft. In addition, the lithology is very complicated. For example, I have readings within the range of 110 micsec/ft to 150 micsec/ft for 15 ft and then suddenly I have very low readings (between 60 and 80misec/ft) for 10 ft. This situation makes it hard for me to draw normal trend line. How do I overcome this issue?

Evren Bektas

Petroleum Engineer- Pore Pressure/Geomechanics

ANSWERS PROVIDED BY INDUSTRY EXPERTS

Michael Drews

Senior Geologist at ConocoPhillips

Evren, usually pore pressure is calculated in shales. Filter your sonic for shales only (the simplest method would be to apply GR cut-offs). Next, pore pressure usually does not change over short intervals, so apply a boxcar filter (e.g. with a 199' window). That should give you a better dataset to work with (unless you work in a shale-poor basin, but then you might like to consider a different pore pressure prediction (PPP) method).

Selim Simon Shaker

Consulting Geoscientist at Geopressure Analysis Services

Evren, follow up on Michael's excellent recommendation on how to draw the Compaction Trend. From shallow to deep DT will gradually decrease (that is, velocity will gradually increase) and at a specific depth this DT trend reverse itself. This is the top of geopressure depth. You can draw this trend and with the help of geopressure software you can calculate PPP. On the other hand, without owning some software, this compaction trend (DTC) can be mathematically calculated and consequently PP can predicted. This link will help you calculate DTC and PPP: http://csegrecorder.com/articles/view/a-new-approach-to-pore-pressure-predictions

Mark Tingay

Senior Geomechanics specialist at Chevron

Evren, further to what Michael and Selim said. You want to make sure you do not have any artifacts in your log data (e.g.55us/ft is faster than casing velocity), and then filter out the log data to leave only the shales. After that, you want to build a NCT founded on the geology. It sounds like the geology is tricky. Can you give us any more details? You said the shales are quite thin - this can make it tricky. Do the shales cover a range of formations, or show variability in age, mineralogy or depositional environment? Is this an area in which the shale compaction may have been affected by uplift and exhumation or high horizontal stresses? These can all make it tricky to develop a reliable NCT. Your PP prediction hangs off of your NCT, so it is important to build that based on a solid geological foundation, and to make sure the inputs into your NCT are shales with reliable petrophysical data and, that are normally pressured and normally compacted.

Selim Simon Shaker

Consulting Geoscientist at Geopressure Analysis Services

Evren, this is an observation we need to be aware of; Predicting PP is from seismic before drilling. However, the exercise of calculating PP during and after drilling from the Sonic is PP Model Calibration utilizing the actual P measurement from an existing well. Consequently, the model's exponents and constants can be used to Predict PP from seismic velocity for a proposed well location in the same basin i.e. pore pressure prediction. Therefore, Evren's exercise will end up creating a PP model with a unique Eaton's exponent. This model will be the basis for any future pre-drilling PPP in the same basin.

Evren Bektas

Petroleum Engineer- Pore Pressure/Geomechanics

Reliable NCT and Eaton's exponent is crucial to getting a good PPP from Eaton's method. As Dr. Shaker said, I need to have a unique Eaton's exponent. In order to determine Eaton's exponent for a specific region, I usually compare PPP with mud weight used to drill an existing well. Then, I try different Eaton's exponents to find the general behavior for that region until the mud weight and PPP match.

Ed Dew

Senior Staff Drilling Engineer at Southwestern Energy

Evren, excellent comments from experts have been offered above. In the GOM & parts of the North Sea, we required shales to be about 50 to 100 ft to ensure that they were adequately sealed to get a good pore pressure from the trend. Onshore in South Texas, I found that in addition to the GR shale cutoff and hole quality cutoffs (hole diameter, rugousity, delta rho, etc.), I would look at the shapes of the sonic and GR in the shale. If the shales were adequately thick and clean enough, there would be a relatively flat profile across the peak. If it was a spike, it was not thick enough to give a good compaction trend. Sometimes a 15 or 20 ft thick zone would work. This helped find the trend through silty or sandy stretches.

Hesham Naiem

Senior Pore Pressure and Geomechanics Engineer

Evren, you have to look at the structure and geology of the area because sometimes in compressional zones which have fractures and minor faults, you will get a lot of sonic spikes and fluctuated readings which makes trend line setting process very difficult. In such areas sonic could be misleading in pore pressure calculations. Also comparing the mud weight while drilling such zone with other places in the area you will see that lowest mud weight used will be in such zone of fluctuated sonic reading.

Odofin David Khulong

Geopressure Specialist

Evren, let me add that calibrating predicted pressure with mud weight used in drilling an offset well, in order to get the appropriate exponential, may be misleading, because most wells are drilled overbalanced. As such, in the absence of other pressure data (RFT, RCI, MDT etc.), we have to resort to using mud weight, then a correction must be made for the margin of error applied to the mud weight when the well was drilled (in order to get the true formation pressure for such well).

Crystal Clemmons

Geomechanics (Pore PressureO consultant and Project Management, Halliburton

Evren, May I ask if you have any evidence in the geological data that would justify breaking the NCTL. Since the line should represent normal compaction we can justify breaking it when we find evidence of abnormal compaction. I agree with Mr. Khulong's comment regarding the use of mud weight data from offset wells. If possible you could find the high and low points for the MW used in each section. Once you bring in those data sets it will help you constrain the model. Once the well is drilled you can enhance your prediction with real-time data feeds into your model to improve accuracy. Also if any additional offset wells are drilled near the prospect well you may bring in the MW used to compare.

QUESTION 120

Can we define Eaton Exponent in a specific area/well or its value set as by default, 3 and 1.2 depending upon log type? If yes then how?

Nasir Mahmood

Assistant Geophysicist at OGDCL

ANSWERS PROVIDED BY INDUSTRY EXPERTS

Steve Hobart

Sr. Pore Pressure Advisor

Nasir, Eaton published values of 1.2 for Resistivity and D-exponent and 3 for Sonic and Velocity. His work was based on US Gulf Coast data from Hottman and Johnson. Those exponents worked for that area. Dr. Eaton, recognizing the empirical nature of his widely used equations, once remarked that "There is nothing magic in those exponents." The published values will probably work well in Tertiary shales where the porosity is not affected by carbonate precipitation and the where the primary mechanism for overpressure generation is undercompaction ("compaction disequilibrium"). In older shales, we sometimes find that reduced exponents work better than the published values. Glenn Bowers suggested that raising the exponent for acoustic data from 3.0 to, say, 5.0 could be a viable means of accounting for pore pressure mechanisms other than undercompaction, namely those related to thermally induced phenomena which create "unloading" (reduction of vertical effective stress).

The best values, of course, are the ones that work. By back-calculating the exponents when you have a good handle on the normal compaction trend (perhaps based on an in situ calibration) and on the shale pore pressures (when sands and shales can be assumed to be in pressure equilibrium). I would recommend using the same exponent in the same formations across a field. When you start changing parameters in an ad hoc, arbitrary manner from well to well within a study area the model loses it predictive value. Making the exponent a function of geologic age might be a worthwhile experiment.

Selim Simon Shaker

Consulting Geoscientist at Geopressure Analysis Services

Nasir, Eaton's and Bowers' methods calculations are derived differently:

- Bowers' method is a modification of the vertical effective stress (ES) that equates the ES in the normally pressure system with the ES of the overpressured system. This can lead to a great disparity of the predicted results since Eaton's X is a product of horizontal effective stress. Please review Terzaghi 3 stages models.
- The exponent 5 that Bowers came up with (2001 OTC paper# 13042) is based on switching from a power law trend to exponential trend based on THREE PRESSURE TESTS (RFT's in ppg values). Choosing the trend type is correctly done IF it is based on the nature of the data NOT just based on the R2 value (least square fit).
- Unloading supported by up-dip transfer of reservoir pressure is geologically UNFIT. If there is a pressure and fluid TRANSFER in the overpressure system, every geopressured reservoir will BREACH. Again, this is supported by Stage A of Terzaghi model.

References: Reservoir vs. Seal Pressure Gradients: Perception and pitfalls.: AAPG 2014 annual convention (Ex.ABS.#1834685) / AADE Annual Convention, April 2012 (AADE-12-FTCE-66).

- Pore pressure (in situ) generation mechanisms are INTERRELATED and the process goes as follow: sediment load, basin subsidence (depth), temperature and dehydration increase, SEAL forms (under-compaction), excess pore pressure is the result. They are in harmony (independent), and cannot be separated from each other. The main reason for generating the over pressure is the SEAL (high stand stratigraphic sequence). A subsurface sedimentary sequence without a seal is a breaching system with a BIG DRY HOLE symbol on it. Reference: New Approach to Pore Pressure Prediction: Generation, Expulsion and Retention Trio, Case Histories from GOM: CSEG (Canadian Society of Exploration geophysicists), RECORDER, January issue 2015, v 40, n 1, pp 44-51/AAPG Sixth Annual Deepwater and Shelf Workshop, Jan.27-28th 2015./GCAGS 2015 Annual Convention, Houston, Sept.2015.

Actually we reduce the exponent going from the old sediments to younger ones e.g. Gulf coast (3), inner and middle shelf (3-2), outer shelf (2-1.5), deep water in mini-salt basins (can range from 2- 0.8) depending how close to the salt mass one is. On the salt Toe it is widely varied due to the salt creep effect. These numbers can be changed due to the age, predominant lithology, and regional tectonic in every sub-basin.

How (the second part of the question): There are several methods, however the easy one is: interval velocity data from seismic gathers close by (preferably on location) of existing wells and sonic logs. The wells should have pressure measurements and drilling event records. Now use Eaton equation with for example exponent 2 and find out how far you are from the measured values. Then adjust the number until you come up with the right X.

PS. Nasir you need to run several iterations with different exponents until you reach a consensus as to the right exponent without swaying and breaking the compaction trend.

Morteza Azadpour

Petroleum Engineer at National Iranian Oil Company

Nasir, the exponent constant x is originally 3 in Eaton's study but requires modification to be implemented in tight unconventional reservoirs. Here is some example of Eaton exponent less than 3 showing this modification in different reservoir formations:

x = 1, Contreras et al, 2011, A case study for pore pressure prediction in an abnormally sub-pressured western Canada sedimentary basin.
x = 1-1.5, Yully P. Solano et al, 2007, A modified approach to predict pore pressure using the d-exponent method.
x = 2.6, Jeff C. Kao et al, 2010, Estimating pore pressure using compressional and shear wave data from multicomponent seismic nodes in Atlantis field, deepwater gulf of Mexico.
x = 0.1-0.3, Kadyrov and Tutuncu, 2012, Sonic log-derived pore pressure prediction in a west Kazakhstan dolomite field.

Mark Tingay

Senior Geomechanics specialist at Chevron

Nasir, as Steve says, these were only the originally published values from Eaton, but they are often modified. There are numerous examples of this, as Morteza has shown. As Bowers indicated, the Eaton exponent can also be raised in places where disequilibrium compaction is not the only overpressure generating mechanism. For example, I used x=6.5 in vertically transferred overpressures in Brunei (Tingay et al., 2009, AAPG Bulletin).

As Steve says, you can also back-calculate the Eaton exponent if you have sufficient PP data and a good handle on the normal compaction trend. I would be very cautious about doing this for just a few wells, or even just one field. However, I have done it using hundreds of direct pressure measurements in overpressured sequences over a large area. This allows you to look at the statistical distribution of 'perfect Eaton exponents' (the Eaton exponent that perfectly fits the pore pressure; Tingay et al., 2009; Tingay et al., 2013 - both in AAPG Bulletin). You can usually get a good idea of the overall overpressure system by looking at what the mean value, and standard distribution of values for perfect Eaton Exponent is.

Also, as Steve say, do not change the Eaton exponent arbitrarily from well to well or field to field - not unless there is clear geological reasoning for doing it. This is part of what I like to call 'Cheatin' with Eaton' - you can basically always fiddle with things to make Eaton fit. But, it is better to get a reasonable fit of PP for all wells using one standard set of parameters, than it is to get a perfect fit for all wells, all using different parameters (otherwise, which parameters do you use for the next well you are going to drill?).

Ekwereubong Ini Jonah

Geophysicist/Field Supervisor at EBRUWASSA

Mark Tingay, the proposed exponential modification of Eatons method by Glenn Bowers, where it is supposed to exceed X= 5, as used by you in the case of vertically transferred overpressure may not be necessary. I used the modified Eatons method without exceeding X=5. As seen in literatures the higher level under estimated overpressure generated by unloading, but Bower's unloading method predicted perfectly (Ini E. J., 2015. Detection and prediction of overpressure zones Niger Delta Basin: A case study of "GRINI" field, Central Swamp Depobelt. Unpublished Thesis Centre for Petroleum Geosciences (CPG), Institute of Petroleum Studies(IPS), University of Port Harcourt).

Jorunn Haver Figenschou

Senior Geomechanics Specialist at Baker Hughes

Jonah, in my work in different regions globally, I have defined regional Eaton's exponents that I use in my work. Often it is a case of having numerous wells with structural control and calibration data that is the basis for setting a regional exponent. As Mark says, the strength of your Eaton's exponents is tested when you get a reasonable result in your interpretations using similar trends and parameters in multiple wells in a region or depositional basin.

QUESTION 121

For pore pressure prediction using Eaton's equation, can one use Katahara's correction by subtracting the velocity at the mud line (Katahara, 2003)? I have found that term adds too much uncertainty if you want to invert for the exponent n, as per Sayers, 2002.

Alejandro Garcia

Geophysicist and Geomechanics at Total E&P

ANSWERS PROVIDED BY INDUSTRY EXPERTS

Mark Herkommer

Geopressure Technology Consultant at Semicircle

Alejandro, I have tried to do this a number of times but with limited success. Although it successfully lowers the Eaton Exponent to the 1.0 - 1.5 range, it causes extreme variation in the predicted pore pressure when the pore pressure is close to hydrostatic.

Keith Katahara

Senior Geophysical Advisor at Hes Corporation

Alejandro, I still use it when I use an Eaton-type method with a normal depth-velocity trend, although most of the time now I use a direct velocity-stress function rather than a normal trend. A place where my modification of Eaton is useful is on the Gulf of Mexico shelf in hard pressure due to smectite-illite transformation. When the effective stress is very low, Eaton's formula does not work. You can see that by considering the limit of zero stress. To get zero stress, Eaton's formula needs to have zero velocity, but in water-saturated shales, the velocity is never going to be much lower than water velocity. So if you need to fit very low effective stress with Eaton, you will find that you need to raise the exponent very high, or you may find yourself looking for ways to increase the overburden.

I like to use formulas that conform to what I think of as physical limits. One limit is that when sediments first settle out of suspension. The velocity is given by Wood's equation (close to water velocity in practice). Bowers' power-law formula honors that limit. Another limit is that at very high effective stress the porosity should approach zero and velocity should approach a notional solid grain velocity. Miller's equation incorporates both limits. Eaton, as used in practice generally violates one or both limits. For that reason, Eaton is not my choice for a starting point for improving our understanding of pore pressure and how it affects velocity. Of course it can still be useful as means for fitting data. In a sense, Eaton is ideally suited to that purpose. I just don't learn much from using it. The way it is used on the GOM shelf is a good example. Folks (including me at times) use the normally pressured section with smectite-rich shales to determine the normal trend, and then use Eaton's formula to compute pressure in deeper overpressured illite-rich shales. Eaton can work very well in fitting such data, even though it makes no sense conceptually to use a normal trend for one type of shale to predict pressure in another type of shale.

It is really a curse that Eaton works as well as it does. That is probably one reason why our discipline has made so little progress. We can get by with adjusting parameters without really understanding what we are doing.

Selim Simon Shaker

Consulting Geoscientist at Geopressure Analysis Services

Keith, I am not questioning the validity of your corrections to Eaton's. However, I support your comment: "That is probably one reason why our discipline has made so little progress. We can get by with adjusting parameters without really understanding what we are doing."

"If you can't reduce a difficult geological (or other) problem to just one 8-1/2 by 11-inch sheet of paper, you will probably never understand it."

- rephrased from quote by Karl Terzaghi (after George Klein)

"Discovery consists of seeing what everybody has seen, and thinking what nobody else has thought"

After Albert Szent-Gyorgyi (1883-1986), Noble Prize in Medicine (1973). Discoverer of Vitamin C. (after Dwight (Clint) Moore)

Gopal Amruthapuri

Geological Advisor – Pore Pressure at Anadarko Petroleum Corporation

I strongly support Keith's last comment on not understanding it. People should stop clicking buttons then use pen & paper to gain understanding.

Kehinde Adesina

Snr. Global Technical Advisor –Drilling & Geomechanics (unconventional & Deepwater), Halliburton

Alejandro, the beauty about what we do is that we are the end users of data and also the people that make sense of it with limited knowledge of what has truly occurred over the geological age of the rock. The exercise of using all available equations to verify what workflow best estimates the realities of the observation seen in a basin is really what we do for living. Knowing that nature is big and complex guarantees that we will remain employed because each basin is just a little dot on the map of the earth. Kudos to all the people that have tried to develop equations - all comes down to ratios of ideal rock compaction history to observations of today or what is expected compared to what is seen as of today. When you can put a trend to that and adjust the magnification to the relative changes in the type of lithology and fluid content, you have done more than enough. Sequences of evaporites, carbonates and sand/shale combination can never be described by an equation. Rather it can be understood when you know the source data that was used in developing the existing equations, the underlining assumptions made and the age of the rock for which it has been tested over time. Today, Eaton equations are still the most consistent for all ages of rock with significant modifications to exponents and definition of where the NCTL is anchored as a starting point for known normal pressures especially when you have several and significant unconformities, and faulting systems that are known to affect the pressure compartments.

For every basin, start with Eaton basics and you can modify. This is really science that makes sense.

Alejandro Garcia

Geophysicist and Geomechanics at Total E&P

Keith, I would be very grateful if you could provide me with more detail on the type of "direct velocity-stress function" you are using. Indeed, I have found your correction to Eaton's equation to give apparently better results for the type of high pressure reservoirs I am working with. However, I have no knowledge/data of the value this parameter should have (I will look into the Wood's limit you mentioned). Thus, I was trying to invert for it, as well as the exponent n, using Colin Sayers Method (Sayers, 2002) but found this to be non-unique. This explains why Eaton is so popular. Any more a physical model has too many parameters, in which case you can fit any data you are given.

Keith Katahara

Senior Geophysical Advisor at Hes Corporation

Alejandro, the "direct" methods that I use are primarily Miller's equation and sometimes Bowers, 1995 power-law equation. By direct I just mean that they relate effective stress directly to velocity. Incidentally Miller's equation was mentioned to me by Glenn Bowers a few years before Terry Miller presented it publicly, so I call it the Miller-Bowers equation. Everyone seems to know about the Miller method from the Knowledge Systems user manual for Drillworks. I found out from Mark Herkommer that the description in the manual is a little distorted from Miller's presentation at the pressure workshop in Galveston in 2003. I think it is worth seeking out the abstract distributed at that conference. Mark H. sent a copy to me a while back but I lost it, for the second time. Keep in mind when you think about these calculations that vertical effective stress is not the only variable that affects velocity. Temperature and thermal history, diagenesis, brine salinity, mineralogy, etc., can all have effects. Lots of folks believe that mean effective stress is a better variable to use than vertical effective stress. Ultimately I think it is unlikely that a single stress variable is going to be sufficient. Matt Hauser gave an interesting talk on compaction in thrust-fault environments at the pressure meeting in San Antonio this year, so folks are starting to think seriously about such issues.

Wood's equation is just the lower bound or Reuss limit for the bulk modulus. The idea is that if you took a rock with a given porosity and broke it up into a suspension of the same density, but with non-contacting grains suspended in water, then the bulk compressibility would be the volume-weighted average of the bulk compressibilities of the water and grains. That is a physical lower bound to the velocities at zero effective stress. Eaton's velocity-stress equation clearly violates that lower-bound.

Fernando Enrique Ziegler

Geophysicist:Pore Pressure/Wellbore Stability

Alejandro, find below the aforementioned abstract by Terry Miller of Knowledge Systems from the SEG/EAGE 2002 Summer Workshop "Geopressure: conceptual advances, applications, and future challenges".

The title of the abstract is "A Velocity/Effective Stress Methodology that Predicts Finite Sediment Velocities at Infinite Effective Stresses". I took the liberty of uploading the abstract to my Dropbox account. You can find the document here: https://db.tt/d6WkZ15t

Dan Ebrom

Lead Geophysicist at Statoil

Alejandro, one of the problems (or solutions) of pressure practice is that we routinely use the easiest available inputs, that is, P-wave velocities. P-wave velocities include fluid effects that are absent from S-wave velocities. S-wave velocities have the pleasing feature of starting at near zero velocity at the mudline (for clastics), and also being more sensitive to pressure variations than are P-wave velocities (as evidenced by an Eaton exponent of 2 for S-wave velocities rather than 3 for P-wave velocities). I showed in the 2007 SEG abstract (Can P-waves be made more sensitive to pore pressure?) that converting P-wave velocities to S-wave velocities by use of Castagna's mudrock regression gives a result similar to Bowers but more stable when inverting for Eaton's exponent (Figure 6). But this has practical issues, since no one (as far as I know) has published the misfit of Castagna's regression regarding clay mineral type and compaction state. Still, it's good to recognize that S-wave velocities (ignoring anisotropy for the moment) carry a purer signal of effective stress than P-wave velocities precisely because S-wave velocities are a function only of the frame properties.

Jon Zhang

Geomechanics/Pore Pressure Expert, Shell

Alejandro, I used Katahara's modified Eaton method. One of the improvements is that when you have hard overpressure and slight overpressure existed in the same region, you may be able to use a single exponent for PPP.

QUESTION 122

Have you tried Eaton exponent equal to 1.5 or 2 for pore pressure prediction from sonic wave velocity? What was your experience?

Vahid Atashbari

Petroleum Industry Specialist

ANSWERS PROVIDED BY INDUSTRY EXPERTS

Shundong He

Development Geologist at Chevron

Vahid, I have modified Eaton exponent number to come up reasonable PP model. There is nothing magic about Eaton's exponent number. He got these numbers by fitting the exponential curve for US Gulf coast region data. But the exponential curve is a good model. In areas other than US gulf coast region, it is Ok to change the exponent number for a better PP model.

Kehinde Adesina

Snr. Global Technical Advisor –Drilling & Geomechanics (unconventional & Deepwater), Halliburton

Vahid, yes, the exponent is just a control factor for the normal vs observation. You need to iteratively define it based on what is consistent for a basin. Age of the rock, and evidence of secondary pressure mechanisms are reasons why the exponent cannot remain 3. GOM exponent 3 is only consistent for rocks younger than Miocene as they are still undergoing under compaction with some of unloading in certain areas depending on depth of burial.

Morteza Azadpour

Petroleum Engineer at National Iranian Oil Company

Vahid, I am agreement with Shundong and Kehinde comments. This is originally 3 in Eaton's study and requires modification to be implemented in unconventional reservoirs. Here is some example of Eaton exponent less than 3, not just for sonic but show this modification in different lithology:

X=1, Contreras et al, 2011, A Case Study for Pore Pressure Prediction in an Abnormally Sub-Pressured Western Canada Sedimentary Basin.

X=1-1.5, Yully P. Solano et al, 2007, A modified approach to predict pore pressure using the D Exponent method.

X=2.6, Jeff C. Kao et al, 2010, Estimating Pore Pressure Using Compressional and Shear Wave Data from Multicomponent Seismic Nodes in Atlantis Field, Deepwater Gulf of Mexico.

And X=0.3, Kadyrov and Tutuncu, 2012, Sonic Log-derived Pore Pressure Prediction in a West Kazakhstan Dolomite Field.

QUESTION 123

It has been said that if you "break" Eaton's NCT line (forming a total of two lines), the lines must be parallel to each other. Is there an exception to this rule? If so, what is it based upon?

Phillip Grainger

Geomechanics Consultant

ANSWERS PROVIDED BY INDUSTRY EXPERTS

Colin Sayers

Scientific Advisor, Schlumberger

Phillip, Eaton is a simple scaling relation that relates two dimensionless quantities:

(i) The ratio of the effective stress to the effective stress that would occur if the sediments were normally pressured.
(ii) The ratio of the velocity (or resistivity, etc.) to the velocity (or resistivity, etc.) that would be observed if the sediments were normally pressured.

The so-called NCT (Normal Compaction Trend) is an approximation to the variation in velocity (or resistivity, etc.) with depth if the sediments were normally pressured. Now the only reason to "break" a trend line would be if you have a good reason to believe that the relation between velocity (or resistivity, etc.) and effective stress is different in different depth ranges (for example, if you move from one type of sediment to another, pass through the smectite -> illite transition, etc.). Since you expect the relation between velocity (or resistivity, etc.) and effective stress to be different in the two different depth ranges, there is no reason to expect that the variation with depth of the velocity (or resistivity, etc.) of normally pressured sediments would have the same gradient.

Phillip Grainger

Geomechanics Consultant

Colin, it is still not 100 percent clear to me. If all your logs (DT, RHOB, RES, GR, NPHI, etc.) have a sudden shift that continues untill the logs' final depth, and have a geologic discordance at that depth of the shift, this suggests a break in the NCT trend. Do you agree? And in this case, wouldn't the trends need to be parallel?

Colin Sayers

Scientific Advisor, Schlumberger

Phillip, If the sediments above and below the "break" are different they will have different relations between velocity (or resistivity, etc.) and effective stress. If this is the case, the variation of velocity (or resistivity, etc.) of normally pressured sediments with depth (the so called "normal trend") will be different, and the trends will not be parallel.

Phillip Grainger

Geomechanics Consultant

Yes, that does make sense. I appreciate the explanation, Colin.

Mark Herkommer

Geopressure Technology Consultant at Semicircle

Phillip, in addition to Colin's comments, one breaks Normal Compaction Trend because the normal compaction sequence has been ostensibly broken by faulting, unconformity, lithology, geologic age, tool measurement change, etc. I suppose it should be said that the NCT should always be constructed in TVD, not MD. Also, the idea that the two segments should have the same slope is not correct. The slope of the NCT is, in the most general terms, related to the rate of sedimentation, at least in the normally compacted section. One can easily imagine a change in the slope of the NCT without an offset in the NCT, simply due to changing geological environments for two different formations. The NCT "break" can be a change is slope, an offset in the NCT, or both. An offset in the NCT is more commonly related to maintaining or adjusting a pressure calibration, such as maintaining normal pressure even though there may be a shift in the tool measurements. A change is slope more often related to a stratigraphic discontinuity or clay diagenesis.

Always keep in mind what the tool is measuring. We may be thinking in terms of shale porosity, but resistivity measures formation fluids, sonic measures acoustic wave velocity, d-exponent measures rock penetration, and so on. Thus, it is possible to break one NCT without breaking all of the NCTs.

Julien Martineau-Fabre

Sr. RT Geomechanics Eng.

Phillip, I agree with Mark's comment: make your NCT in TVD, not in MD. Also, the "rule" of a "unique NCT" has been proven not so regular. As an example, if you use DXC for your PP, then you may have to break it with bit change, diameter, mud system, BHA. However, you can expect to keep parallel slope after the breaks. For other measurement - petrophysical, I would tend to avoid breaks unless driven by lithology and formation. As an example, an unconformity or a formation/age change can justify a break and a change in slope, assuming you know more on the geological side. That is, changes in sedimentation process/speed or, a change in marine environment. For those, I already broke NCT and changed slope for RES or DT or RHOB. Often, my NCT in log scale for DT is not a unique line but a broken NCT with various slope. Eventually, and again I agree with Mark, a break in one NCT does not necessarily mean a break on others.

Pore pressure prediction using Bowers method

Bowers (1995) calculated the effective stresses from measured pore pressure data of the shale and overburden stresses (using Eq. 9.1) and analyzed the corresponding sonic interval velocities from well-logging data on the Gulf of Mexico slope. He proposed that the sonic velocity and effective stress have a power relationship as follows:

$$V_p = V_{ml} + A\sigma_e^B \tag{9.13}$$

where V_p is the compressional velocity at a given depth; V_{ml} is the compressional velocity in the mudline (i.e., the sea floor or the ground surface, normally $V_{ml} \approx 5000$ ft/s, or 1520 m/s); A and B are the parameters calibrated with offset velocity versus effective stress data. Rearranging Eq. (9.13) and considering $\sigma_e = \sigma_v - P$, the pore pressure can be obtained from the velocity, as described in Eq. (9.13), as:

$$P = \sigma_v - \left(\frac{V_p - V_{ml}}{A}\right)^{\frac{1}{B}} \tag{9.14}$$

For Gulf of Mexico wells, $A = 10-20$ and $B = 0.7-0.75$ in the English units (with p, σ_v in psi and V_p, V_{ml} in ft/s). Eq. (9.14) can also be written in terms of transmit time simply by substituting $10^6/\Delta t$ for V_p and $10^6/\Delta t_{ml}$ for V_{ml}

$$P = \sigma_v - \left(\frac{10^6\left(\frac{1}{\Delta t} - \frac{1}{\Delta t_{ml}}\right)}{A}\right)^{\frac{1}{B}} \tag{9.15}$$

where Δt_{ml} is the compressional transit time in the mudline, normally $\Delta t_{ml} = 200\,\mu s/ft$ or $660\,\mu s/m$. The effective stress and compressional velocity do not follow the loading curve if formation uplift or unloading occurs, and a higher than the velocity in the loading curve appears at the same effective stress. Bowers (1995) proposed the following empirical relation to account for the effect of unloading curves:

$$V_p = V_{ml} + A\left(\sigma_{max}\left(\sigma_e / \sigma_{max}\right)^{1/U}\right)^B \tag{9.16}$$

where σ_e, V_p, V_{ml}, A and B are as before; U is the uplift parameter; and

$$\sigma_{max} = \left(\frac{V_{max} - V_{ml}}{A}\right)^{\frac{1}{B}} \tag{9.17}$$

where σ_{max} and V_{max} are the estimates of the effective stress and velocity at the onset unloading, in absence of major lithology changes; V_{max} is usually set equal to the velocity at the start of the velocity reversal. Rearranging Eq. (9.16), the pore pressure can be obtained for the unloading case:

$$P_{ulo} = \sigma_v - \left(\frac{V_p - V_{ml}}{A}\right)^{\frac{U}{B}} \sigma_{max}^{1-U} \tag{9.18}$$

where P_{ulo} is the pore pressure in the unloading case. The Bowers method is applicable to many petroleum basins (e.g., the Gulf of Mexico). However, this method overestimates pore pressure when shallow formation is poorly consolidated or unconsolidated, because the velocity in such a formation is very slow.

QUESTION 124

How can I determine A and B parameters in Bowers method for a specific region?

Evren Bektas

Petroleum Engineer- Pore Pressure/Geomechanics

ANSWERS PROVIDED BY INDUSTRY EXPERTS

Yogesh Kulkarni

Pore Pressure Analyst/Field Engineer, Geoservices, Schlumberger Company

Evren, If you are not using any commercial software, a spreadsheet will be helpful provided that you have enough of a dataset for the region with calibration data to plot on cross plot. A & B can be optimized by finding out the difference in computed and in situ effective stress (based on available calibration data). Use a least square method to compute sum of error. Then use add-in analysis tools like solver to minimized this error by optimizing value of A & B with conditional constraints like A<=15 and/or B<=1. A & B values will have better match depending upon the amount of data (from different wells) covering the region.

Odofin David Khulong

Geopressure Specialist

Evren, when I asked the same question of Glenn Bowers himself, the response I got from him is as stated below:

I use a trial-and-error process. I'm assuming you have a well that you want to use to derive values for A, B, U, and V_0 (I also treat V_0 as a calibration parameter). First, I create a plot in which I display all available information about how that well behaved as it was being drilled. That would include: 1) measured pore pressures, 2) drilling fluid densities (mud weights), 3) pore pressures estimated from formation fluid influxes (kicks), 4) wellbore problems (tight hole, stuck drill pipe, pack-offs), and 5) high mud gas incidents. Next, I make an initial guess for V_0, A, and B. My typical first guess would be V_0=4930 ft/s, A=14.2, B = 0.724. I then use these parameters to compute the pore pressure, and compare the result with the data on the plot. If there is an interval in which I think unloading has occurred, then I only consider data from outside that interval at this stage of the calibration process. If the resulting pore pressures appear consistent with the well data, then I'm done. If not, I try a different set of parameters, and keep repeating the process until I'm satisfied with the match

V_0 is the least sensitive parameter, while B is the most sensitive. So, I try adjusting V_0 first, while leaving A & B fixed. Next, I try changing V_0 & A, while leaving B fixed. And if I still don't like the match, then I'll adjust B. I go through a similar iterative process to find U. My first guess for U is always 3.13. I have adopted this trial-and-error-process because: 1) in most of the areas in which I've worked, there are not enough "known" pore pressures, and therefore, velocity-effective stress data points, to apply standard curve-fitting techniques, and 2) in sands with significant vertical structure, say several thousand feet, the pore pressures measured in those sands may not reflect the pressures in adjacent shales.

Selim Simon Shaker

Consulting Geoscientist at Geopressure Analysis Services

Odofin, are these numbers A, B and U related to any subsurface geology processing? Or just curve fitting numbers. Some geological facts:

1. What is the validity of comparing the laboratory Load and Unload tests to the actual geological process? Unless procedure goes from deposit, bury, subside, heat to later on uplift and emerge (expose) the whole section. This is applies only in structural uplift (Zhang SPE 145964).
2. Fluid expansion due to heat occurs gradually in the whole section with subsidence and not in a specific part of the sedimentary column unless it is fed by volcanic or geyser activity (Tingay, 2015).
3. What is the validity of the vertical Effective Stress (ES) theorem for PPP? Terzagi 3 stages model does not support the concept of equating the ES above and below the geopressure top (vertical effective stress model).

Odofin David Khulong

Geopressure Specialist

To the best of my knowledge, A, B, and U don't represent any physical parameters, as such, just curve fitting parameters. According to the word of Selim Simon Shaker on another trend on this group, it is a curse that some of this practices work, despite being not so grounded in geologic principles. That Bowers model prediction does match reservoir pressure data (RFT, RCI, etc.) is beyond any reasonable doubt.

Selim Simon Shaker

Consulting Geoscientist at Geopressure Analysis Services

Odofin, calibration of PP profile is more than matching the reservoir pressure data (RFT, RCI, etc.) to the velocity - effective stress trend. Reservoir with hydrocarbon charge (that is where most of the RFT's take place) pressure data is completely unrelated to PPP in shale. There are several tools for calibrating PPP besides comparing the PPP in shale with measured pressure in sand. Check this out: Causes of disparity between Predicted and Measured Pore Pressure, The Leading Edge. August 2002 pp.756-760. I am assuming that you plot your RFT's in ppg mwe (based on the method you use). Check this out also: The Controversial Pore Pressure Conversion Factor: psi to ppg mwe. TLE, December 2003. Nothing is beyond any reasonable doubt when you dealing with mother earth.

Philip Heppard

Geologist at ConocoPhillips

Evren, I noted with some satisfaction that in Odofin's comment above he reported that Glenn's "first guess" is A = 14.2 and B is 0.724 with a V0 intercept of near water velocity of 4930 ft/sec. This is consistent with what I have determined as a mixed clay shale trend from a number of basins where clay mineral composition is less than 20% smectite with the rest dominated by illite and kaolinite. My average for these global data sets is A = 14.0 and B = 0.72. For smectite-rich rocks which compact much slower the average is still A = 14.0 but B is 0.67 to 0.68. The data I am referring to was published in the bulletin of the Houston Geological Society, February 2010, D. Ebrom and P. Heppard. Getting far from these values means you are just making stuff up. The other comments above cover the process of working out the details. In a new basin is one of the first things I do is find wells that define what normal shale compaction is.

Tim Rolph

Pore Pressure Analyst at BP

Evren, first you need pressure data that you are confident represent the shale pressure (eg. connection gas data, pressure in enclosed, flat sands etc). Then, in the shales adjacent to each pressure point, obtain the value of shale velocity (sonic logs; VSP etc) and an estimate of the overburden pressure. When you have sufficient data, for each point calculate VEL-5000 (if working in ft/s) and VES (PP-overburden). Next take natural logs of each calculated parameter and plot LN (VEL-5000) against LN (VES). If you fit a linear regression to the data, the slope of the line gives you the Bower's B parameter and the intercept gives you LN (A). You can also get uncertainty estimates.

Jerome Asedegbega

QI Geophysicist/Pore Pressure Analyst

Evren, Tim Rolph's way of determining A & B is more scientific and such results would reflect the geology of the immediate vicinity and could be confidently used for any undrilled prospects nearby. Note: A & B represent your compaction curve and U for Unloading case (and are not the same everywhere) so adjusting the parameters to match the measured pressure data in a particular well may introduce a high level of uncertainty for the next well (I called it cheating). Thus, you need sufficient data to be able to derive robust Bower's parameters that can be used for accurate pre-drill PP in the prospect area.

Tim Rolph

Pore Pressure Analyst at BP

I was asked, how much data do you need for the regression? Well, it is not just a game of numbers. Ideally you would want a set of data (VES: velocity pairs) that represent as broad a range as possible in terms of VES/velocity. The rigor of the linear regression is improved by, and justifies the use of, a single set of parameters over the entire well. Uncertainty of the regression parameters provides you with an uncertainty envelope for your real-time PP prediction. If you are able to acquire VES and velocity data pairs from a number of your offset wells, then the uncertainty in the regression parameters will be a more robust reflection of the impact of lateral and vertical heterogeneity in shale the mineralogy.

Ngurah Beni Setiawan

Geomechanics (petro Technical Expert) at Schlumberger

Evren, As Tim mentioned in his post, it is not about quantity of the data. It is true that (ideally) what we want is to have good quality data with a wide range of pressure and VES. I have been in a situation where we have "enough" valid data but still it is not "enough" for me to be able to derive Bowers' parameters with confidence. The data points are just clustered in a narrow cloud where a wide range of Bower's parameters seemed to give me good prediction. Another thing is to (always, if possible) have a blind-test well. The blind-test well will ensure the predictability of our parameters QC'd for at least the study area.

QUESTION 125

What is Vmudline when using Bowers' equation for pore pressure prediction? Is it velocity of the water column?

Shraddha Chatterjee

Project Geophysicist at CGG

ANSWERS PROVIDED BY INDUSTRY EXPERTS

Sankhajit Saha

Geomechanics Consultant at Baker Hughes

Shraddha, It is velocity of shallow sediment, and it considered to be near 1500m/sec.

Odofin David Khulong

Geopressure Specialist

Shraddha, It is the velocity of particles settling out of suspension, which have been generally accepted to be close to velocity of water. The value 1524 m/s is assigned to it. But experience has shown me that using that value result in Bowers method over estimates the pressure in the shallow section. As such, I do iterate it to get a good match with the measured pressure at such depth (sometimes, 1200 m/s does give me good result).

Phillip Clegg

Global Training Manager, Ikon Science (Geopressure Technology)

Shraddha, In his original equation Bowers suggested 5000 ft/s as V_0, but as with many of these types of equations there is room for local calibration. V0 is usually taken as seabed velocity. In reality it's the velocity where the vertical effective stress is zero.

QUESTION 126

I used Bowers method for pore pressure prediction from sonic logs and seismic velocity at well location. I got good result when I used Bowers unloading parameter (U) for PPP from the sonic log. I didn't see any effect of "U" on PPP from seismic velocity. Is Bowers' method only useful for pressure prediction for shale only, as seismic velocity contains effects of both shale and sand?

Ifikhar Satti

Research Scholar at Petronas University, Malaysia

ANSWERS PROVIDED BY INDUSTRY EXPERTS

Mark Herkommer

Geopressure Technology Consultant at Semicircle

Iftikhar, If you do not see any effect of "U" then something is wrong. It is a very potent parameter that can create very wild fluctuations in your PPP. Here is what I would check:

(1) Is U = 1? This means no effect. In order for U to work, it must be greater than 1.
(2) Vmax not reached. For most software applications, Vmax is a trigger that causes the PPP to move from the loading limb (Type 1 geopressure) to the unloading limb (Type 1 geopressure).
(3) Other geopressure Type triggers may not be reached.

Bowers is basically 2 cross plots, the loading limb and the unloading limb. If for some reason you are not switching to the unloading limb, which is controlled by "U", then you will not see any effect from using or changing "U". Beyond that, it is hard to guess why the parameter is not changing your interpretation. As for different lithologies, you should consider creating a separate cross plot for each lithology, or better yet, each formation.

Iftikhar Satti

Research Scholar at PETRONAS University, Malaysia

Mark, I tried to change all the parameters (Maximum Velocity depth, Values of A and B etc.) but didn't see any effect of "U" on PPP from seismic velocity, while value of "U" showed significant effect on PPP from sonic logs. I am also trying to solve this puzzle but not successful yet.

QUESTION 127

Can I have the steps or workflow on how to calculate PPP by Bowers method using seismic velocity and sonic/checkshot/vsp?

Ronald Siregar

Operation Geologist, Joint Development Area Malaysia - Thailand

ANSWERS PROVIDED BY INDUSTRY EXPERT

Mark Herkommer

Geopressure Technology Consultant at Semicircle

Ronald, The biggest step in the whole process is creating the Bowers cross plot. After that, you simply enter the cross plot from the velocity (Y-axis) until you get to the correct curve (loading or unloading), and then follow the line down to the effective stress on the X-axis.

Then, PP = OBG - EFF_STRESS

So, how to create that darn cross plot? If you are not using commercial software that has a built in ABU determination function, you will probably need to use a spreadsheet. When I first experimented with this I used a least-squares method for finding A and B, afterwards finding U more or less by trial and error. Unfortunately the least-squares method does not produce the optimal values for A and B. You need a method that will minimize total error, not just error along one axis. The solution is to use MS Excel Solver add-in function. To understand the Solver, it's important to understand the basic concept of what it does and how it works. There are 3 primary components you should become familiar with.

Target Cell. This is the cell that represents the goal or objective of the problem. In our example, Bowers parameter "A" and "B" will be varied to create the lowest error; the error is the target cell.

Variable Cells are the cells that can be modified to arrive at the desired outcome. In our example, that will be Bowers parameter "A" and "B"

Constraints. These are restrictions or limitations to what Solver can do to solve the problem. For example, "A" cannot exceed 15 and "B" cannot exceed 1

To Prepare to use Solver:

Create the lines that evaluate Bowers loading formula using the observed effective stress

Calculate the error squared (or abs error) between the Bowers formula velocity and the observed velocity

Sum the error in some cell

Once you start Solver A and B will be calculated so as to minimize the error. When it is done, you will have found the optimal A and B.

QUESTION 128

Can you explain the reason and solution (if any) for the mathematical flaw in Bower's regression between acoustic velocity and effective stress? As per Bower's power law equation for loading/DC mechanism: Acoustic Velocity = Mudline (or water) velocity + A* Vertical Effective Stress^B.

Anjanava Das Purkayastha

Associate Subject Matter Expert - Pore Pressure Prediction at Shell

ANSWERS PROVIDED BY INDUSTRY EXPERTS

Steve Hobart

Sr. Pore Pressure Advisor

Anjanava, I will recommend you read Bower's paper "Pore Pressure Estimation from Velocity Data: Accounting for Overpressure Mechanisms Besides Undercompaction". This is where the Bowers equation was introduced to the industry. In it, Bowers explains that the equation only applies to sediments with velocity-effective stress relationships that are on the virgin compaction curve.

While on the virgin compaction curve, with zero effective stress, the sediment is by no means a "rock". There is no mathematical flaw. Bowers' equation is an empirical relationship, not derived from first principles. It is consistent with

Terzaghi's effective stress definition, which in turn derives from Newton's principle regarding the equilibrium of forces. However, it is one of several empirical relationships which are consistent with Terzaghi's concept and which attempt to model a relationship between (vertical) effective stress and some proxy for porosity. In this case, the proxy for porosity is the speed of sound through given sediment (generally a claystone or shale) which is subjected to no overpressure generation mechanism other than undercompaction (compaction disequilibrium). There are numerous papers which acknowledge the existence of a "critical porosity" with respect to the speed of sound in the sediment (e.g. Raymer et al.). Above the critical porosity, the compressional waves essentially "see" the sediment as a slurry having a speed of sound essentially equal to that of the liquid phase. Sediment which has not yet compacted to less than 50% porosity (some authors use lower values) is still above the critical porosity.

Above the critical porosity, there is no practical relationship between speed of sound and porosity, between speed of sound and effective stress or between speed of sound and density (Gardner equation). The actual mud-line velocity is a reasonable approximation for the velocity in claystones having porosities greater than the critical porosity. The place to expect the velocity to be equal to the mud-line velocity is at the mud-line. Once buried, sediment will have a lower porosity and, if below the critical porosity, a higher velocity. There is no reversal of porosity under gravitational compaction (the virgin compaction curve). Mechanisms that create gas chimneys and geysers violate the assumption that the sediment is still on the virgin compaction curve. If there is any pore pressure in excess of the lithostatic, it is due to at least one mechanism other than gravitational compaction, in which case the equation does not apply.

Selim Simon Shaker

Consulting Geoscientist at Geopressure Analysis Services

Anjanava, I am always speculative and in doubt that equating the effective stress (which having the same ES using the velocity-depth cross plots) above and below the top of geopressure is scientifically incorrect. The velocity increases from - 5000' ft/sec at the mudline to a maximum value (V max) and reverses its trend at a depth where stress and lithology are adequate to form the top seal (TOG). Velocity increases above the TOG due to compaction (NCT) and reverses slope deeper in the section to values reflecting compartmentalization entrapment rather than compaction (Shaker, 2012). Therefore, equating the ES above TOG (hydrodynamic gradient in shale and sand) and below TOG (Geopressured Hydrostatic gradient in sand and exponential in shale) is physically unlikely.

However, Bower's calculated the effective stress at point B as follows:

Sigma B= Sigma max (sigm A/sigma max)^U (Bower's 1995) .... (1). It is a power law curve fitting of RFT data (RFT's) where is U=3.13 in GOM.
This is the backbone of his analytical calculation: V=Vo A sigma^B .... (2) where Vo,A and B are curve fitting parameters.

Here is my opinion:

Sigma A = OB at A – pressure at A (Hydrodynamic pressure gradient in sand and shale due to dewatering).
Sigma max = OB at TOG – pressure at Transition pressure gradient (between hydrodynamic and hydrostatic geopressured)
Sigma B = OB at B – Geopressured pressure gradient (linear in sand and exponential in shale).

I question if you can compare the Sigma A and B despite the fact that they are under different pressure mechanisms (regime) and also use three parameters for calibration fitting? Keep in mind the measured PP in sand (RFT) was the calibration tools for the curves fitting.

On Steve's comments above:

"While on the virgin compaction curve, with zero effective stress, the sediment is by no means a "rock" If effective stress at point A is zero therefore, the resultant of equation (1) is zero and the whole theory falls apart? Thanks Steve, this makes me more doubtful about these hypothetical methods. I believe we need to revisit all the vertical effective methods after a thorough understanding of pore pressure mechanism. What are the driving mechanisms of PP build up? Is it generation, expulsion or retention of the excess pressure? You can generate excess pressure from burial, heat, diagenesis etc., but the overpressure symptom occurs only if retention takes place (TOG) otherwise the excess pressure vent to the sea floor? I am still suspicious about the validity of all the vertical effective stress methods.

Phil Holbrook

Chief Scientist and Educator, Force-Balanced Petrophysics

Pore fluid pressure is a scalar load sharing relationship expressed by the effective stress theorem. The loading limb for any lithology is Bowers' "virgin compaction curve". This curve can then be revisited by subsequent compaction. The variations of vertical and horizontal stresses are encompassed by the effective stress theorem. I have two US patents that cover both Bowers' and Eaton's viewpoints in my "state-of-the-art" sections. Both Bowers and Eaton substitute depth (a coordinate) for average confining load, S(average) which is the heart of the mathematical and physical problem. This is dimensionally incorrect for starters. We're in the woods until we address this problem.

Eric von Lunen

Senior Geophysical Advisor & Team Lead at Nexen Inc

I think a little pragmatism is in order as Colin Sayers mentioned. For example, p-wave velocity is sensitive to bulk modulus and rigidity, and shear velocity is sensitive to rigidity only in the simplistic sense of mechanical properties when material is isotropic. An anisotropic media such as over pressured shale with microfractures would have distinct shear wave or rigidity modulus changes in my opinion.

Mark Herkommer

Geopressure Technology Consultant at Semicircle

Colin has pretty much nailed it. The Bower's loading limb is directly related to the Athy porosity decay equation (1931):

$$PHI = PHI0 / e \wedge bz$$

and time-average equation:

$$PHI = PHI0 * (RHO\_MAX - RHOB) / (RHOB\_MAX - RHO\_FLUID)$$

Phil Holbrook

Chief Scientist and Educator, Force-Balanced Petrophysics

Anjanava, Mark & Colin, First of all, the Dvorkin and Nur paper cites an ancient 1931 compaction function. It is not a regression. What you refer to as the time average equation is correct only for pure water and pure isotropic minerals.

Neither has very much to do with loading, unloading, or measurements on porous sedimentary rocks. Bowers stitched these together as a depth function if you can imagine that. The effective stress theorem does not mix concepts like this and is dimensionally correct. It is correct for the full range of porosity if you use solidity (1 - porosity) as a measure of strain.

For everyone's information, the "first fundamental in situ stress/strain relationship" was first published in 1995 and is still true today. Therein, each of the 5 most common lithologies has a mineral specific compaction function which is power-law linear. Their zero porosity intercept (Sigma max) is closely related to average mineral hardness which is mechanically sensible. Athy, Bowers and Eaton have nothing to do with it.

Anjanava Das Purkayastha

Geomechanist-Pore Pressure Prediction, Shell

I would like to thank everyone for responding to the question with such elaborate answers. I think Colin's point can be a solution to this very confusion. I'll work on that.

Phil, Sorry, but I could not get the point in your last comment. I may be missing something.

Mark Herkommer

Geopressure Technology Consultant at Semicircle

Phil, Let's get back to basics… Athy states:

"Definite relations have actually been observed from a laboratory study of more than 2,200 samples taken from wells in northeastern Oklahoma and parts of Texas." Please see Athy's Figure 2 for the compaction curve together with the data. That it was not obtained by statistical regression may be true, but there are literally hundreds of examples since then that have been produced.

The statement "first fundamental in situ stress/strain relationship" being published in 1996 is not entirely accurate. In 1959 Hubbert and Rubey published dimensionally correct relationships for isotropic and (importantly) anisotropic rocks, including stress reductions owing to fluid buoyancy effects.

With that said, I think we are getting far afield from the original question regarding a "mathematical flaw" in the Bower's loading limb. I think we can all agree that all offshore wells have at least three intervals: the water column, a water+sediment interval, and a sediment+water interval. Hedberg (1936) recognized this and named the putative stages of clay compaction: mechanical, dewatering, deformation, and recrystallization. To my thinking the question really is what is the nature of the water/sediment interface? Clearly as we move from 100% water to say 30% water+70% sediment the acoustic response will go from a fluid ITT (189 usec/ft) to a sediment ITT. If there is a flaw, it is in the simplification of the water/sediment interface where mechanical compaction begins, but not in the principle itself. Raymer, Hunt, and Gardner took a stab at the problem in 1981 in their SPWLA conference paper.

Owing to the inherent error bars on velocity values, I think Colin's point is where we can basically begin and end. I would only add that V_0 is frequently used, not as an interpreted value, but as a convenience. It is often manipulated to ensure that the first non-water time (i.e, the sea floor) accurately corresponds to bathymetric surveys. In any event, whatever the value, in the Bowers' method it must correspond with the zero effective stress state.

Pore pressure prediction using Miller's method

The Miller sonic method describes a relationship between velocity and effective stress that can be used to relate sonic/seismic transit time to formation pore pressure. In Miller's sonic method an input parameter "maximum velocity depth," d_{max}, controls whether unloading has occurred or not. If d_{max} is less than the depth (Z), unloading has not occurred, and the pore pressure can be obtained from the following equation (Zhang et al., 2008):

$$p = \sigma_v - \frac{1}{\lambda} In\left(\frac{V_m - V_{ml}}{V_m - V_p}\right) \tag{9.19}$$

where V_m is the sonic interval velocity in the matrix of the shale (asymptotic travel time at infinite effective stress, $V_m = 14{,}000 - 16{,}000$ ft/s); V_p is the compressional velocity at a given depth; λ is the empirical parameter defining the rate of increase in velocity with effective stress (normally 0.00025); d_{max} is the depth at which the unloading has occurred. If $d_{max} \geq Z$, then unloading behavior is assumed, and the pore pressure in the unloading case is calculated from the following equation:

$$P_{ulo} = \sigma_v - \frac{1}{\lambda} In\left[a_m\left(\frac{V_p - V_{ulo}}{V_m - V_{ml}}\right)\right] \tag{9.20}$$

where a_m is the ratio of slopes of the virgin (loading) and unloading velocities in the effective stress curves σ_{ul} (normally $a_m = 1.8$) and $a_m = V_p/V_{ulo}$; σ_{ul} is the effective stress from unloading of the sediment; V_{ulo} is the velocity where unloading begins.

Pore pressure prediction using Tau model

A velocity-dependent pore pressure prediction method was proposed by Shell through introducing a "Tau" variable into the effective stress equation (Lopez et al., 2004; Gutierrez et al., 2006):

$$\sigma_e = A_s T^{B_s} \tag{9.21}$$

where A_s and B_s are the fitting constants; T is the Tau variable, and $\tau=(C-\Delta t)/(\Delta t-D)$; Δt is the compressional transit time either from sonic log or seismic velocity; C is the constant related to the mudline transit time (normally $C=200\,\mu s/ft$); and D is the constant related to the matrix transit time (normally $D=50\,\mu s/ft$). Then, the pore pressure can be calculated from Eq. (9.21) using Eq. (9.1), i.e.,

$$P = \sigma_v - A\left(\frac{C-\Delta t}{\Delta t - D}\right)^{B_s} \tag{9.22}$$

The best-fitting parameters for the Gulf of Mexico are $A_s=1989.6$ and $B_s=0.904$ (Gutierrez et al., 2006). The Tau model and Miller's method are similar to the Bowers method. The advantage of Miller's method and the Tau model is that both the effects of the matrix and mudline velocities are considered on pore pressure prediction.

PORE PRESSURE PREDICTION USING SEISMIC DATA AND WELL DATA

QUESTION 129

How can I use seismic velocities to predict pore pressure? Can Petrel produce a pore pressure volume?

Bright Osuagwu

Petroleum Geophysicist at Afren

ANSWERS PROVIDED BY INDUSTRY EXPERTS

Erik Johnson

FugroGeoscience Division

Bright, fluid pressure is a function of Effective Stress, which in turn, affect acoustic velocity.

$$\text{Effective Stress} = \text{Overburden Pressure} - \text{Fluid Pressure.}$$

If you have well log data, you can use a linear regression to solve for OBP as a function of water depth + burial depth. After calibrating the seismic velocities to the sonic log(s), you can then solve for Effective Stress as a function of depth and velocity in the log domain using multi-linear regression. Once you have the OBP and the Fluid pressure, you can calculate the fluid pressure at any depth, and compare the predictions to the measured well data as a check on the estimates. You can calculate required mud weight as MW(ppg) = 6.3*FP/Z.

If you want to do this in 3D, you will need to calibrate the seismic velocities in a program that works not in cells but in regular samples such as a geophysical forward modeling program with a trace and sample calculator to apply the corrections and formulas in 3D. You will also need a seismic-to- well tie program to calibrate the velocities in time. You can then convert the relevant cubes to depth and transfer them to Petrel. Since the cubes will be fairly low frequency, you may not have to worry about sample to cell up-scaling too much. I am not a Petrel expert, but I believe that it has sufficient calculation functions so that the final calculations and calibrations can be done there.

José Adauto DE Souza

Senior Geophysicist,

Bright, you may retrieve "pseudo rock densities" from seismic velocities. These densities may be used to compute overburden (geo-pressure). With a 3D interpolation/extrapolation algorithm a "low precision" 3D overburden cube may be obtained. Well logs may be used to calibrate and enhance cube final data quality. This type of 3D data may help you understand/predict HC (hydrocarbon) migration paths to the reservoir. Pore pressure is basically associated to the presence HC and water. In a confined ambient pore, pressure is increased by overburden. In bottom of www.exploratoryvision.com homepage there is an example of a Geo-pressure 3D cube. In this case only well logs where used (with densities retrieved from sonic logs).

Phil Bailey

Global Geomechanics Focal Point, Rock Mechanics Well Engineer

Bright, I would also recommend checking that the velocities have been processed with PPFG prediction in mind. This can be a manual or calibrated process, to avoid over prediction and lithological effects. CGG and other providers can assist with this in more detail.

QUESTION 130

It is well known that one can calculate pore pressure via density logs (obtaining overburden gradient) and then - by Bowers' and Eaton's approaches. In my case I have only a 3D seismic cube, an interval velocity cube and resistivity/sonic well logs (which were recorded not from the surface datum and which contain at least ~1000 m (3300 ft) blind/deaf zones). In general there is a lot of printed literature related to this issue but most of them are only theoretical discourses without practical solutions. So, the question is: is it possible to calculate pore pressure with only a 3D seismic cube, an interval velocity cube and some limited coverage resistivity and sonic well logs?

Oleksandr Okrepkyi

Geologist I category - Nadra Integrated Solutions

ANSWER PROVIDED BY INDUSTRY EXPERTS

Emanuel Flores

Geophysical Team Leader, Seismic Characterization Services en CGG

Oleksandr, actually in deep water in the Gulf of Mexico, we just have seismic data for PP (pore pressure) analysis at a well location to drill (specifically seismic velocity from HD picking analysis or tomography and seismic stacks). We obtain a density from seismic velocity via Gardner's equation. For 500 m below water bottom we use Miller equation using porosity variations with depth, and these are input data as overburden gradient, computing the sum of values. For trend velocity we have to analyze some points in the cube to check the variability of trends with water depth. From this analysis we obtain constant values for each point and we create a new cube krigging those constants point to point. This produces a new cube of values coming from points (pseudo wells) and using an approximation for velocity trend with depth you will have a velocity trend cube. With interval velocity (input data), velocity trend and overburden gradient, you can estimate a Pp cube expressed in equivalent density, and using a density constant for Hp (1.03 g/cc water density) you have the input data for Eaton equation.

Oleksandr Okrepkyi

Geologist I category - Nadra Integrated Solutions

Emanuel, your comment is really useful and it looks even like a manual with a lot of details, hence I definitely will have some questions: If I understood you correctly, we can "draw" such a sequence of needed iterations (I will generalize them):

1. We have seismic velocities which can be used as input data for Gardner's and Miller's equations. As a result it is now possible to calculate overburden gradient (OBG).
2. Next step is velocity cube checking and creating a new one while checking variability of trends. It seems to be an important moment in the offshore areas. What about onshore? I suppose that creating an interval velocity cube is a necessary step in both offshore and onshore?
3. Having a trend and calculated OBG, now it is possible to calculate Pp/Peff etc.

Well, must say, I really appreciate your help, it makes me understand this issue quite better: thank you.

Emanuel Flores

Geophysical Team Leader, Seismic Characterization Services en CGG

Oleksandr, you will create a new velocity cube, this has the variability of trends, and for example, if you have some wells inside the area, you can interpret the normal compaction trend just taking the interval that you consider as normal. If you have deviations from the trend those values are "over-pressured" because generally, you may have overpressure or some gas sands, because in both cases you will have decreases in velocity. But if you have some AVO or another data you will have support for your interpretation. Anyway, when you take the velocity values that you consider as normal, the equation of the regression versus depth will give you the necessary constants at each point. Also you can create some pseudo wells in the same exercise where you have interval velocity. This step is important onshore and offshore, even more if you have a very complex geology, for example, faults that are separating regions, or a very strong deformation on water depth.

QUESTION 131

If pore pressure prediction can be obtained using seismic data, is it practical to do it in the early stages of exploration? What's the importance of it and how can I predict it?

ANSWERS PROVIDED BY INDUSTRY EXPERTS

Noemi Ordonez Perez

Sr. Pore Pressure Geologist at Repsol

One in many reasons to do Pore Pressure predictions at early stages of exploration is its application to calculate maximum HC column heights trough top seal capacity analysis. This helps to avoid overestimation on the reserves calculation.

Julien Martineau-Fabre

Sr. RT Geomechanics Eng.

Another point of performing PPFG prediction at the early stage is to be able to draw the draft of the well design. Like this, should some potential issue rises (shallow gas, HP, fluid transfer, potential seal breach etc), then dedicated teams would have time enough to consolidate the well design and optimize it, reducing uncertainties before the well is being drilled.

Ayman Khorshid

Subsurface Geologist team Leader & Senior Operation Geologist

The pore pressure prediction helps in well architecture and mud program. P shale derived from seismic data helps in the pre-well planning and knowing what type of problem can be expected while operating and helps in preparing early solution.

Dilip Kumar Mahata

Production Seismologist at Shell

Pore Pressure Prediction in early stage will help:

1. Well and mud design
2. Understand the drill-ability
3. Estimate and optimize cost of exploration
4. Calculate maximum HC Column
5. Predict over pressure zones
6. Understand pressure related fault behavior
7. Exploration risks and IP Volumes
8. Define what is a valid hydrocarbon trap
9. Predict fluid behavior in trap

Selim Simon Shaker

Consulting Geoscientist at Geopressure Analysis Services

We have forgotten to mention sequence stratigraphy function in collaboration with the seismic interval velocity. Seismic reflection surfaces where velocity is picked are the lithological boundaries. Therefore, we should eliminate the velocities related to NON shale and very fine clastics from our calculations.

Scott Cintula

Senior Partner at Integrated Surface Petrophysics

I agree with Selim's comments. In addition, I would state that "Velocity based pore pressure prediction 'Pre-drill' analysis of a proposed well location, postmortem of relative analogue well(s) and surrounding intersects is in a word "is vital" for a safe, efficient, as well as relatively confident, execution of a drilling program. Without a pre-drill analysis of any validity, a pore pressure analyst can really only confirm 2 of (in my experience) 5 primary objectives of a PPP workflow and would be at the wellsite:

- Depth of Onset into an abnormally pressured interval
- Magnitude of change in pressure through an interval would be,
  1. Rate of Change.
  2. Pressure generation mechanism.
  3. Compartmentalization - (feel free to add to this list). This is just how I personally prioritize these objectives.

A multi-disciplinary, integrated approach would vary slightly upon geologic setting. And also consider the other side of this type of work. From seismic data we derive rock properties from direct and empirical relationships to predict fracture gradient/rock strength which is also of vital importance in the early stages of planning.

QUESTION 132

Which applications are used for pore pressure prediction from 3D seismic data, well log data and integrate both seismic and well data?

Nasir Mahmood

Assistant Geophysicist at OGDCL

ANSWERS PROVIDED BY INDUSTRY EXPERTS

Rob Dudley

Geoscientist

Nasir, The Ikon software (RokDoc) is robust, detailed and full of QC.

Fernando Enrique Ziegler

Geophysicist:Pore Pressure/Wellbore Stability

Nasir, Personally, I use a variety of different software, mostly depending on what is needed. For a three-dimensional pore pressure analysis, from scratch (i.e. from seismic gathers):

- Geolog: for data conditioning of well logs
- Drillworks Predict: for pore pressure analysis of offset wells
- Hampson-Russell: for data conditioning of seismic gathers
- VelPRO: for semblance-based velocity picking

- Permedia: for integrating pore pressure analysis of offset wells with the velocity model and also, great for pore pressure (seismic and well-based) analysis with respect to conventional basin modeling.

Victor Botelho

Geomechanics Consultant III at Halliburbon

Nasir, If you want do geopressure calculations based on interval velocities, Drill works 3D is perfect. Drill works Presage would be better for basin analysis using seismic and real data from offset wells.

PORE PRESSURE PREDICTION IN DEVIATED WELL

QUESTION 133

What would be the best way to calculate the overburden for a planned deviated well?

OBG at well location vertical and at TD may indeed be very different, hence both have to be modeled. However, how do I handle the transition from kick off point to TD? Assuming of course there is a significant departure and a different sediment load at the spud location and the TD target.

Bertrand Bocquet

Wellsite Geologist for BP Egypt

ANSWERS PROVIDED BY INDUSTRY EXPERTS

Marcel F. Castro

Geomechanics Engineer at Schlumberger

Bertrand, I am going to answer your question in two different ways "easy and difficult".

Easy: You will need to get the density log from surface to TD from any offset well and propagate this property to your projected trajectory. Integrating the density profile you will get a very good 0.5 ft resolution approximation. As you don't have a better resolution density log profile no matter how accurate your geological model is this always will be your limitation, also if you are using any interpolation method to fill the gaps in your density profile that is a limitation. With this methodology you will get very good overburden estimation for your well and you will do this very fast.

Difficult: A 3D model will not necessarily give you the best answer, but for sure this approach will consume much more time and effort. For an accurate 3D model you will need many offset wells to be incorporated in your model and again you will be constrained due the population method, your cells size and vertical log resolution. In many cases you don't have all the density profiles in your offset wells so, at some point you are using interpolation methods which will give you uncertainty in your density cube.

Finally I think the most important decision for you is to decide how important your planned well is. How good and complete is your offset data and how much time do you have to complete your project to provide an answer. Always you can explain the limitations of your methods and the improvements that you can do in order to improve your first approach, including of course the time and efforts to do that.

Nitin Bhardwaj

Senior Operations Geologist (Geopressure and Geomechanics)

Bertrand, This is indeed a tricky situation because deviated wells essentially pose a 3D problem, whereas in PPFG prediction we are mostly looking at 1D solution. This is a particular problem in cases where wells are drilled in location close to offshore shelf-slope transitional areas. The best solution for this problem is to use a 3D modeling tool. In this method you first need to build a 3D density cube or 3D overburden cube which covers the full well trajectory. Then you can extract OBG along the well path in MD or TVD. If you don't have the 3D modeling tools then you may requires 3 separate sets of velocity profiles, one going from well surface location, one going from TD and one going from the middle point between TD and kick off. They can then be transformed from velocity to density to OBG. Once transformed you need to splice them at appropriate levels to make a composite profile. This may not be the best solution but it is better than using a profile based at well surface location.

Bertrand Bocquet

Wellsite Geologist for BP Egypt

Nitin, That was more or less my idea, but with more than 3 sets of velocity, to reduce the 'uncertainty'. However, the more I dig into deviated wells the more questions pop up. A traditional best practice is to integrate data from the well to check/correct our OBG. (Either DT transform, Rhob from LWD or wireline, etc.). However, when you reach your deviated TD, the real time data you integrated, lets say near your kick off interval becomes irrelevant, hence your OBG is wrong, with all implications. (Again, this assumes a significant sediment load difference between your surface location vertical and the TD).

Tebis Llobet

Geomechanics Advisor

Bertrand, note that, even for a vertical well, overburden from simple vertical integration of formation density is probably wrong, anyway, if there is substantial lateral variation. It is indeed a 3D problem, but there are other methods, besides FE and FD (e.g.: point-load methods based on Boussinesq's model), that allow one to make reasonable estimates without the pains of laterally extending the model.

At any rate, most overburden estimation methods completely ignore the deposition history and the frictions/creep usually associated with it.

Bertrand Bocquet

Wellsite Geologist for BP Egypt

Tebis, It is probably true that even for 1D case, OBG based on density integration may be different from the one estimated by FE techniques. However as FE techniques are relatively cumbersome, integration methods are usually preferred both in 1D and 3D domains.

Tebis Llobet

Geomechanics Advisor

Bertrand, In the 1D case (say, vertical - or deviated - well in horizontally-layered cake) except for really poor discretizations, the results should be the same. In 2 or 3D, however, if you have substantial lateral variation in the density (as in near a salt dome, under it, or where topography varies a lot), vertical integration of the formation density is most certainly a poor estimate.

Ekene Okoye

Geomechanics Specialist at Baker Hughes

Bertrand, Calculating Overburden in deviated, horizontal well, extended wells, and surface-onshore to target-offshore is a 3D problem. As mentioned earlier, this issue has to be addressed by a 3D subsurface modeling tool which can handle complex structure covering the entire trajectory. This has been taken care of in the Jewel Suite 3D Geomechanics application.

Matt Czerniak

Rig-based PPFG Real-Time Operations, Independent Geopressure Consultant

I have had to do that on multiple occasions for sub-salt deviated wells. Extracting a 1D projected overburden along a deviated well path from a 2D or 3D volume in Gocad is fairly simple and straightforward. But time was short, I have resorted to constructing end-member 1D profiles and then doing a linear interpolation where the salt thickness changes along the well path. Salt thickness can change gradually or abruptly, but my approach has successfully captured this effect in both cases.

OVERPRESSURE AND WELLBORE INSTABILITY

Abnormal pore pressures, particularly overpressures, can greatly increase drilling time and cause serious drilling incidents (e.g., well blowouts, pressure kicks, fluid influx) if the abnormal pressures are not accurately predicted before drilling and while drilling. A study of 2520 gas wellbores drilled on the Gulf of Mexico shows that more than 24% nonproductive time was associated with incidents of kicks, shallow water flow, gas flow and lost circulation (Dodson, 2004), which were caused by improper pore pressure and fracture gradient prediction. In the deepwater of the Gulf of Mexico (water depth over 3000 ft), incidents associated with pore pressure and wellbore instability took 5.6% of drilling time in nonsubsalt wells, and 12.6% of drilling time in the subsalt wells (York et al., 2009). Furthermore, an investigation of drilling incidents shows that there are 48 kicks for 83 wells drilled in the deepwater Gulf of Mexico, and at least 21% of those kicks resulted in loss of all or part of the well (Holand and Skalle, 2001). The abnormally high pore pressures not only caused the kicks and blowouts, but also induced geologic disasters, such as mud volcano eruptions (Davies et al., 2007; Tingay et al., 2009). Therefore, accurate pore pressure prediction is of crucial importance for operators to reduce borehole trouble time and avoid drilling incidents.

To optimize drilling decisions and well planning in overpressured areas, it is essential to carry out pore-pressure predictions before drilling. Knowledge of pore pressures implies knowledge of the effective stress, which is a key input for several geomechanics applications, such as fault slip and fault seal analysis, and reservoir compaction studies. It is also a required input for 3D and 4D seismic reservoir characterization. Because the seismic response of shales and sands depends on their compaction history, the effective stress will govern the sedimentary seismic response. This is in contrast to normally pressured regimes, where the depth below mudline (or overburden stress) is typically used to characterize the compaction effect.

QUESTION 134

How does the coupling effect of (increasing or decreasing) pore pressure and under-compaction affect overpressure? Furthermore, what are the indications to watch out for in order to prevent wellbore instability during drilling; and excessive porosity loss, thus failure of the formation surrounding the wellbore during production?

Oluwasesan Ayo Alalade

SLS Offshore Data Analyst at Weatherford

ANSWERS PROVIDED BY INDUSTRY EXPERTS

Fernando Enrique Ziegler

Geophysicist:Pore Pressure/ Wellbore Stability

Ayo, The minimum required mud weight needed to prevent mechanical wellbore instability can be estimated by calculating shear failure using rock properties derived from compressional sonic data. Before you drill, this may be calculated using any data from nearby wells. As you are drilling, one assumes that you are drilling with a mud weight greater than your estimates of (pore pressure and) shear failure to prevent mechanical wellbore instability. Your estimates can be adjusted if and when you encounter signs of mechanical wellbore instability such as a tight hole, hole size enlargement not associated with wash-outs, packing-off, increased volume of cuttings, fill on bottom after tripping, "non-drilled" cuttings or caving, or when problems are more noticeable when you stop circulating (changes from equivalent circulating density to equivalent static density).

Oluwasesan Ayo Alalade

SLS Offshore Data Analyst at Weatherford

I recently learned pore pressure calculations and prediction do not work for sand. Can anyone explain why we cannot determine the pressure in sand formations? What about cemented sand, for instance SS with interbedded limestone or dolomite (as its cement); or very compacted sandstone? And I'm basing these instances on bulk density determination as the problem for the inconsistencies of pore pressure prediction in SS formations. Are there basically any easily accessible papers, research, or suggestions why pore pressure prediction can't be used in SS formations?Basing sand pressures primarily on its overlying shales and assuming the sand isn't faulted at its top and is conformable to the shales, is a pretty big assumption in my thinking.

Keith Katahara

Senior Geophysical Advisor at Hes Corporation

Ayo, I think you are talking about estimating formation pressure from other measured properties such as sonic velocity, density or resistivity. In shales, such properties tend to depend primarily on effective stress; basically overburden minus pore pressure, which governs mechanical compaction. That dependence on effective stress allows us to compute pressure from velocity or other properties. In sands or other formations such as limestones, velocity and other properties tend to not be strongly related to effective stress. They depend more on temperature and cementation/diagenesis, except for shallow unconsolidated sands. Generally, the greater the extent of diagenesis and chemical compaction, the harder it is to estimate pressure in this way. As shales get hotter and more cemented it eventually becomes impractical to infer pressure in shales from velocity. Of course there are other ways to get at pressure. You can directly measure pore pressure in permeable formations like sandstones, but you can't measure it in shales. Sands will flow during drilling if the wellbore pressure drops below formation pressure, whereas shales will not flow. And so on.

Yogesh Kulkarni

Pore Pressure Analyst/Field Engineer, Geoservices, Schlumberger Company

Ayo, In addition to the method mentioned above by Keith, PP in sand can be estimated based on various gas signatures (hydrocarbons recorded on surface while drilling) and drilling events as compared with respective annular mud weights.

J. Grant Brydle, P. Geoph

Interpretation Geophysicist

Ayo, As part of your research on pore pressure prediction (PPP) you can look at these papers:

AAPG Memoir 76, Pressure Regimes in Sedimentary Basins and Their Prediction, 2002, AAPG, edited by Alan Huffman & Glenn Bowers.

SEG Distinguished Instructor Series No. 13, Geophysics Under Stress: Geomechanical Applications of Seismic and Borehole Acoustic Waves, 2010, SEG, by Colin M. Sayers.

RT PORE PRESSURE ANALYSIS

QUESTION 135

A client has an important well and decides they need RT PP analysts (offshoreobviously). The client has a pre-drill(s) analysis (either in-house or by a 3rd party). What information should the guys going out to the rig be given and why?

Julian Coker
WSG/PP

ANSWERS PROVIDED BY INDUSTRY EXPERTS

Yogesh Kulkarni
Pore Pressure Analyst/Field Engineer, Geoservices, Schlumberger Company

Julian, it would be preferable the crew going for RT analysis should be part of pre-drill study/basin analysis, even if client got its in-house pre-drill done. Other than pre-drill, the information to carry to offshore should also include: set flags and communication protocol with Client for RT operation, operation intervention scope of PP analyst (if new client). Generally, these points are part of pre-drill preparations. Thorough pre-drill improves probability of successful Job.

Julian Coker
WSG/PP

Kulkarni, I am assuming communication protocols and paperwork are in place. I am only interested in the data. But why do you think the crew involved in the RT work should also be involved in the pre-drill?

Ekene Okoye
Geomechanics Specialist at Baker Hughes

Julian, As RT modeling and analysis is usually done to reduce the uncertainties and avoid surprises, you sure need to have the pre-drill model. It's important to have the right personnel who understand the pre-drill model with its details but also have knowledge of the basin being drilled. It's important that the pre-drill sensitivity analysis is included to indicate the type of data or logs to be acquired while drilling to reduce the uncertainties in the pre-drill model.

Yogesh Kulkarni
Pore Pressure Analyst/Field Engineer, Geoservices, Schlumberger Company

Julian, If RT PP analysts do the pre-drill analysis/offset well study and understand regional geology in terms of origin and mechanism of pressure regime and build a model for RT, it would boost the confidence of the crew if they do pre-drill on their own to device strategies for RT job to tackle uncertainties. I guess details of the pre-drill study cannot be conveyed completely just by forwarding PPT or pre-drill model files with trend lines.

Julian Coker
WSG/PP

Ekene, by 'sensitivity analysis' do you mean checking out the fudge and bs factors one puts in one's analysis? Do you work onshore or offshore? But both Ekene & Kulkarni make the point that the best guys to do the pre-drill are the same people (or company that we send offshore). My opinion is that when we do a pre-drill (or anything to do with geology) the earth is too complex, so we have to make assumptions. If the same company is used for the Pre-drill and RT, they will have no problems sharing those assumptions. However, what happens if the folks who are going off-shore have not been involved in the pre-drill? What datasets should they be given, and why do they often not get what they think is required?

Hatem Alnajar

Pore Pressure & Geomechanics Independent Consultant

Julian, I worked both Pre-drill and RT for a long time and realize that Pre-drill is not necessarily the same as RT. There might be a big difference between RT and Pre-drill. For example, I worked a well in NA (Mauritania) where the Pore pressure Ramp (pre drill) was showing 13.4 ppg MW while RT showed Hydrostatic PP. So the suggested MW was reduced and the well was drilled to 10.2ppg and npt as suggested. A casing point was ignored and the well was drilled below suggested Pre-drill mud weight and a casing point was neglected. I suggest that not only the pre-drill model be transferred to RT monitoring people but also all pre-drill information has to be given to the RT people so they can build their own pre-drill models. That includes pre-drill Res, DT, RHob, Seismic Vel, LOT and FIT.

Alan Morrison

Senior Business Development Manager at GeoLog International

Julian, as scientists working up pre-drill models, we deal with ranges of uncertainty, based on the data available. Your average rig based drilling crew (and I do mean the entire crew) probably feel safer given something more rigid to work with (possibly a single number!). A company can act on at least 2 philosophies –

(A) The rig based RT crew is there to identify all manifestations of pore pressure and communicate that back to people to interpret it (the pre-drill team).

(B) The rig based crew is also there to interpret and collaborate with the shore based team that derived the pre-drill. The information supplied in each case varies hugely. Team A might be kept purposely in the dark so as not to cloud their analysis with pre-conceived ideas! Team B might be given everything, included in the pre-drill analysis. I've seen both applied, and would generally err in favor of the latter rather than the former myself. When drilling departments see differing opinions, they tend to lose trust of somebody and that is never a good thing.

Tim Sheehy

Senior Manager Real Time at Ikon science

Julian, the best situation is where the RT team is very involved in the pre-drill and if that is not possible they should get all the data from the pre-drill. The switch from PP prediction to PP detection is critical as others have already stated. And while there can of course, only be one mud weight (putting aside dual gradient for now) nailing down the best number to report at the rig is always informed by a good pre-drill. A key point behind all the data is that this can never be just a LAS file. It must also include contextual type info and notes, assumptions (already mentioned above) etc. Basically the RT crew should be involved in any kind of Peer Review/DWOP type meeting concentrating on the geology.

Julien Martineau-Fabre

Sr. RT Geomechanics Eng.

Julian, As an RT specialist, I can do a pre-drill as well. The RT operators should be given not only the pre-drill, but also the data used and given explanations on resulting model (regional knowledge, ovp mechanism, methods). I know clients can be reluctant in providing such (independence analysis, tight data), but when presenting my model (pre-drill or RT), I keep insisting on the fact that a model is a model, and not the truth. Even the post-drill is another model since we estimated pp in shale and measured it in sand.

DENSITY EXTRAPOLATION IN PORE PRESSURE PREDICTION

QUESTION 136

I have assumed sediments density at the surface in order to calculate overburden stress and gradient. Now I have to extrapolate it to the first measurement point on the density log (95ft, density 1.878). How can I extrapolate the density?

Izhar ul Haq

Geologist

ANSWERS PROVIDED BY INDUSTRY EXPERTS

Mark Tingay

Senior Geomechanics specialist at Chevron

Izhar, If you first density data point is only at 95ft, then you really don't need to extrapolate at all. I would just assume that the average density from the surface to your first data point is the same as the density at your first data point.

It's really not that important when the density data you have starts so shallow. The biggest source of error in overburden gradient calculation is related to the depth of that first density data point (I did a detailed analysis of method and errors sources in Tingay et al., 2003, Marine and Petroleum Geology, Vertical stress variation in the Baram Basin, Brunei). The further from the surface you need to estimate data to the top of the density log, the more likely you will have errors. However, when your first data point is at only 95ft, then even a huge error in average density from the surface to 95ft will make almost no difference to your vertical stress. For example, let's take an extreme case and say your estimate of density from the surface to 95ft was a huge error - like 1.0 sg error (8.345 ppg error - a simply ridiculous error that would never happen). Even with such a massive error, this equates to an error in your overburden gradient of about 0.2 ppg - which is not much at all.

Basically, if you have a density log starting at 95 feet, you are extremely lucky - no need to extrapolate, just use that value from the surface to 95ft. It is really when you start needing to extrapolate more than 1500m or so of density data that you are at risk of significant errors. Also, make sure that you check your density log for artefacts caused by borehole rugosity, and be sure that the data is in TVD - other common errors.

Dariush Javani
Mining Engineer
Mark, Your answer to Izhar was very useful. But, what is your suggestion for off-shore wells? I mean if there is no data for first sections (usually there is no available logs for 24" section) of a well at off-shore. What should be our assumptions to get a reasonable extrapolation?

Waqar Hussain
Mud Logging Technical and Support Engineer, Sagemines Diamoco Group France
Dariush, For offshore, you need to take the water depth and sea water weight into account, plus you need to see what was in 24 inch section. As for offshore south pars you have argillaceous limestone which is totally fractured. That is why you do not get return, but geologist do tell you what is there so you need to calculate it with that weight. For **Izhar:** if you do not have any data on density logs but you can get this data from the mud log, you can go ahead with your project.

Dariush Javani
Mining Engineer
Waqar, If we have access to the field which is going to be studied then we can sort out these issues. But my question is if we do not have access how are we to make a reasonable estimation?

Waqar Hussain
Mud Logging Technical and Support Engineer, Sagemines Diamoco Group France
Dariush, Without data access, it will be not correct, but still with knowledge of the surrounding area you can get a rough idea and go ahead with your calculation. But as it is overburden stress that you want to calculate, the lack of good data will affect the calculation to the end. So in my view for research purpose you need to request more accurate data, to add more cream to your research work.

QUESTION 137

I am doing a post-drill pore pressure analysis in an offshore basin. I have a couple of questions. Firstly, I have density log data from 600 m to 5000 m and water depth is 60 m. So for the over burden calculation, I would need the density from 60 m to 600 m. What is the general industry practice for assuming the density where no log is present (in this case from 60-600m)? Secondly, is the normal compaction trend a least square best fit trend (I mean is it the L-2 Norm best fit curve) assuming we use only the shale data points?

Sumeet Mavani
Geoscientist

ANSWERS PROVIDED BY INDUSTRY EXPERTS

Steve Hobart

Sr. Pore Pressure Advisor

Sumeet, One common technique is to assume a mudline density for the area in which you are drilling. Depending on the age of the sediments and other environmental factors, this may range from less than 1.5 g/cc (deepwater Gulf of Mexico) to maybe as high as 2.2 g/cc (for example, wherever Somenath is used to working). Continental shelf mudline density in the Gulf of Mexico is often assumed to be between 1.9 and 2.0 g/cc). During the last Ice Age, the mean sea level was significantly lower than today, resulting in erosion of softer sediments from what is now the continental shelf.

Another technique is to simply extrapolate from logged density data to the mudline using a least-squares technique. Some may apply this technique to the logarithms of densities.

I am partial to converting the density log to density porosity and then extrapolating the porosities to the mudline using a least squares fit on a logarithmic scale. This would conform to Athy's equation, which relates porosity to TVD using an exponential relationship. Of course, hard data trumps algorithms. Therefore, if there were cores taken nearby of the near mudline sediments, such as for a hazard survey, you may be able to obtain actual values that you can use for the density profile just below the mudline. The normal compaction trend-line (NCTL) is whatever the expected value is in a normally pressure sediment of the porosity-sensitive measurement. Eaton methods for Resistivity/Conductivity, Sonic/Velocity, Dx/Dxc use trendlines that are linear with depth on a logarithmic basis. You may want to consider non-linear trends for sonic/velocity data. The Bowers, Miller and Hamouz-Mueller methods come to mind in this regard.

The NCTL is applicable only to the lithology for which the underlying porosity - effective stress relationship holds. Restricting its use to a consistent shale mineralogy is considered a best practice by many.

COMPACTION TREND

One of the major causes of abnormal pore pressure is abnormal formation compaction (compaction disequilibrium or undercompaction). When sediments compact normally, formation porosity is reduced at the same time as pore fluid is expelled. During burial, increasing overburden pressure is the prime cause of fluid expulsion. If the sedimentation rate is slow, normal compaction occurs, i.e., equilibrium between increasing overburden and the reduction of pore fluid volume due to compaction (or ability to expel fluids) is maintained (Mouchet and Mitchell, 1989). This normal compaction generates hydrostatic pore pressure in the formation. Rapid burial, however, requires faster expulsion of fluids in response to rapidly increasing overburden stress. When the sediments subside rapidly, or the formation has extremely low permeability, fluids in the sediments can only be partially expelled. The remaining fluid in the pores of the sediments must support all or part of the weight of overlying sediments, causing the pressure of pore fluid to be abnormally high. In this case porosity decreases less rapidly than it should with depth, and formations are said to be undercompacted or in compaction disequilibrium (Fig. 9.6). The overpressures generated by undercompaction in mudrock-dominated sequences may exhibit the following characteristic: the abnormal pore pressure change with depth is subparallel to the lithostatic (overburden) pressure gradient (Swarbrick et al., 2002). The compaction disequilibrium is often recognized by higher than expected porosities at a given depth and porosities deviating from the normal porosity trend. However, the increase in porosity is not necessarily caused merely by undercompaction. It could also be caused by other reasons, such as microfractures induced by hydrocarbon generation. Therefore, porosity increase and abnormal pressure may be caused by both undercompaction and hydrocarbon generation. Thus, by knowing the normal porosity trend and measured formation porosities (which can be obtained from well log data), one can calculate the formation pore pressure.

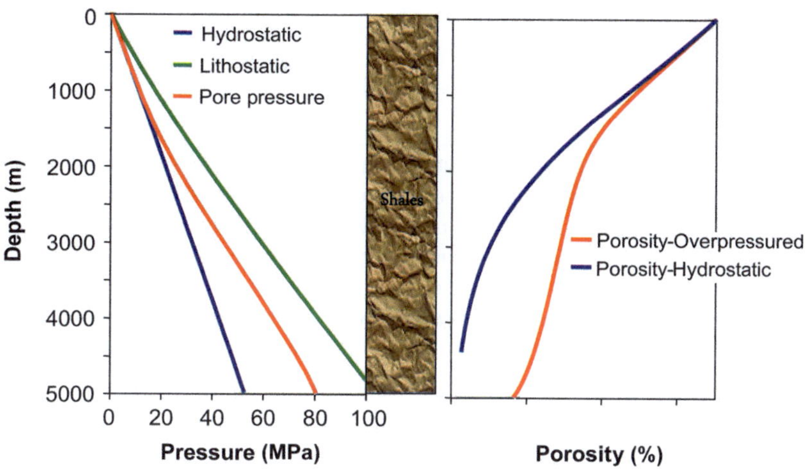

FIG. 9.6

Shows compaction disequilibrium curves.

Eaton's Resistivity Method With Depth-Dependent Normal Compaction Trendline

In Eaton's original equation, it is difficult to determine the normal shale resistivity or the shale resistivity in the condition of hydrostatic pore pressure. One approach is to assume that the normal shale resistivity is a constant. However, the normal resistivity (R_n) is not a constant in most cases, but a function of the burial depth, as shown in Fig. 9.7.

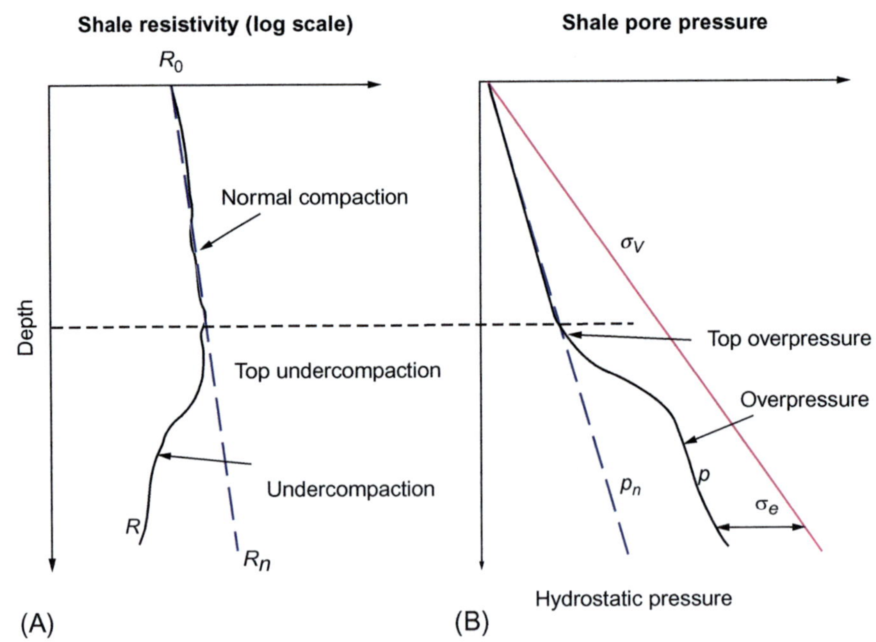

FIG. 9.7

Schematic resistivity (A) and pore pressure (B) in an undercompacted basin. The inclined line in (A) represents the resistivity in a normally compacted formation (normal resistivity, R_n). In the undercompacted section the resistivity (R) reversal occurs, corresponding to an overpressured formation in (B). In the undercompacted/overpressured section, resistivity is lower than that in the normal compaction trendline (R_n). In the figure, σ_v=lithostatic or overburden stress; σ_e=the effective vertical stress; p_n=normal pore pressure; p=pore pressure.

Thus a normal compaction trendline needs to be determined for pore pressure prediction. Based on the relationship of measured resistivity and burial depth in formations with normal pressures, the following equation of the normal compaction trend of resistivity can be used (refer to Fig. 9.7):

$$\ln R_n = \ln R_0 + bZ$$
or
$$R = Re^{bZ} \tag{9.23}$$

where R_n is the shale resistivity in the normal compaction condition; R_0 is the shale resistivity on the mudline; b is the constant; and Z is the depth below the mudline. Substituting Eq. (9.23) into Eq. (9.11), the Eaton's resistivity equation can be expressed in the following form:

$$P_{pg} = OBG - \left(OBG - P_{ng}\right)\left(\frac{R}{R_0 e^{bz}}\right)^n \tag{9.24}$$

where R is the measured shale resistivity at depth of Z; R_0 is the normal compaction shale resistivity in the mudline; b is the slope of logarithmic resistivity normal compaction trendline. A case study is examined to verify the adapted Eaton's resistivity method with depth dependence.

Case Application

The studied basin is located in a deepwater field in Green Canyon of the Gulf of Mexico, United States. The water depth is 5000 ft, and the Tertiary formations are mainly shales (mudstones) with some sandstones. The target reservoir is located in the Miocene sandstones. Several offset wells are analyzed to examine pore pressures in this field.

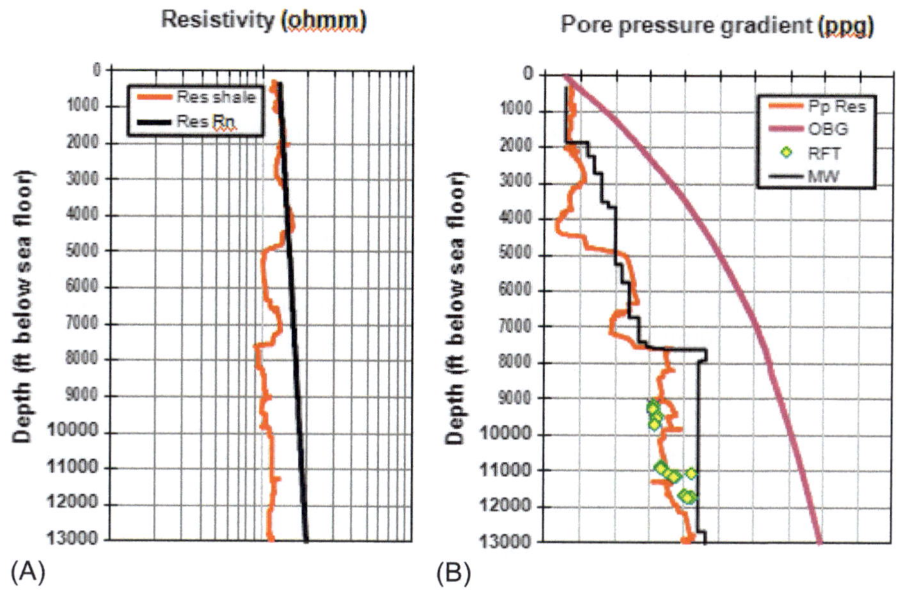

FIG. 9.8

Pore pressure calculated by adapted Eaton's resistivity method with depth-dependent compaction trendline in a deepwater postwell analysis in the Gulf of Mexico. (A) plots the resistivity in shale and the normal resistivity calculated from Eq. (9.23), and the resistivity is plotted in logarithmic scale. (B) shows the overburden stress gradient (*OBG*), mud weight used while drilling (MW), measured pore pressure gradient (RFT) and pore pressure gradient (PP Res) calculated from resistivity using Eq. (9.24).

Fig. 9.8 shows the pore pressure calculation in an oil well using the modified Eaton's resistivity method in this basin. Prior to the pore pressure calculation, the normal resistivity compaction trend is first analyzed based on Eq. (9.23), as shown in Fig. 9.8A. With calibration of the measured pore pressure data, the normal compaction trendline is obtained with the following parameters in this basin: $R_0 = 1.28\,\Omega\text{m}$, $b = 0.000034$. Pore pressures calculated from the adapted Eaton's method (Eq. 9.24) with $n = 1.2$, $P_{ng} = 8.7$ ppg) are compared to the measured pore pressures from the repeat formation tests (RFT) and mud weight. Fig. 9.8 indicates that the formation is in normal compaction when depth is <4900 ft below the sea floor. Deeper than this depth (from 4900 to 7600 ft), the formation is slightly undercompacted with a lower resistivity than the normal compaction trend (Fig. 9.8A), implying that the pore pressure increases, as shown in Fig. 9.8B. From 7600 to 13,000 ft, the formation is further undercompacted and more elevated pore pressures exist. Fig. 9.8 demonstrates that the adapted Eaton's resistivity method gives a fairly good result in the pore pressure calculation. It should be noted that the pore pressure in the formation near the wellbore is affected by drilling-induced stresses (Zhang and Roegiers, 2005). Therefore, in order to obtain the formation pore pressure, the deep resistivity is needed for the pore pressure calculation.

Eaton's Velocity Method With Depth-Dependent Normal Compaction Trendline

Slotnick (1936) recognized that compressional velocity is a function of depth, i.e., velocity increases with depth in the subsurface. Therefore, the normal compaction trendline of travel time should be a function of depth. The oldest and simplest normal compaction trend of seismic velocity is a linear relationship given by Slotnick (1936) in the following form:

$$V = V_0 + kZ \tag{9.25}$$

where V is the seismic velocity at depth of Z; V_0 is the velocity in the ground at the surface or at the sea floor; k is a constant. Sayers et al. (2002) used this relationship as the normally pressured velocity for pore pressure prediction. A normal compaction trend for shale acoustic travel time with depth in the Carnarvon Basin was established by fitting an exponential relationship to averaged acoustic travel times from 17 normally pressured wells (van Ruth et al., 2004):

$$\Delta t_n = 225 + 391e^{-0.00103Z} \tag{9.26}$$

where Δt_n is the acoustic transit time from the normal compaction trend at the depth of investigation (μs/m); Z is in meters. A similar relationship was used for a petroleum basin in Brunei (Tingay et al., 2009):

$$\Delta t_n = 176.5 + 461.5e^{0.0007Z} \tag{9.27}$$

Based on the measured sonic transit times in the formations with normal pore pressures, as illustrated in Fig. 9.9, the following general relationship of the normal compaction trend of the transit time is:

$$\Delta t_n = \Delta t_m + \left(\Delta t_{ml} - \Delta t_m\right)e^{-cZ} \tag{9.28}$$

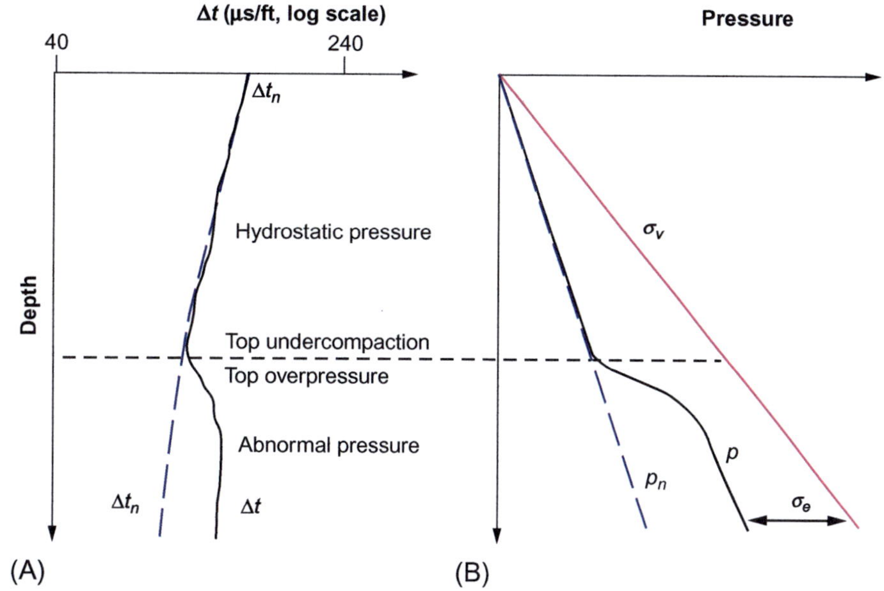

FIG. 9.9

Schematic plots showing sonic transit time (Δt) measured in shale, (A) the normal compaction trend of the transit time in the normal pressure condition (Δt_n) and (B) the pore pressure response to the transit time (Δt).

where Δt_m is the compressional wave transit time in the shale matrix (with zero porosity); Δt_{ml} is the mudline transit time; and c is the constant. Substituting Eq. (9.28) into Eq. (9.12), the modified Eaton's sonic equation can be expressed in the following form:

$$P_{pg} = OBG - \left(OBG - P_{ng}\right)\left(\frac{\Delta t_m + \left(\Delta t_{ml} - \Delta t_m\right)e^{-cZ}}{\Delta t}\right)^3 \tag{9.29}$$

The same case study, as presented in Eaton's resistivity method with depth-dependent normal compaction trendline, is used to examine the adapted Eaton's sonic transit time/velocity method. Fig. 9.10 shows the determination of normal compaction trend of sonic transit time from Eq. (9.28) and pore pressure calculation using the Eaton's sonic method with depth-dependent compaction trendline from Eq. (9.29).

By calibrating the measured pore pressure data, the normal compaction trend is determined from Eq. (9.28), which results in the following parameters: $\Delta t_m = 70\,\mu s/ft$, $\Delta t_{ml} = 200\,\mu s/ft$ and $c = 0.000245$. Compared to measured pore pressure data, the pore pressure calculation from Eq. (9.28) (Fig. 9.10) gives a very good result.

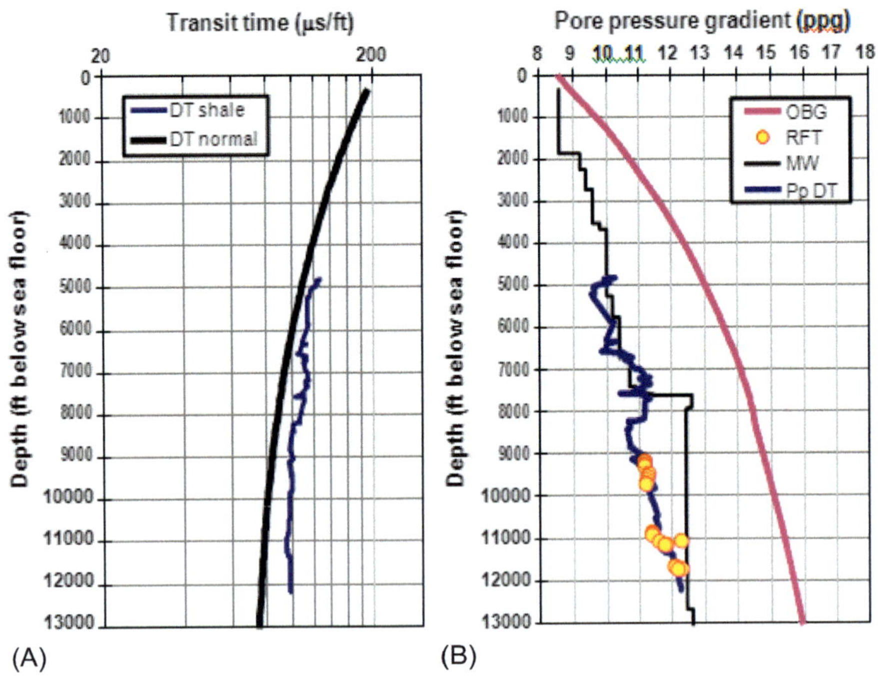

FIG. 9.10

Pore pressure calculated by adapted Eaton's sonic velocity method with depth-dependent compaction trendline in a deepwater well in the Gulf of Mexico. (A) Plots the sonic transit time in shale and the normal transit time calculated from Eq. (9.28), and the transit time is plotted in logarithmic scale. (B) The overburden gradient (*OBG*), mud weight used while drilling (MW), measured pore pressure (RFT), and pore pressure gradient (PP DT) calculated from sonic transit time using Eq. (9.29).

QUESTION 138

Ever since the seminal paper by Hottmann and Johnson, people in the industry tend to create normal compaction trends as functions of depth. For instance, the logarithm of sonic slowness or of resistivity is assumed to be linear in depth. Most folks also think they are deriving trends that really relate slowness or resistivity to vertical effective stress. So it seems logical to create normal compaction trends as functions of overburden stress rather than depth, especially where density varies a lot as in a section with salt. I have often thought about trying that but never had the energy, especially since I basically stopped using such normal trends. But it would certainly make it easier to compare normal trends from one area to another or to other vertical-effective-stress methods for estimating pore pressure. So has anyone tried it, and if so, did it work well?

Keith Katahara

Senior Geophysical Advisor at Hess Corporation

ANSWERS PROVIDED BY INDUSTRY EXPERTS

Gopal Amruthapuri

Geological Advisor – Pore Pressure at Anadarko Petroleum Corporation

Keith, I have tried few times where a petroleum basin system has multiple pressure mechanisms present. My answer to your question is neutral. One point I would like to emphasis on is Overburden characterization is the most important task in this exercise as in understanding all principal stresses.

Alan J Cohen

Chief Technology Officer at Headwave Inc.

Keith, why not just plot against vertical effective stress, rather than depth?

Keith Katahara

Senior Geophysical Advisor at Hess Corporation

Alan that is basically what some pore pressure analysts do, including me. We relate velocity directly to effective stress. But many other pore pressure experts establish a normal trend in depth first, and then use something like Eaton's equation to estimate the departure from normal VES as a function of the departure from normal velocity or resistivity. The latter method does not in general give VES as a single-valued function of velocity. This method does have the virtue of allowing the analyst to implicitly account, to some extent, for temperature, age, varying horizontal stress, and other depth-dependent variables that may be affecting velocities. My thought is that using overburden rather than depth for normal trends might make it a little easier to compare different methods and results from different areas.

Frans Kets

External Consultant Geophysics and Quantitative Interpretation, Wintershall

Keith, we are using velocity and resistivity as proxies for porosity, and then porosity as a proxy for effective stress. We use vertical effective stress as a proxy for total effective stress, which is fine for most extensional regimes, but not very much so in strike slip and extensional regimes. It is hard to get a handle on the total stress field (and of all the other factors you mention), so that is why I consider regional calibration still to be the most "reliable". It is nice though to have some case studies at hand where we see individual effects affecting the NCT so that we can "kind of" calibrate, for example, a faster increase in velocities as function of depth in compressional regime areas.

Christian Clark

Senior Pore Pressure Advisor – Deep Water Gulf of Mexico – Freeport-McMoRan O&G

Keith, one of issues is how to effectively use Normal Compaction Trends in relation to deposition and burial depth. We can make the assumption that normal compaction is relatively constant. However, changes in compaction, which would indicate it is not a constant principal, can cause issues in properly modeling PP development. If we look at Mississippi Canyon for example, one basin development is not necessarily the same as another basins development. Add the complexity of salt movement in relation to sediment deposition and it becomes easier to see how compaction may not be constant through geological time. We will assume the models are based on Miocene sediments, not Lower Tertiary or Cretaceous morphology.

When comparing the result of one basin development to another, even though relatively close (distance wise), you can easily observe how normal compaction models can be very different from one another. Depending where you are in the GOM, I have found through repeated trial and error, in an area such as Mississippi Canyon, the comparative use of NCT's can be misleading. Additionally, would we want to create normal compaction trends as a function of overburden stress rather than depth based models? Some geological assumptions might have to be made of course, such as compressional / extensional effects, rebounding, offloading. Moreover, we know that velocity data can at times be misleading or inaccurate enough to under predict as well as over predict shale based and sand based pressures. With this in mind, would we want to minimize our use of velocity data due to increased risk and uncertainty or make it a larger component of the workflow?

Selim Simon Shaker

Consulting Geoscientist at Geopressure Analysis Services

Keith, let us identify the NCT. It is the trend of sediments' porosity reduction after deposition due to compaction. It starts when dehydration begins and ends where top seal forms at the fluid retention depth. Several papers attempted to define the EXTENT and SLOPE of the compaction trend:

Burst John (1969 AAPG Bulletin, vo. 53), Smith J.(1970 mathematical geology), Shaker S (2007, AADE 07- NTCE -51 and CSEG 2007,#1 pp.29-35), Wessling et al. (2013, SEG Geophysics).

Compaction trend (CT) does not carry normally pressured values and we should not call it (NCT). It starts when PP exceeds the shallow normal hydrostatic (e.g. 0.465 psi.ft/ 8.9 ppg GoM) and bottoms at the FRD (fluid retention depth) ranges from 10 to 11.5 ppg (contingent on stratigraphy). The CT is a unique value for each CDP/well location and cannot be generalized or used on a regional scale. It is the result of OB, sedimentation rate, lithology and the stress perturbation of any deep seated structural features at specific location. The slope and extent of CT represents the petrophysical data trend where sediments are subject to post sedimentation dewatering.

Gopal, Christian and Frans point of view is what we should consider. Christian excellent comment sums it all. However, Keith's proposal could be a great endeavor in a relaxed basin. Give it a shot.

Michael Drews

Senior Geologist at ConocoPhillips

Keith, I found that using overburden (below mud line) vs. shale effective stress or overburden (below mud line) vs. shale velocity does a pretty good job when comparing offset well pressures in environments with significant variance on overburden. For example, when comparing sub-salt offset well shale pressures, I usually get much better alignment using this method than a VES vs. depth or velocity vs. depth method.

QUESTION 139

When we are determining the normal compaction trend line, should we use a single line for the whole field, or should it vary from well to well? I am trying to use a common linear trend line for a field, but a common line does not fit to some wells in the area.

Tuba Evsan

Geologist at Turkish Petroleum

ANSWERS PROVIDED BY INDUSTRY EXPERTS

Yogesh Kulkarni

Pore Pressure Analyst/ Field Engineer, Geoservices, Schlumberger Company

Tuba, ideally a single compaction trend line should work for field provided the origin of overpressure is compaction disequilibrium only. If the compaction trend line does not match for some wells in field then probably revisiting well data to look for geologic differences would give insight into overpressure mechanism (ex. unloading/ load transfer). Personally I find cross plot helpful (velocity-density, velocity-ES).

Mark Tingay

Senior Geomechanics specialist at Chevron

Tuba, It is always better to get a good fit with one consistent method than a whole bunch of perfect fits with different methods. It is always possible to tweak your NCT and other inputs and aspects to 'make' your prediction fit an offset well. However, if you need to use a different method for each offset well, how can you possibly have confidence in predicting a planned well? As Yogesh says, unless there is a valid geological reason to vary your NCT, then you should use a single NCT for all wells in the field. Changing your NCT and other inputs just to 'make the model fit' is what is sometimes referred to as "Cheatin' with Eaton", and is not good industry practice.

Better to look at why a constant model is not giving you good results - changes in geology, issues such as lateral transfer and buoyancy, faulting, etc.

Selim Simon Shaker

Consulting Geoscientist at Geopressure Analysis Services

Tuba, Great observations by all. However Tingay's comments: "Cheatin' with Eaton", goes to the core of PPP modeling pitfalls because:

1. Compaction trend is a unique value for each CMP/ CDP (before drilling) and should not be used as a common value unless you have flat beds without any structural or stratigraphic impact.
2. In a field where wells are already drilled "it is calibration rather than prediction". That is mean finding the right common exponent rather than swaying and breaking the NCT to fit MPP (measured) vs. PPP (predicted).
3. MPP (RFT's and MDT's) should not be used in ppg for calibrations, since it shows negative slopes on these kinds of plots.

4. Assuming that the MPP is equal to PPP at the calibration intervals is a major hurdle. Because, we observe kicks (MPP>PPP) and we also observe loss of circulations (MPP<PPP), assuming MW pressure is derived from the prediction model.
5. You can find a lot more on: AAPG 2014, Annual Convention, extended abstract ID # 1834685; AADE 2007-NTCE-51; TLE 2003, pp. 1223-1225; and TLE 2002, pp. 756-760.

Nitin Bhardwaj

Senior Operations Geologist (Geopressure and Geomechanics)

Tuba, let me add few points from my experience. There can be two situations:

(1) Cases where you may need to change the NCT in the same well itself.
(2) Cases where you keep a single NCT in the well but the NCT parameter may need to be change spatially in the basin from area to area.

In the first type of case, this is usually required when you have large unconformities present (break in the sedimentation) or as earlier commented above, change in the dominant pressure mechanics from mechanical compaction related to chemical compaction or secondary mechanisms.

In second type of case: this is usually required when there is large variation present in the spatial patterns of sea bed topography and sedimentation. In large basins you can expect some change in the NCT pattern, however in such cases the change of NCT parameters are generally range bound, for example you may have to use slightly different values of mudline sonic velocities in areas on shelf (low water depth) in comparison to areas in deep waters.

In general you should try to use a single NCT with appropriate exponents. If you change the NCT then you must back that up with geological reasons for doing so. Additionally, if in your case the use of single NCT is resulting in higher pressure than the MW, then you should be doubly conscious about reducing them by changing the NCT.

Morteza Azadpour

Petroleum Engineer at National Iranian Oil Company

Tuba, I've had your problem before. It was due to lateral transfer and buoyancy effect. Do you have inclined aquifer? It might be due to existence of additional overpressure such as buoyancy force within the hydrocarbon accumulation or some excess pressure by lateral transfer due to extension of inclined aquifer in the studied area. Inclined aquifers can create an increase in pore pressure at the crest of a structure. Using a common linear trend line for a field is not a suitable estimation which would not fit with the measured data in some wells especially at the crest or down dip.

QUESTION 140

How can I develop 3D a normal compaction trend using the Eaton method for 3D pore pressure prediction using seismic velocity? In Drillworks we can develop 3D normal trend for Bowers' Method. Bowers' method can also be applied using values of A, B and U without normal trend but for Eatons' method NCT is necessary.

Iftikhar Satti

Research Scholar at PETRONAS University, Malaysia

ANSWERS PROVIDED BY INDUSTRY EXPERTS

Colin Sayers

Scientific Advisor, Schlumberger

Iftikhar, when you ask how to develop a 3D Normal Compaction Trend, then I assume that you are referring to a normal trend defined by parameters that vary laterally due, for example, to a variation in water depth, sedimentation rate, etc. Is that correct? If so, then you need to choose a suitable parameterization and determine the parameters defining the lateral variation by inversion of data at a number of offset wells having sufficient data for the inversion.

Iftikhar Satti

Research Scholar at PETRONAS University, Malaysia

Colin, I have many offset wells that can be used for this purpose.

Fred W Schroeder

Seismic Interpreter

Iftikhar, I am assuming you want to generate a 3D normal compaction trend volume for seismic based Eaton PP prediction in Drillworks. From my experience it is necessary to write an UDM with a few of line code for "single trend line"(refer to one location) over the field; or UDP for 'multi trend line"(refer to one location).

Iftikhar Satti

Research Scholar at PETRONAS University, Malaysia

Fred, Yes. I want to generate the 3D trend volume but I don't know how to do it. Have you done it?

Fred W Schroeder

Seismic Interpreter

Yes. What version of Drillworks you are using is? Last time I did it using Drillworks 3D version11.6 (very old) in 2013. To write an UDM "single trend line" you need to find the mudline depth and velocity of the seismic volume, and to define a linear Eaton trend line equation.

Iftikhar Satti

Research Scholar at PETRONAS University, Malaysia

I am using Drillworks 5000.8.1.1. I know the Mudline depth and velocity also. Can you share UDM you wrote?

Phil Holbrook

Chief Scientist and Educator, Force-Balanced Petrophysics

Iftikhar, The particular version of Drillworks has no effect on the governing physics. Eaton and Bowers both involve depth functions that are dimensionally incorrect. If this were true you wouldn't need anything other than depth to make your prediction. Average seismic velocity includes all lithologies and assumes that they are the same. This is a lithologic mismatch that you are also ignoring.

Hatem Alnajar

Pore Pressure & Geomechanics Independent Consultant

Ifitkhar, you can easily construct an Eaton NCT either in Predict or in 3D in Drillworks. Just go to the NCT icon and select "semi log Interval velocity" in 3D. Or you can select the same in Predict. If you want Eaton resistivity you can select "Semi-log resistivity", again both in 1D (Predict) or in 3D, for Eaton sonic select "semi log sonic".

Iftikhar Satti

Research Scholar at PETRONAS University, Malaysia

Hatem, I tried this option but we can use only one value of A & B. So if we have multiple wells and NCT varies from well to well due to complex geology (i.e fault compartments) then how can we incorporate this variation?

Dayo Adebayo

Geomechanics Consultant at Halliburton

Iftikhar, based on my experience you don't need to develop a 3D NCT model for seismic velocity transformation. All you need is a working regional overburden and NCT model to establish your well pore pressure prediction to seismic velocity transformation which is a function of your calibration technique. Several techniques can be employed for seismic based pore pressure modelling such as trend kriging or linear velocity shift at depth. The basic idea of trend kriging is to predict the value of a function at a given point by computing a weighted average of the known values of the function in the neighborhood of the point. That is, the prediction at the NCT location would be distributed across the seismic velocity cube. So, in this case you only need one NCT model.

Mark Herkommer

Geopressure Technology Consultant at Semicircle

Iftikhar, you should consider using a product called pQube which performs a global 3D analysis on a volume of seismic velocity functions. Each NCT is unique and fitted to the velocity function at its location. Parameters that are used for the automatic construction of the NCTs include water depth, mud line velocity, velocity inversion sensitivity, min/max velocity and depths for NCT fitting. NCTs types that are supported are polynomial (1st - 5th order), exponential, hyperbolic, as well as ad hoc user defined NCTs hung off the water bottom.

Selim Simon Shaker

Consulting Geoscientist at Geopressure Analysis Services

Mark, you hit the nail on the head in your post: "Each NCT is unique and fitted to the velocity function at its location." Iftikhar, also consider reading "Calibration of geopressure predictions using the NCT" CSEG, RECORDER, Jan. 2007.

Colin Sayers

Scientific Advisor, Schlumberger

Iftikhar, note that it is not necessary to fit a normal compaction trend to velocity data. The normal compaction trend can be obtained by inversion of pressure, density and velocity information. This approach avoids the need to assume that sediments are normally pressured at shallow depths, for example. Shallow sediments can be overpressured, resulting in errors if the conventional approach of fitting a normal compaction trend to velocity data is used. This problem is avoided by using an inversion-based method. For more details, see http://library.seg.org/doi/abs/10.1190/1.1500391

Phil Holbrook

Chief Scientist and Educator, Force-Balanced Petrophysics

Colin, you are mixing apples with oranges following this inversion approach. The p wave velocities of the primary lithologies are vastly different. They reach their terminal velocities in different ways due to consistently force balanced physics. Clay minerals have an inter-particle repulsive layer normal to their "c" axis. No other minerals for example, quartz, calcite, anhydrite, dolomite or halite do. The matrices of these non-clay lithologies are not sensitive to the scalar pore fluid pressure. Further the transit time, that is the average of reflected seismic impedances, is not either. True you can invert these. But in the end, what sort of animal do you have?

I know we'd like to solve for pore fluid pressure if we could. But there are no downhole or reflected seismic instruments that do this. The scalar pore fluid pressure is the difference between average confining load (S ave) minus the average effective stress (Sigma ave). The only correct way to accomplish this is through the effective stress theorem. Overburden can be inverted from a bulk density RHOB or pseudo bulk density log. Sigma max is very close to average mineral hardness. That's how rocks respond. Apparently this still bears repeating. The first fundamental in situ stress/strain relationship was first published in 1995. The earth has no brain. It resists compaction with the constituents at its disposal. These are average mineralogy and pore fluids. Each of these has a bulk density, bulk modulus, hardness and stiffness. These are the active parts of the earth's constitutive mechanical system. It's not hard if you take the direct approach.

Colin Sayers

Scientific Advisor, Schlumberger

Phil, each different lithology has its own velocity (or resistivity, etc.) vs. effective stress relation which needs to be inverted separately using available velocity, density and pore pressure data.

Phil Holbrook

Chief Scientist and Educator, Force-Balanced Petrophysics

Colin, I'm glad to hear that we agree. Let's call this a scalar forced balanced inversion.

Colin Sayers

Scientific Advisor, Schlumberger

Phil, not necessarily a scalar: that is just an approximation. Effective stress is a tensor, and so are conductivity and elastic stiffness.

QUESTION 141

What will be the effect of an unconformity (if it is present in a normal compacted zone) on normal compaction trend development?

Iftikhar Satti

Research Scholar at PETRONAS University, Malaysia

ANSWERS PROVIDED BY INDUSTRY EXPERTS

RM Iman Argakoesoemah

Senior Petroleum Exploration Geologist

Satti, If the age gap of the sediments of the overlying and underlying sequences is not significant even though they are separated by unconformity (SB), the effect should be irrelevant. Most of the effect might be observable if the unconformity occurred due to large tectonics uplift followed by erosion of a significant amount of thickness of the underlying sediments as in the Sunda Fold, thrust belt, etc.

Iftikhar Satti

Research Scholar at PETRONAS University, Malaysia

RM Iman, It is an erosional unconformity due to uplifting and the age gap is almost 2 Ma. The erosion thickness is not exactly known but whole lithology group is eroded, approximate eroded thickness is 150-200m. I tried two compaction curves (above and below the unconformity) and I noticed that the one below unconformity gave better results as compare to the one which includes both zones (above and below unconformity).

David Rajmon

Consulting Geologist, GeoSophix

Iftikhar, modern compaction models (in basin modeling tools or Touchstone) usually describe porosity change during burial as a function of vertical effective stress plus diagenesis. Taking mechanical compaction only, you can use an equation like this:

$$\text{PHI} = \text{PHImin} + (\text{PHIdepo} - \text{PHImin}) * \exp(\text{ComCoeff} * \text{VES}).$$

Decompression does not increase PHI substantially. Play with it in Excel and you'll see that porosity decreases significantly at low VES. The higher the VES (deeper) the rock compacts less and less. So the effect of erosional unconformity on porosity depends on how deep the rocks were buried before erosion and how deep they are now (neglecting lithological changes). If the rocks were buried to 6 km uplifted and reburied to 5 km, you will hardly notice any difference. However, if they were buried to 6 km and reburied to 1 km the difference in porosity will be huge.

Phil Holbrook

Chief Scientist and Educator, Force-Balanced Petrophysics

Iftikar and David, clearly the concept that was programmed was uni-axial. The two horizontal loads, (maximum and minimum) are entirely neglected. These are not zero. In fact gravity is the intermediate principal stress in the strike-slip regime and the minimum principal stress in the thrust fault regime.

The way to do this that captures all natural variability is to use the effective stress theorem symmetrically coupled to gravity. This will solve your fundamental problems of which there are many. All of these forces are scalars meaning they are equal in all directions. The first fundamental in situ stress/strain relationship describes in detail the lithologic effects which you also choose to ignore. These are all detailed in "The Earth's Constitutive Mechanical System", Holbrook, 2001. The Earth can't ignore any of these. Why should you?

David Rajmon

Consulting Geologist, GeoSophix

Phil, yes, what I described is a simplified concept sufficiently accurate in an extensional tectonic regime. VES dependent porosity models are currently coded in all basin modeling tools I know and I think they are common in pore pressure prediction workflows too. Many exploration experts do not realize these VES dependent models are grossly inadequate for a compressive regime and may be off in a strike-slip regime too.

Steve Hobart

Sr. Pore Pressure Advisor

There are three classes of "normal compaction trend-line" (NCTL). Each class of NCTL attempts to model the normal compaction behavior of porosity or one of its proxies (e.g. Resistivity, Velocity, Dxc) as plotted versus depth.

1. The NCTL used with Eaton-style analysis relates the actual shale value to the value it is expected to have at the same depth in order calculate pore pressure. If there is a major unconformity, it is generally acceptable to have a break in the continuity of the NCTL in order to reflect the missing compactible section. The sections above and below an unconformity may have had different burial histories which could have different implications for the porosity - effective stress relationship.
2. Trend-lines used in an Equivalent Depth approach assume that the shale porosity proxy value at the depth of interest also represents the same vertical effective stress value at the shallower "normal pore pressure" depth. Since salinity and temperature change with depth, this technique is not expected to work with unnormalized shale resistivities, even if there is no unconformity. If an NCTL for the Equivalent Depth method is calibrated above and below the unconformity such that it becomes discontinuous at that depth, values from the depth of interest may in principle reference different porosity - (normal) effective stress relationships, rendering this method as potentially inaccurate in the presence of an unconformity or fault.
3. NCTL values for what Alixant and Desbrandes called "explicit methods" can be constructed by substituting (OB - PPnormal) for the vertical effective stress. The trend-lines are not (or at least should not be) used for

> pore pressure estimation, but rather illustrate how the input parameters for methods like the Baldwin-Butler, Bowers or the Miller methods would predict the normally pressured shale velocity or DT. "Explicit" methods do not rely on depth as an input parameter. If the formations above and below the unconformity are characterized by the same input parameters, the only impact of the unconformity on this class of normal compaction trend-lines will be if there is a density contrast across the unconformity which affects the input vertical stress (OBG). For example, by using a sonic - porosity transform of choice, (DT = f(porosity), one can create an NCTL for DT for the Baldwin-Butler relationship with DTnormal = f(porosity) = f(1 - ((OB - PPnormal) / A)^(-B)), wherein, if one so chooses, A is Holbrook's sigmamax and B is the exponent, both of which are claimed to be functions of mineralogy.

In summary, the impact of an unconformity on the NCTL is dependent on the application.

I would like to, hopefully, clarify the following statement by David, which is startling when taken out of the context of basin modeling. "So the effect of erosional unconformity on porosity depends on how deep the rocks were buried before erosion and how deep they are now (neglecting lithological changes). If the rocks were buried to 6 km uplifted and reburied to 5 km, you will hardly notice any difference. If they were buried to 6 km and reburied to 1 km the difference in porosity will be huge." This is not a statement about actual physical behavior of the rocks, but rather a deficiency in basin modeling software which regards the Lahann equation as a reversible virgin compaction stress relationship. David is talking about porosity resulting from a basin modeling algorithm. As we all should know, even if the software doesn't recognize it, compaction is to a very large degree an irreversible, plastic deformation.

David Rajmon

Consulting Geologist, GeoSophix

Steve, let me clarify my comment. I wrote: "Decompression does not increase PHI substantially." In other words, mechanical compaction is mostly irreversible. The comparison referred to porosities across the erosional unconformity, that is, "If they were buried to 6 km and reburied to 1 km the difference in porosity (across the unconformity) will be huge".

Iftikhar's question is broad and leaves room for many geological scenarios. Rather than jumping on trends fitted to well log curves, I prefer thinking about geological processes affecting sediment porosity first. I chose to limit my discussion to a simple VES driven mechanical compaction to illustrate a basic concept for siliciclastic sediment to get Iftikhar thinking about processes that control sediment porosity. Other processes need to be taken into account in a specific geological scenario such as deviatoric stress or diagenesis. Lithologic variation is another issue (e.g. different types of sands). After considering these, then one can predict what would be the expected unconformity signature in well logs.

Phil Holbrook
Chief Scientist and Educator, Force-Balanced Petrophysics
Steve, NCTL is a dimensional mis-match. As you said earlier depth is a coordinate not an answer. These things vary with depth, but what does this imply much less prove? Any attempts to model the normal compaction behavior of porosity or one of its proxies (e.g. Resistivity, Velocity, Dxc) as plotted versus depth is suspect as you pointed out,

1. The NCTL used with Eaton-style analysis relates the actual shale value to the expected value at the same depth in order calculate pore pressure. If there is a major unconformity of significant fault throw, it is generally acceptable to have a break in the continuity of the NCTL in order to reflect the missing compactible section. The sections above and below an unconformity may have had different burial histories which could have different implications for the porosity - effective stress relationship.
2. Trend-lines used in an equivalent depth approach assume that the shale porosity proxy value at the depth of interest also represents the same vertical effective stress value at the shallower "normal pore pressure" depth. Since salinity and temperature change with depth, this technique is not expected to work with unnormalized shale resistivities, even if there is no unconformity. If an NCTL for the Equivalent depth method is calibrated above and below the unconformity such that it becomes discontinuous at that depth, values from the depth of interest may in principle reference different porosity - (normal) effective stress relationships, rendering this method potentially inaccurate in the presence of an unconformity or fault.
3. NCTL values for what Alixant and Desbrandes called "explicit methods" can be constructed by substituting (OB - PPnormal) for the vertical effective stress. The trend-lines are not (or at least should not be) used for pore pressure estimation, but rather illustrate how the input parameters for methods like the Baldwin-Butler, Bowers or the Miller methods would predict the normally pressured shale velocity or DT. "Explicit" methods do not rely on depth as an input parameter. If the formations above and below the unconformity are characterized by the same input parameters, the only impact of the unconformity on this class of normal compaction trend-lines will be if there is a density contrast across the unconformity which affects the input vertical stress (OBG). For example, by using a sonic - porosity transform of choice, (DT = f(porosity), one can create an NCTL for DT for the Baldwin-Butler relationship with DTnormal = f(porosity) = f(1 - ((OB - PPnormal) / A)^(-B)), wherein, if one so chooses, A is Holbrook's sigmamax and B is the exponent, both of which are claimed to be functions of mineralogy.

More so, the compaction of any lithology IS a function of that mineralogy. Please re-visit "The first fundamental in situ stress/strain" relationship for the details. This feeds into the effective stress theorem which is a dimensionally correct scalar force balanced relationship. That relationship is EXPLICIT as are those of Alixant and Des

Brandes. For hopefully everyone's comfort, sigma average is almost directly proportional to average mineral hardness.

These are both measures of compaction resistance. Unloading has an elastic limit. This may appear to be vertical, but it has a finite residual stress/strain. I agree with you when you said "As we all should know, even if the software doesn't recognize it, compaction is to a very large degree an irreversible, plastic deformation". The book, "The (Earth's Constitutive Mechanical Systems) is a symmetrical and force balanced material Science" summarizes this viewpoint.

NORMAL COMPACTION TREND IN CARBONATE RESERVOIR

QUESTION 142

Which techniques can I use to determine the normal compaction trend (NCT) for shale acoustic travel time in carbonate environment? According to my research I have noticed that NCT determination differs from shale to carbonate environment. In carbonate it is more spontaneous and non-linear. Please correct me if I am misled.

Myrat Charyyev
Wireline Field Engineer at Schlumberger

ANSWERS PROVIDED BY INDUSTRY EXPERTS

Mark Tingay
Senior Geomechanics specialist at Chevron

Myrat, the whole point of using a NCT is based on the assumption that porosity changes (and thus changes in sonic velocity, resistivity, etc) are related to changes in effective stress (stress minus pore pressure). The prediction methods then try to equate deviations from the NCT to variations in effective stress and then values for pore pressure. However, this entire approach is based on the assumption that rocks compact due to changes in effective stress. This assumption holds true for clastics and some carbonates. However, some rocks (including some carbonates) can be considered to be 'stress insensitive' - they have very stiff frameworks that do not show significant porosity response to changes in effective stress. Furthermore, carbonate porosity and petrophysical properties can be extensively influenced by diagenesis. All this adds up to make it very difficult to use classic porosity based prediction methods in some carbonate rocks.

My recommendation is to start with the geology. What type of carbonates are they? How would you classify them? What do you know about their burial and diagenetic history? Is there any evidence that your carbonates show 'stress sensitive' behaviour? Are there other lithologies you could use (e.g. if you have alternating shales and carbonates, then try doing your PPP just on the shales)? Are there overpressures in your carbonates? If so, what is the likely generation mechanism (does not have to be disequilibrium compaction)?

PPP in stress insensitive rocks remains one of the great challenges in pore pressure prediction.

Steve Hobart

Sr. Pore Pressure Advisor

Myrat, I could quibble with the phrase "stress insensitive" in Mark's comments above since all material have some degree of sensitivity to stress, but I think he did convey the idea that porosity in carbonates is often a function of factors other than vertical effective stress. The mineralogical content of "carbonates" varies, as well as the texture, geochemical history and other factors which can affect porosity. There are several classifications of porosity within carbonates.

Mark's recommendation to look for adjacent shales is often a good one. However, even these may become tainted with carbonates, either from the original depositional environment or subsequent to deposition through movement of subsurface water and accompanying dissolution and (re-) precipitation of the relatively soluble calcium carbonate. This dissolution and (re-) precipitation can impact the porosity of shales and marls over geologic time as well. If you can pick shale points which have similar, and relatively low, concentrations of carbonates, you will have a shot at using compaction/undercompaction models (models that use normal compaction trend lines) for effective stress and pore pressure.

The normal compaction trendline is specific to a mineralogical mix for which there is a consistent relationship between porosity and effective stress. This relationship assumes a consistent compactable mineralogy for each NCT segment and that mechanical compaction is the primary mechanism for reduction in porosity. When using petrophysical measurements, we also need to remain cognizant of and alert to factors other than porosity which can affect the measurements.

Tom Arnold

R&D, Training Director for Paladin Geo. & Petroskills SME/Instructor

The bottom line here is, 'the answer is in the rocks'. There is no model that will work without calibration in all circumstances and rock types. NCT trends fall apart in sediments that resist compaction. You must find some shale to work with.

Mark Tingay

Senior Geomechanics specialist at Chevron

Myrat, Tom is right - understanding the geology is the key. However, there simply isn't shale to use sometimes. I've seen many examples, particularly in the Middle East, of overpressures in thick carbonate sequences without any shale to use. The NCTs in these carbonates is just a straight vertical line - no sonic or resistivity response at all in the overpressured rocks because there is no porosity change.

Carbonate reservoirs are a major challenge for our industry. It may take a paradigm shift to figure out good methods. However, I think you first need to understand your geology and figure out why and how your rocks are overpressured. Those two bits of knowledge will take you a long way towards building some sort of prediction strategy.

Kehinde Adesina

Snr. Global Technical Advisor –Drilling & Geomechanics (unconventional & Deepwater), Halliburton

Myrat, other important factors to consider apart from the carbonates type and geology would be depth of burial, source rock potentiality and geological history.

Phil Holbrook

Chief Scientist and Educator, Force-Balanced Petrophysics

Kehinde, when carbonates and claystones make up the entire stratigraphic sequence, mineralogy controls porosity and pore fluid pressure. I'm assuming a calcite - marl sequence.

The fluid pressure in any lithology is a load bearing problem wherein Average load (S ave) = average effective stress (Sigma ave) + pore fluid pressure (Pp).

This is the effective stress theorem which many experts choose to ignore. It's a dimensionally correct, force balanced conservation law that the earth cannot ignore.

Finn Engstrøm

Principle Petrophysics at Maerst Oil

Myrat, I know that many people assume that net effective stress is equal to weight-of-overburden minus the pore pressure and that the effective stress, defined in this way, defines the normal compaction trend (porosity vs depth). Unfortunately this is often a too simplistic model for carbonates. You need, as a minimum to include the Biot factor in the net effective stress equation, and you also need to consider average grain size (which defines the minimum water volume that cannot be removed by compaction), shaliness, strain rate, saturation-history and local stress model (ratio of local net effective horizontal and net effective vertical stress).

If you include all the above you can start understanding the mumbling from nature. That you need to include e.g. the Biot factor to reduce the effect of the PP is relatively clear from the often seen rapid variations in porosity for carbonates around 20 pu in overpressured environments. The rapid variations can be seen as a consequence of small variations in the porosity decay constant coupled with the rapid change in the Biot factor around 15-20 pu. Here carbonates often display layers with around 15-20 pu between layers with hardly any porosity at all.

Phil Holbrook

Chief Scientist and Educator, Force-Balanced Petrophysics

Finn, the assumption that all lithologies compact exactly like shales and have the same velocity is too simplistic. Also the effective stress theorem (first sentence), S ave = Sigma ave + pore fluid pressure is a dimensionally correct conservation law. All of this worked very well with a Biot coefficient of 1.0 for Michel Krief et al. (1990).

Finn Engstrøm

Principle Petrophysics at Maerst Oil

Phil, I agree that from a dimensional point of view that Save = Sigma-ave + Pf is correct, but so is Save = Sigma-ave + B*Pf, where B is the Biot factor, a pure number between 0 and 1. With respect to your reference to Michel Krief, then I'm not sure I understand your reference as he has delivered many Biot-models to me as input to porosity-vs-stress models.

I did not write that all lithologies compact exactly like shales. In fact I tried to write that many carbonates have a very complex compaction models that include saturation history (not relevant for shales), stress history (likely relevant for shales) and grain size (perhaps relevant to shales, but it should likely be replaced by rheology).

Michel Krief

Senior Petrophysicist at OMV Exploration & Production GmbH

Finn, The assumption that Biot's factor equals to 1 is made implicitly, sometimes explicitly, in most published rock physics/rock mechanics models. Those models are actually based on the Terzaghi's theory which is intended to describe the behavior of soils.

It can be shown that the Biot'slow frequency model reduces to the Wood's equation for suspensions when the Biot factor is made equal to 1. This is, not exactly representative of all carbonate rocks, or any other rock for that matter.

Due to the complexity of carbonate rocks, only local studies of the variations of Biot's factor can be made. By local, we need to mean a single reservoir (possibly layer) in a given field. This is only modeling work for practical purposes, not theoretical work which remains to be done.

NEW THEORETICAL MODELS OF PORE PRESSURE PREDICTION

Pore Pressure Prediction From Porosity

As introduced before, undercompaction is the primary cause of formation overpressuring, which occurs primarily in rapidly subsiding basins and in rocks with low permeability. The indicators of undercompaction are higher pore pressures and larger formation porosities than in the normal compaction condition. It is commonly accepted that porosity decreases exponentially as depth increases in normally compacted formations (e.g., Athy, 1930):

$$\varphi = \varphi_0 \, e^{-cZ} \tag{9.30}$$

where φ is porosity; φ_0 is the porosity in the mudline; Z is the TVD below the mudline; c is the compaction constant in 1/m or 1/ft. The same relationship exists in porosity and effective stress (e.g., Dutta, 2002; Flemings et al., 2002; Peng and Zhang, 2007):

$$\varphi = \varphi_0 e^{-a\sigma_e} \tag{9.31}$$

where a is the stress compaction constant in 1/psi or 1/MPa. As discussed previously, porosity is an indicator (a function) of effective stress and pore pressure, particularly for the overpressures generated from undercompaction and hydrocarbon cracking. Therefore, pore pressure can be estimated from formation porosity. Fig. 9.11 illustrates how to identify undercompaction and overpressure from a porosity profile.

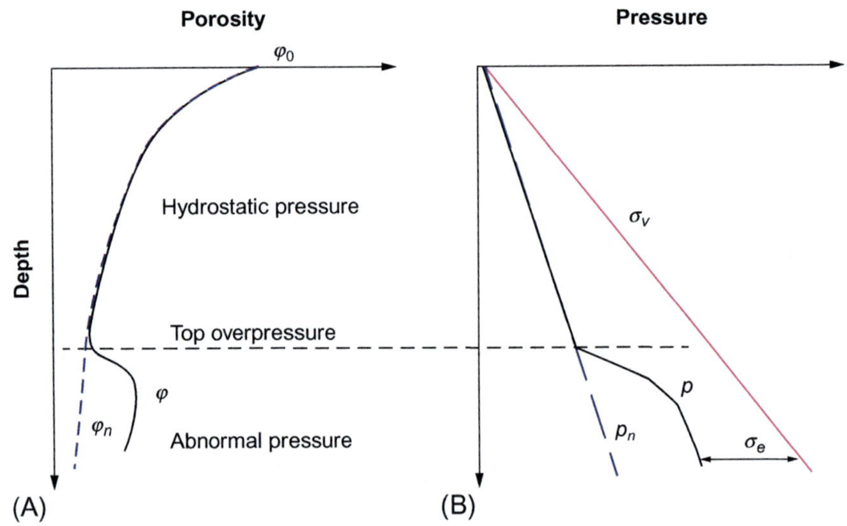

FIG. 9.11

Schematic porosity (A) and corresponding pore pressure (B) in a sedimentary basin. The dash porosity profile in (A) represents normally compacted formation. In the overpressured section the porosity reversal occurs *(heavy line)*. In the overpressured section, porosity is greater than that in the normal compaction trendline (ϕ_n).

When the porosity trend is reversed, the undercompaction occurs and overpressure is generated. The starting point of the porosity reversal is the top of undercompaction or top of overpressure. In the formation with undercompaction, porosity and pore pressure are higher than those in the normally compacted formation. Efforts have been made to use porosity data for predicting pore pressure in shales and mudstones. For instance, Holbrook et al. (2005) presented a porosity-dependent effective stress for pore pressure prediction. Heppard et al. (1998) used an empirical porosity equation similar to Eaton's sonic method to predict pore pressure using shale porosity data. Flemings et al. (2002) and Schneider et al. (2009) also applied porosity-stress relationships to predict overpressures in mudstones. Another author derived a theoretical equation for pore pressure prediction from porosity according to normal compaction trend of porosity (Zhang, 2008). The pore pressure gradient can be calculated from the following equation (refer to Appendix A for derivations):

$$P_{pg} = OBG - \left(OBG - P_{ng}\right)\frac{In\,\varphi_0 - In\,\varphi}{cZ} \tag{9.32}$$

where φ is the porosity in shale at depth of Z, which can be obtained from sonic or density logs; φ_0 is the porosity in the mudline (in the ground surface or sea floor); Z is the depth below the mudline; c can be obtained from the normal compaction porosity trendline in Eq. (9.34). From Eq. (9.32) the pore pressure, overburden stress and porosity have the following relationship:

$$P = \sigma_v - (\sigma_v - P_n)\frac{In\varphi_0 - In\varphi}{cZ} \tag{9.33}$$

The primary difference between Eq. (9.33) and other existing pore pressure-porosity equations is that the pressures calculated from Eq. (9.33) are dependent on depths. In other words, the normal compaction trendline of porosity is not a constant, but a function of depth. When porosity (φ) at the depth of interest is greater than the normal porosity (φ_n) at the same depth, the formation is overpressured. The normal compaction porosity trendline can be determined from the following equation (re-written from Eq. 9.30):

$$\varphi_n = \varphi_0 e^{-cZ} \tag{9.34}$$

Pore Pressure Prediction From Transit Time or Velocity

Porosity (φ) in Eqs. (9.32), (9.33) can be obtained from density logs, sonic logs or from seismic interval velocities. The following equation represents one of the methods to calculate porosity from compressional velocity/transit time proposed by Raiga-Clemenceau et al. (1988):

$$\varphi = 1 - \left(\frac{\Delta t_m}{\Delta t}\right)^{\frac{1}{x}} \tag{9.35}$$

where x is an exponent to be determined from the data. In a study of compaction in siliciclastic rocks, predominantly shales, in the Mackenzie Delta of northern Canada, Issler (1992) obtained laboratory measurements of porosity in shales and determined a value of 2.19 for the exponent x and used a fixed value of shale transit time, $\Delta t_m = 67$ μs/ft. Substituting Eq. (9.35) into Eq. (9.32), the pore pressure can be predicted from the compressional transit time/velocity, i.e.,

$$P_{pg} = OBG - (OBG - P_{ng})\frac{C_m - In\left[1 - (\Delta t_m / \Delta t)^{1/x}\right]}{cZ} \tag{9.36}$$

where C_m is the constant related to the compressional transit time in the matrix and the transit time on the mudline, and $C_m = \ln[1 - (\Delta t_m/\Delta t_{ml})^{1/x}]$. Porosity is often estimated by the empirical time average equation presented by Wyllie et al. (1956):

$$\varphi = \frac{\Delta t - \Delta t_m}{\Delta t_f - \Delta t_m} \tag{9.37}$$

where Δt_f is the transit time for the interstitial fluids. Substituting Wyllie's porosity-transit time equation into Eqs. (9.32), (9.33), we obtain the following equations to calculate pore pressure gradient (P_{pg}) and pore pressure P:

$$P_{pg} = OBG - (OBG - P_{ng})\frac{In(\Delta t_{ml} - \Delta t_m) - In(\Delta t - \Delta t_m)}{cZ} \tag{9.38}$$

$$P = \sigma_v - (\sigma_v - P_n)\frac{In(\Delta t_{ml} - \Delta t_m) - In(\Delta t - \Delta t_m)}{cZ} \tag{9.39}$$

The normal compaction trendline of the transit time in Eqs. (9.38), (9.39) can be obtained from the following equation (refer to Appendix B for derivations):

$$\Delta t_n = \Delta t_m + \left(\Delta t_{ml} - \Delta t_m\right)e^{-cZ} \tag{9.40}$$

This normal compaction trend allows the normal transit time to approach the matrix transit time at a very large depth, which is physically correct, as Chapman (1983) pointed out. Unlike other methods, this proposed sonic method (Eqs. 9.38, 9.39) uses a normal compaction trendline that is asymptotic to matrix transit time and therefore better represents the compaction mechanism of the sediments. The other advantage of this model is that the calculated pore pressures are dependent on depth, and both effects of the matrix and mudline transit time are considered.

CASE APPLICATIONS—GULF OF MEXICO

The proposed methods (porosity model in Eqs. 9.26, 9.27 and velocity model in Eqs. 9.32, 9.33) have been applied in several petroleum basins (e.g., the Gulf of Mexico, North Sea and onshore conventional gas and shale gas formations) to verify their applicability. Three examples of deepwater wells in the Gulf of Mexico are presented in the following section.

Pore Pressure From Porosity Model

The same case study as shown in Figs. 9.8 and 9.10 is used in this section for comparison purposes. An important step in performing pore pressure analysis is to select the clean shales in well log data. Prior to applying the porosity model, one needs to obtain the shale porosity from the density log or sonic log. Clay minerals have a crystalline structure that can contain more radioactive elements than sandstones. Therefore, gamma ray data in well logs can be used to differentiate shale intervals from those of other lithologies. Drawing shale base lines in gamma ray data is the first step to identifying and separating shale from other rocks, as shown in Fig. 9.12. The high gamma ray values (N 75–110 API) are assumed to be pure shale, whereas points where gamma ray values are less than the shale base lines are not used for analysis. Shale points defined on the gamma ray log are then transferred to the corresponding sonic log or other logs used for pore pressure analysis (KSI, 2001).

Fig. 9.12 demonstrates how to obtain shale porosity from gamma ray and sonic logs. Porosity is calculated from the shale transit time and Wyllie's equation (Eq. 9.37) using Δt_f=200 μs/ft and Δt_m=70 μs/ft. Using this porosity data, the normal compaction trend in porosity is analyzed based on Eq. (9.34) with the parameters of φ_0=0.8 and c=0.00024, as shown in Figs. 9.11A and 9.13A. Then, pore pressure gradient is calculated from the porosity model (Eq. 9.32) with parameters obtained from Fig. 9.13A (φ_0=0.8, c=0.00024) and P_{ng}=8.7 ppg. The calculated pore pressure gradient is compared to the measured pore pressure from the RFTs and mud weight, as shown in Fig. 9.13B. Compared to the pore pressure obtained from the

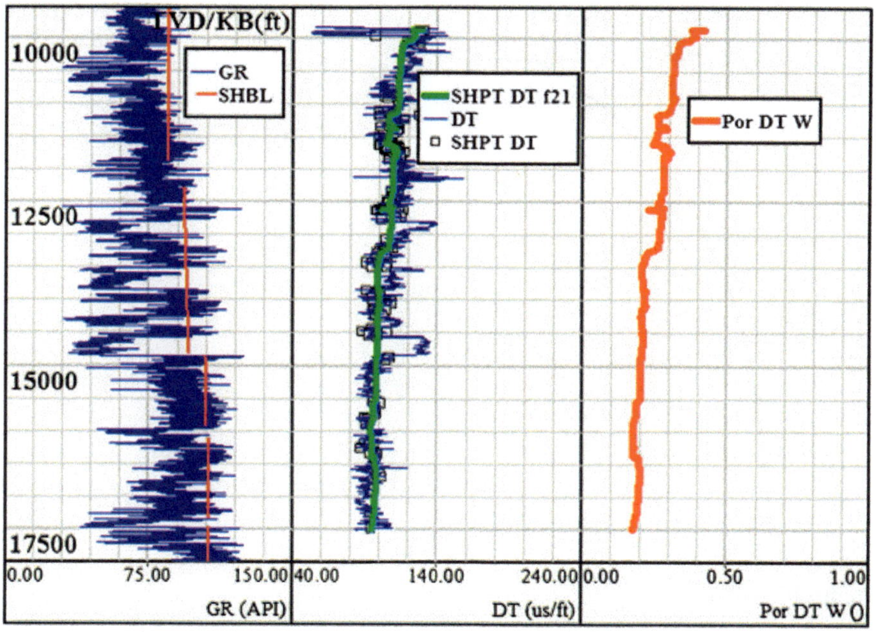

FIG. 9.12

Log data and the calculated porosity from sonic transit time by Wyllie's equation (Eq. 9.37) in a deepwater well of the Gulf of Mexico. In this figure, the gamma ray and shale base lines (SHBL) are shown in the left track; the sonic transit time (DT), shale points of the transit time (SHPT DT), and filtered shale transit time in shale (SHPT DT f21) are plotted in the second track; and the calculated porosity from the filtered shale transit time is shown in the right track.

modified Eaton's sonic method (Fig. 9.10B), the proposed porosity method (as shown in Fig. 9.13B) gives a better result in terms of matching measured pore pressures.

Pore Pressure From Sonic Transit Time Model

This case study presents another postdrill pore pressure analysis in a Gulf of Mexico deepwater oil field with water depth of 3850 ft. The formations are Tertiary shales and sandstones, and the target zone is located in the Middle Miocene sandstones. Fig. 9.13 shows a postdrill pore pressure analysis to examine the proposed sonic model. Pore pressure gradient is calculated from the proposed sonic equation (Eq. 9.38) using $\Delta t_{ml}=200\,\mu s/ft$ and $\Delta t_m=70\,\mu s/ft$, $P_{ng}=8.75$ ppg, and $c=0.00028$. The pore pressure gradient is also estimated using the Bowers method (Eq. 9.15) with $\Delta t_{ml}=200\,\mu s/ft$, $P_{ng}=8.75$ ppg, $A=14$, $B=0.745$. Compared to the result obtained from the Bowers method, the proposed method (Eq. 9.38) gives a better result in pore pressure calculation, particularly in the shallow section, where other sonic methods overestimate the pore pressures. The proposed sonic method uses a normal compaction trendline and

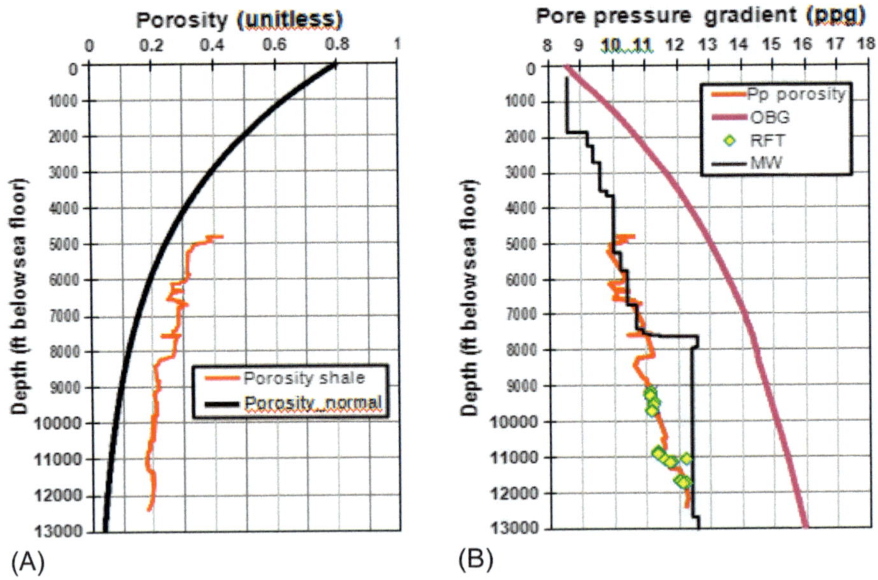

FIG. 9.13

Pore pressure prediction from the porosity method (Eq. 9.32) using the data shown in Fig. 9.10. (A) Presents porosity of shales obtained from Fig. 9.12 and normal compaction trendline of porosity calculated from Eq. (9.31). (B) Plots overburden stress gradient (*OBG*), mud weight used while drilling (MW), measured pore pressure (RFT), and pore pressure profile calculated from the porosity by Eq. (9.32) (PP porosity).

the calculated pressures are dependent on depth. Therefore, the pore pressures at both shallow and deep depths can be better predicted.

Pore Pressure From Sonic Transit Time Model in Subsalt Formations

Pore pressure prediction is very challenging in subsalt formations because the formation water may be affected by salt and has high salinity, thus a much lower resistivity. In addition, the salt tectonics may affect the formation compaction status, causing changes to the normal compaction trend. Furthermore, the surface seismic survey cannot provide reliable interval velocities in the subsalt formations. Therefore, a sonic log is critically important for pore pressure prediction while drilling. We examined a borehole drilled in a deepwater subsalt formation of the Gulf of Mexico. The field description can be found in Zhang et al. (2008). Pore pressure gradient is calculated from the proposed sonic equation (Eq. 9.38) using $\Delta t_{ml}=120\,\mu s/ft$ and $\Delta t_m=73\,\mu s/ft$, $P_{ng}=8.75$ ppg and $c=0.00009$. Compared to the measured formation pressures, the proposed method (Eq. 9.38) gives an excellent pore pressure calculation in subsalt formations, as shown in Fig. 9.14. It should be noted that the mudline transit time needs to be adjusted ($\Delta t_{ml}=120\,\mu s/ft$, instead of the $200\,\mu s/ft$ used in conventional cases) to make a better pore pressure estimation in subsalt formations.

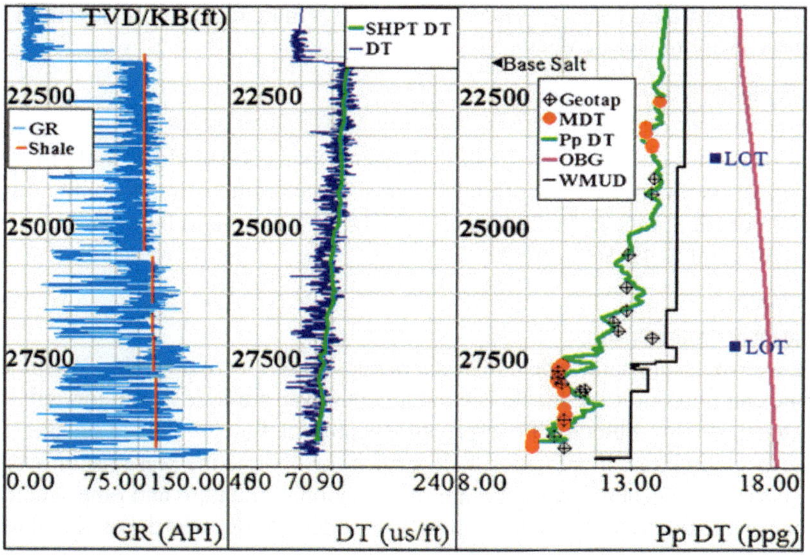

FIG. 9.14

Pore pressure calculation from the sonic transit time proposed in this paper (Eq. 9.38) in deepwater subsalt formations of the Gulf of Mexico. In this figure, the gamma ray and shale base lines are shown in the left track; the sonic transit time (DT) and filtered shale points of the transit time (SHPT DT) are plotted in the second track; and the calculated pore pressure from the filtered shale transit time (PP DT) is shown in the right track with comparison to the measured formation pressures (MDT and Geotap).

QUESTION 143

I have mapped a series of isolated sand bodies (mid slope setting) and amplitude and isochore match very well with available accommodating space. Dominant class 1. Compaction seems not to have affected porosity for the most part. I have near and far offset-gathers fairly flattened. The well result shows very high amplitude and high gradient tight sands. The sand bodies I mapped have much lower amplitude and gradient, most likely still class 1 AVO. Rock properties show they can be interpreted as lower elastic impedance bodies relative to the tight sand penetrated. Some sands that kicked while drilling (over pressure) are mapped in I/G space very close to shale background too. The interval seems to match with lower interval velocities (FWI velocities). How can I discriminate high porosity from over-pressure? Or are both associated?

Binh Nguyen

Geophysicist at Bp

ANSWERS PROVIDED BY INDUSTRY EXPERTS

Giorgio Cavanna

Geoscientist at ENI E&P

Binh, If the fluid is a brine, overpressure should cause Vp/Vs to increase (gradient becomes more positive) as well as if porosity increases. In gas sands the effects are opposite: overpressure causes Vp/Vs to decrease (gradient gets more negative) while higher porosity increases Vp/Vs.

Binh Nguyen

Geophysicist at Bp

From what Giorgio explained I can understand that you should be able to see the fluid effect. Is this in well consolidated sands, where compaction is dominating factor, and porosity range is 18-20% or even higher? Or it is valid for old rocks, tighter, and PHI in real of 12-16% only?

Gautam Sen

Consultant at Sahara Group

Binh, over-pressure cannot be a regional phenomenon. If we do a bit of basin analysis, one can find out if the locale is conducive to over pressure. Inversion will surely show locales of low velocity if overpressure is present. Lateral plots of seismic velocities, especially interval velocities, will further corroborate the presence of overpressure. Porous areas on the contrary, are more regional in nature and other geological evidences like outcrops of the formation, age of deposition, post depositional tectonics, and diagenesis can help in understanding the genesis and seismic like velocity, inversion, and AVO can further support their presence. If well data are present, logs and sedimentological studies on cores/cuttings can further corroborate the picture. Thus the answer is integration of all possible data.

Binh Nguyen

Geophysicist at Bp

Guy, the question comes from all you mentioned above, we have plumbers in the team, AVO and RP is just one way to integrate the story.

Gautam Sen

Consultant at Sahara Group

Thus, the answer lies in integrating basin analysis, sedimentological studies, petrophysical studies, elastic inversion, AVO and then rock physics. That is, establish cause and effect, the latter from the seismic expression.

Binh Nguyen

Geophysicist at Bp

From the very start I did colored inversion for intercept and gradient after having cleaned up the near and far offsets for AVO, and could map where the elastic impedance was relative softer than the tight zones in each sand body (we have mapped top and base of cca 10-12 bodies). The next step, is trying to see some discrimination between and inside of each body. There is hint from a well (log and sidewall core) that porous and pressured sand (only one thin interval though) mapped very close to shale above it. Investigating the velocity is in progress. From Giorgio's comment at first, if pressured gas (or light HC) moves towards negative gradient (negative AI too I believe) and pressured brine sands moves towards positive gradient (and positive AI), this means a fluid effect should be seen. We ran fluid substitution but for much tighter sands (8-10%) and saw the effect was very small. I did map the fluid zonation in AIGI (intercept gradient impedance in seismic) but did not see them following the depth contour; they were very random throughout the sands bodies.

Gautam Sen

Consultant at Sahara Group

Overpressure in unconsolidated rocks with negative AI is understandable.

Gradient is negative with overpressure when the fluid is gas is also Ok.

I'd like to see some plots where the gradient (Vp/Vs) is positive with increasing pressure in the case of brine filled pores. We would also like to know a cause of overpressure in the case of consolidated sediments other than tectonic uplift.

PORE PRESSURE PREDICTION IN CARBONATE RESERVOIR

QUESTION 144

MDT shows that there are scattered pressure points in a calcite limestone reservoir that make it difficult for interpretation and we get invalid pressure guardians about 3-6 psi/ft. There are dry and tight pressure points in the over pressure intervals. These gradients are gained from high quality pressure points. Petrophysics shows that there are some anhydrite sub layers in the reservoir but according to lithology data, the overpressure zones have been detected in some interval where there is low percent of anhydrite. Is that overpressure related to calcite limestone or anhydrite sub layers? What is the best method for evaluating over pressure, NCT or some ratio method that uses a drilling exponent? Or is there any good method for pore pressure prediction in carbonate reservoirs.

Hani Asaei

Sr. Petroleum Geomechanics and Hydraulic Fracturing Engineer

ANSWERS PROVIDED BY INDUSTRY EXPERTS

Sam Green

Geopressure Technical Lead (EAME) at Ikon Science

Hani, Pore pressure is typically generated by the rock's response to an applied load, typically burial related (vertical stress) but it could be horizontally generated in the appropriate tectonic regime. NCT models for pore pressure prediction rely on the assumption you can link porosity to pore pressure but there are many reasons the porosity of a carbonate may not relate to its pore pressure, i.e. multiple cycles of dissolution and cementation as well as any fracturing. It's for these reasons that you can get good tests and dry tests in nearby wells within the same interval.

Most, best practice approaches to pressure prediction in carbonates rely on the knowledge/estimation of pressure in the shale and reservoirs above and below the carbonate layer. Combine that with geology of the carbonate (high energy vs. low energy deposits, likely diagenesis, cutting descriptions) and you can begin to build a pressure model. High porosity carbonates behave a lot like other reservoir units (form internal fluid gradients) and low porosity carbonates behave a lot like shales (form either overburden parallel trends or form pressure transition zones between the pressures in the overlying and underlying clastics).

Hani Asaei

Sr. Petroleum Geomechanics and Hydraulic Fracturing Engineer

Sam, you mean we should consider low porosity carbonate as shale and use the empirical models such Eaton? What is your idea about Zamora's method? It uses a drilling exponent to estimate pore pressure. Have you tested it on carbonate reservoirs? Is that applicable for over-pressure zones?

Sam Green

Geopressure Technical Lead (EAME) at Ikon Science

Hani let me clarify my comment. I meant that low porosity carbonates behave like shales in terms of low permeability and therefore can form non-fluid gradients within themselves but you cannot treat them like shales for PPP. Eaton will not work. The fast velocities of carbonate rich units result in very low PPP which is not appropriate. I haven't tried using the d-exponent in carbonates.

Hesham Naiem

Senior Pore Pressure and Geomechanics Engineer

Hani, I'd like to share with you some ideas that will, I hope be useful to you. First the d-exponent method was designed specially to estimate pore pressure in shale. Characteristics of shale are different from those of carbonates in many ways, so it is best to avoid using d-exponent in carbonate because it leads to errors. A while back, Agip Company invented a parameter called Sigma log to estimate pore pressure in carbonates. It is also used to estimate the carbonate total porosity as a reflection of a rock's pore pressure. Pore pressure in carbonates as far as I know depends on quantity of fluids in the rock and the rock confining stresses applied on the fluids inside the carbonates. So it is useful to use Sigma log to estimate pore pressure and at the same time monitor the well stability conditions specially gas analysis and abnormal gas indications while drilling as a tool for refining pore pressure values.

Selim Simon Shaker

Consulting Geoscientist at Geopressure Analysis Services

Hani, the first sentence on your post started with "MDT shows that there is scatter amongst the pressure points in calcite limestone reservoir that make it difficult for interpretation and we get invalid pressure guardians about 3-6 psi/ft."

My comment follows and it might let you think about the next step in your project.

1. What type of Carbonate environment? AAAP Memoir 33 classified the carbonate depositional environment to 12 categories. That is where to start to find out if your carbonate sequence is subject to compaction during deposition.

2. Were the scattered MDT pressure points obtained in the reservoir where hydrocarbons exist? If this is true, your MDT measurements represent HC column and are not related to formation pore pressure build up.
3. Is the scattered distribution of the MDT in one body reservoir or different mini-compartments within the overall pay zone?
4. Is the 3-6 psi/ft gradient calculated from the RKB? If so, this is one of the infamous pitfalls in pore pressure gradient calculations. Gradient is usually measured between two points in the same reservoir.

Hani Asaei

Sr. Petroleum Geomechanics and Hydraulic Fracturing Engineer

Selim,

1. Actually I do not know details about the type of carbonate. It is the Dalan formation in Persian Gulf and is composed of limestone, dolomite and anhydrite. Petrophysics shows that there are some anhydrite sub layers in the reservoir but according to lithology data, the overpressure has been detected in some interval where there is a low percent of anhydrite.
2. Scattered MDT pressure points were obtained in the reservoir, but where there is water. Actually, we get these points exactly after WGC. We have gas gradient and after that in water zone there are scattered pressure points. Is it usual or not?
3. Actually, according to petrophysical studies there is no compartmentalization.
4. Gradients were measured between two points in the same reservoir.

Bill Godsey

Owner, Geo Logic Environmental Services, LLC

Hani, what sediments lie above and below your carbonate? Do you have any structural elements affecting your area of interest?

Hani Asaei

Sr. Petroleum Geomechanics and Hydraulic Fracturing Engineer

Bill,

Dashtak (Sudair) (Caprock): 400 m, Shale with thin layers of clay stone, dolomite limestone, anhydrite, dolomite, limestone.

Kangan (Khuff) (Reservoir):190 m, Limestone and dolomite with layers of anhydrite and clay stone

Dalan (Khuff) (Reservoir): 300 m, Limestone and dolomite with layers of anhydrite and clay stone. We have scattered points in Dalan after gas gradient.

Bill Godsey

Owner, Geo Logic Environmental Services, LLC

Hani, this appears to be a classic sequence of stratigraphy for over-pressured zones. You essentially have confinement above and below, with what I would suspect is minimized porosity on the flanks of your reservoir. If you add structural elements such as faulting or salt based movements to increase dip, the over-pressurization is enhanced. Nonetheless, a thorough assessment of the geology and hydrogeology is essential.

Pritish Mukherjee
Sr. Specialist Geologist, R&T Subsurface Domain Kuwait Oil Company

Hani, I am in Kuwait and we have the Jurassic Hith-Gotnia with alternating layers of anhydrite-salt. Even in a small area of Kuwait, the cycles are ranging from 4 to 2 over the area. Complexity comes because the Anhydrite cycles do contain carbonates and at places, these are HC reservoirs which are not yet fully understood. There is a Najma kerogen rich mudstone which lies below and that has possibly, charged these carbonates through migration channels along faults in the overlying cycles. Seismic is full of multiple interference. Although each cycle of salt and anhydrite is mapable, the carbonate distribution within anhydrite cycles cannot be mapped. It is challenging since having carbonate does not mean there will be a reservoir and hence may not show an overpressure effect. Although Cretaceous Makhul shale can be used for NCT, there is no shale in true sense below these cycles. It is a Catch 22 situation. I think Sam's and Mr Shaker's views on establishing the geology behind these carbonates could be a good idea however, we feel that the NCT needs to be broken up and calibrated with detailed drilling responses well by well rather than over a 3D cube.

QUESTION 145

What would be the best workflow using seismic and well log to generate an abnormal pressure cube within thin limestone layers (gross 7-10m thick) deposited along with anhydrites and alternating salts? Only the Anhydrite cycles contain limestone deposits and those are usually tight but not always. There is no shale in this 300-400m thick salt - anhydrite cycle pack.

Pritish Mukherjee
Sr. Specialist Geologist, R&T Subsurface Domain Kuwait Oil Company

ANSWERS PROVIDED BY INDUSTRY EXPERTS

Hatem Alnajar
Pore Pressure & Geomechanics Independent Consultant

Pritish, it looks like compaction disequilibrium is not the main cause of overpressure in your example. Overpressure in your case might be due to salt / gypsum - anhydrite transform. In such environments; conventional pore pressure prediction methods might not give the right answer. A more robust method must be applied, such as Basin modeling. A well known application in this case might be using Presage (A basin modeling software for pore pressure prediction - a Part of the Drillworks Software). A suggested workflow can be as follows:

1. Construct 1D PP model from log and well data and calibrate to known PP values.
2. Construct a Basin Model utilizing PP from 1D above.
3. Calibrate Seismic to DT from well data.
4. Predict PP from Seismic and then use in Basin modeling creating pseudo wells
5. Re-calibrate your basin model in Presage using all data.

6. Open 3D in Drill works.
7. Construct and transfer 3D PP cube from Presage to 3D application in Drill works.

Yogesh Kulkarni

Pore Pressure Analyst/ Field Engineer, Geoservices, Schlumberger Company

Pritish, as there is no shale in the mentioned interval, normal compaction trend based methods for pore pressure prediction will not be successful. However, understanding origin of the overpressure mechanism – using a basin modeling approach with high resolution velocity modeling that uses relationship between velocity, effective stress and diagenesis factor along with velocity-density crossplots may prove productive.

Medium / high dip, assuming geobodies to be connected, lateral transfer mechanisms may be applied with help of a good seismic profile. If the expected reservoir height and fluid densities are known, pressure calculation via buoyancy techniques, making some assumptions on pressure at the base of the fluid column, can be accomplished. Further it can be checked against gas events experienced in offset wells versus RFT, MDTs.

Pritish Mukherjee

Sr. Specialist Geologist, R&T Subsurface Domain Kuwait Oil Company

Yes, devoid of shale the NCT based approach will not work. Having said this, Presage based models can suffer from a lack of knowledge as to the real cause of geopressure generation. Whether it is salt-anhydrite conversion, or charging of thin carbonate layers from down dip is a matter of assumptions. By using the well data and events, we are able to generate a confident 1D model. However, there is a real lack of confidence attached to stepping out. Incidentally the same cycles have been drilled based on what the well requires with MW ranging from 14.5 to more than 20 ppb in the past. So, clearly the pressure pods are guided by down dip connectivity through the limestone. The real challenge is that it is not possible to pick the limest packs within anhydrite due to lack of contrast. Hence, what Yogesh says about high density velocity picks could be the nearest possible help, but I doubt if that will give good results. Anyways, thanks a lot for the inputs. I will try to combine your steps to get some answers on pseudo wells and test in blind wells with a decent spread over a larger area.

Kehinde Adesina

Snr. Global Technical Advisor –Drilling & Geomechanics (unconventional & Deepwater), Halliburton

Pritish, simple geological understanding of the rocks would suggest that high pressures in non shale and sands intervals must be due to fractures. Vertical, slant and lateral fractures would behave differently in limestones, dolomites or anhydrites depending on the stress regime. I suspect that you have lateral fractures and since they are open and intersect the wellbore, that is the reason for these High PP in thin sections of formation. Vertical resolution of RES or DT cannot see these fractures. If the wellbore is stable enough to run logs, then run image logs and use those to identify beddings, open and closed fractures, orientations and distributions. We have many of these cases in Saudi and it depends on proximity to the tectonic plates of the Arabia Sea and the history of fracture generations / episodes.

QUESTION 146

How can I measure and estimate the pore pressure in evaporate layers, particularly the Gachsaran formation in the Iranian oil fields?

Hossein Talebi

Structural Geologist and Geomechanics at National Iranian South Oil Company

ANSWERS PROVIDED BY INDUSTRY EXPERTS

Morteza Azadpour

Petroleum Engineer at National Iranian Oil Company

Hossein, Evaporate layers (carbonates, sulfates, and chlorides) present particular problems for pore pressure and fracture gradient prediction. This is because of diagenesis, complex pore structure, secondary porosity, highly variable texture and mineralogy and some other problems. But all of them are not created equal and each evaporate type has its own behavior.

Most of the work on PPP from well log data in Iran is in regard to reservoir layers not cap-rock (Gachsaran) because of limitation in full-set well-logs and pressure test in these formations. But PPP from seismic data may help you. I suggest the following thesis by Ehsan Nosrat in Persian:

Ehsan Nosrat, 2007, "Pore Pressure Prediction Using 3D Seismic Attributes", Amirkabir University of Technology (Tehran Polytechnic), Department of Petroleum. In chapter 6 of this thesis you can find the Pore Pressure Prediction at Gachsaran formation using seismic attributes. There are also some other works but not in Gachsaran Formation.

PORE PRESSURE PREDICTION IN AN UNCONVENTIONAL RESERVOIR—CASE STUDY BOSSIER AND HAYNESVILLE SHALES

In the tight gas plays, pore pressure prediction is of critical importance for drilling operations to improve drilling efficiency, reduce borehole trouble time, and avoid risks of well blowouts. However, unlike conventional reservoirs, mechanisms of abnormal pressure in shale gas formations are rarely reported, and accurate pore pressure prediction in such formations can be problematic. This section presents our recent findings in gas plays in Bossier and Haynesville shales in North Louisiana. The gas plays have highly overpressured pore pressure of up to 18 ppg. This hard overpressure, if not correctly predicted, can cause severe drilling risks from gas kicks and possible well blowouts. Fluid kicks and influx in the Bossier and Haynesville shales are common drilling events. However, drilling highly overbalanced is also to be avoided, as the mud weight can exceed formation fracture gradient, causing mud losses and lost circulation. We analyzed drilling and well-logging data in over 50 wells in this project to study the overpressured mechanism in the shale formations. For pore pressure analysis, the rocks' petrophysical properties and geology are basic inputs and are introduced in the following sections.

Geology and Lithology in the Bossier and Haynesville Shales

The gas-bearing formations of Bossier and Haynesville shales are very tight shale with extremely low matrix permeability (0.00007–0.0007 mD). The targets of the gas play are mainly in the Jurassic Lower Bossier and Haynesville at depths of about 11,500–13,000 ft TVD. The lithology in this area is comprised of Tertiary, Cretaceous and Jurassic aged stratigraphic units. The lithologies are characterized primarily as claystones (shales) and sandstones. There are also limestone formations, such as Knowles limestone in Cotton Valley Group and Smackover limestone at the bottom of the Haynesville shale, as shown in Fig. 9.15.

The structure in the studied field is characterized by flat strata in the north part of the field and inclined strata broken by some normal faulting. The rapid change

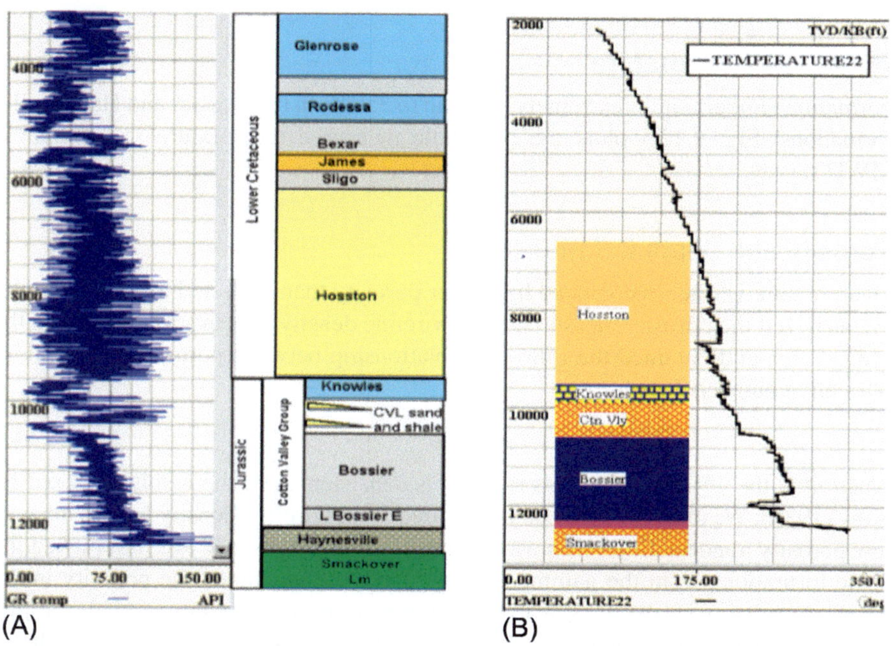

FIG. 9.15

Typical lithology and formation temperature in the studied area. (A) Typical stratigraphic column and gamma ray log; (B) downhole temperature measured from MWD.

in formations tops and inclinations causes changes in rock properties and pore pressures. Regional correlation sections across northern Louisiana reveal a prominent mid-Cretaceous unconformity, suggesting a later uplift event that affected the area.

It now appears that the present-day Sabine High represents a Tertiary uplift event. A maximum missing section due to Tertiary uplift is estimated to be on the order of several thousand feet. Timing of uplift and erosion appears to be post-Wilcox (mainly Oligocene).

Rock Properties Associated With Pore Pressure Prediction

Rock properties are the basic input for pore pressure calculations. In the shale gas formations, however, rock properties are different from the conventional formations. For instance, shale gas formations have very low porosity and the presence of gas in the shales adversely affects the compressional velocity. These present great challenges in pore pressure prediction using conventional log-based pore pressure prediction techniques. In this section, we briefly present some physical and mechanical rock properties that affect pore pressure prediction and fracture gradient analysis.

Formation Temperature

Formation temperature is extremely high, up to 350°F, in the Bossier and Haynesville formations. Fig. 9.15B plots the downhole measured temperature in a borehole MWD survey.

Porosity and Depth Relationship

Overpressure is often recognized by higher porosity than expected at a given depth. We analyzed the porosity measured from wireline density logs in several wells in the project area and obtained the following relationship between porosity and depth in normally compacted shales:

$$\varphi = 0.5\,e^{-0.00036Z} \tag{9.41}$$

where Z is the TVD in feet. This relation is consistent with the result proposed by Athy (1930). We examined porosity measured from wireline density logs to analyze any porosity abnormalities, as a way to quantify pore pressures. We first picked the shale formations from the gamma ray log, and then calculated the porosity in the selected shales from the density log. By doing so, porosity points of shales at different depths were obtained. Then we filtered the porosity points from these shales to get rid of obvious outliers. Finally, we fitted this shale porosity data to depth. Fig. 9.16 presents a typical porosity profile with a normal compaction trend plotted in this shale gas field. Here in the second and third tracks, the red curve is the observed porosity-depth trend and the green curve is the normal compaction trend from Eq. (9.41). It shows that the shale is in normal compaction when depth is <6400 ft. Deeper than that, the formations (at Sligo and Hosston) show as slightly undercompacted with higher porosity than the normal compaction trend. In the gas plays of Bossier and Haynesville, the porosity is significantly higher than the normal compaction trend. This porosity increase may be caused by the undercompaction of the shales or by gas generation.

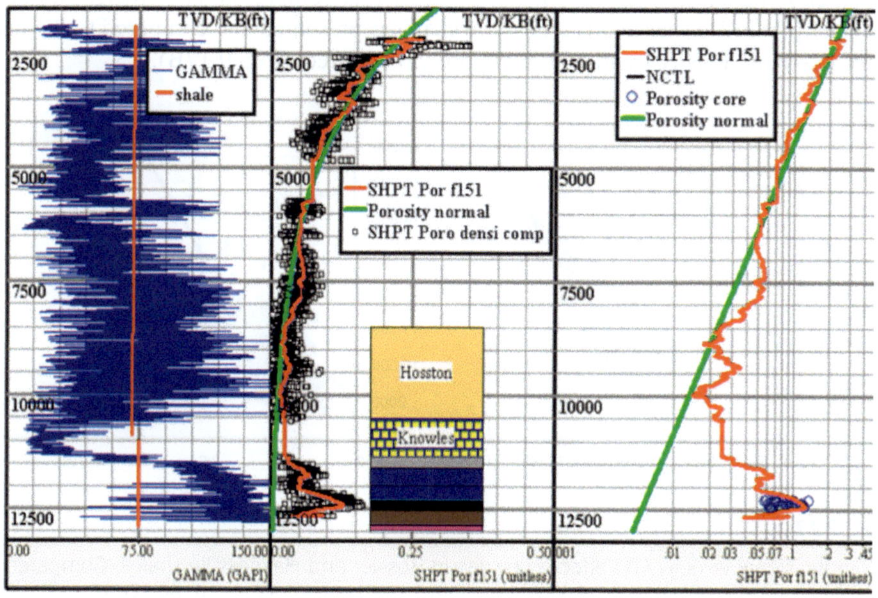

FIG. 9.16

Porosity and depth relationship plotted in both linear and logarithm scales in a shale gas well. The *green line* is the porosity-depth trend in normal compaction case (from Eq. 9.1). Left track shows gamma ray *(blue)*, and shale line *(red)*; middle track presents porosity points in shale *(black dots)*, filtered porosity of shale points *(red curve)*, normal compaction trend line of porosity *(green curve)*, and lithology column; light track plots filtered porosity of shales in logarithm plot *(red curve)*, porosity from core tests *(blue circles)* and normal compaction trend line of porosity in logarithm plot *(green curve)*.

Gas Effect on Compressional Velocity

Sonic velocity or transit time plays a key role in pore pressure and fracture gradient prediction (e.g., Poisson's ratio calculation). For the shale or claystone without gas effect, compressional and shear velocities follow noticeable trends. For Gulf Coast mudrocks, Castagna et al. (1985) presented the following relationship:

$$V_s = 0.8621V_p - 3846 \tag{9.42}$$

where V_s, V_p are the shear and compressional velocities respectively in feet per second. Plotting compressional and shear velocities obtained from wireline logs in the studied area, we found that the observed $V_p - V_s$ trend in the Bossier and Haynesville shales deviates from the normal $V_p - V_s$ trend (Eq. 9.42), as shown in Fig. 9.17A. The $V_p - V_s$ relation shows that V_p is slow compared to the normal $V_p - V_s$ trend. This slowdown is thought to be a Gassmann type of velocity effect due to the presence of gas in the shale, as it occurs mainly in compressional velocity, and because gas apparently has little effect on the shear velocity. A compressional velocity and bulk density plot in Bossier and Haynesville shales is presented and compared to the normal V_p – bulk density trend without hydrocarbon effect (Fig. 9.17B).

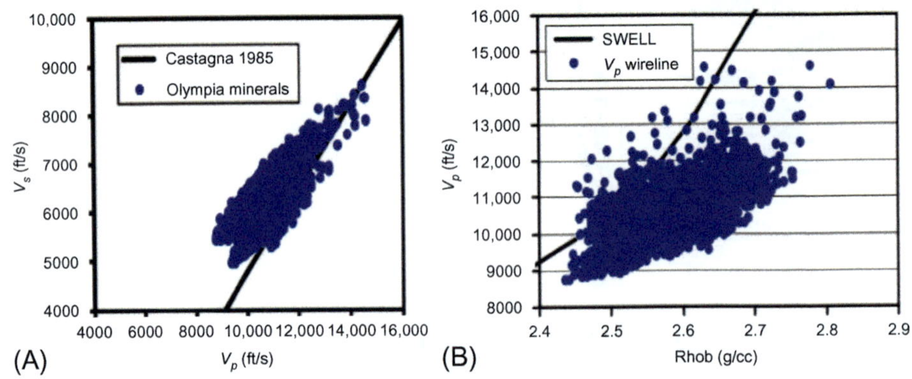

FIG. 9.17

Compressional velocity (V_p) affected by gas in Bossier and Haynesville shales. Compared to established trends V_p is too slow for a given V_s or density. (A) $V_p - V_s$ plot showing V_p reduction compared to the normal $V_p - V_s$ trend; (B) V_p versus bulk density and the baseline of V_p and bulk density in the Gulf of Mexico showing gas effect on V_p.

It again shows that the V_p in Bossier and Haynesville is much slower than the normal trend of mudrocks in the Gulf of Mexico. This V_p and bulk density relationship is useful in identifying gas/HC-bearing formations. The drill gas from mud-logging data also verifies the V_p reduction by gas in shale gas formations (Fig. 9.18).

This slowdown in compressional velocity needs to be corrected for pore pressure and fracture gradient prediction. Given that gas has a small effect on the shear wave velocity, we can therefore calculate the compressional velocity from a shear velocity and avoid the gas effect. Using an appropriate $V_p - V_s$ relation (e.g., Eq. 9.42), we can estimate V_p from the downhole measured V_s. Fig. 9.19 demonstrates an example of the compressional transit time calculated from the shear transit time taken from a wireline log and its comparison to the measured compressional transit time. It

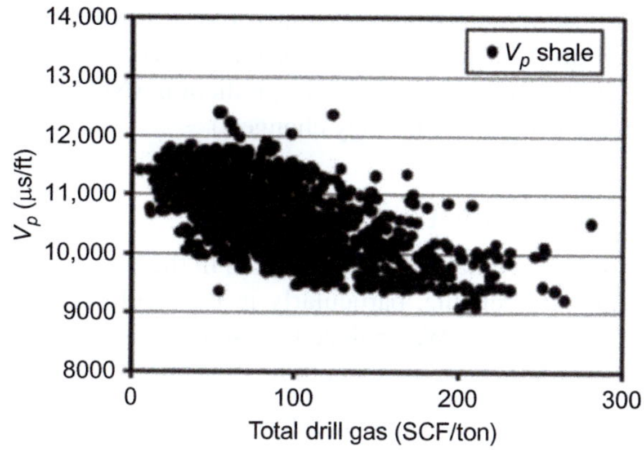

FIG. 9.18

V_p versus drill gas in a well demonstrating that V_p decreases as drill gas increases.

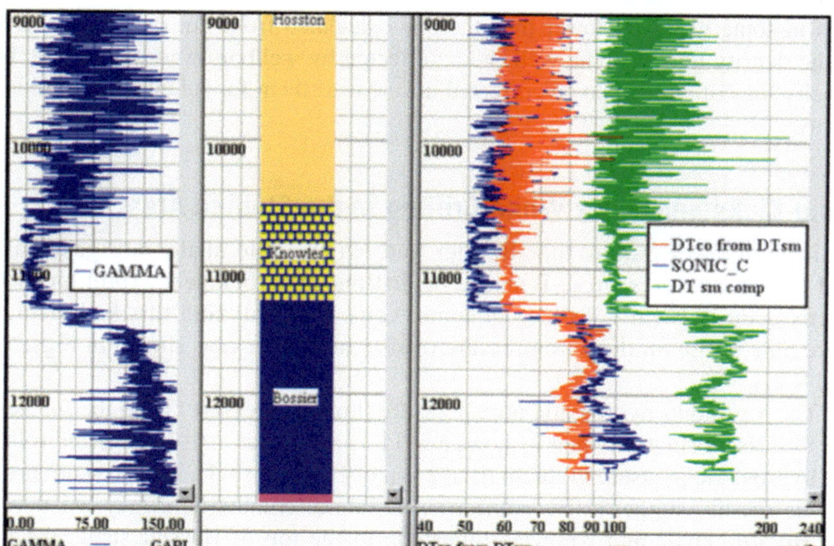

FIG. 9.19

Correction on gas effect on compressional transit time (DT) by using $V_p - V_s$ trend (Eq. 9.42) and compared to the one from wireline logging in a well (the depth in feet). The right track presents downhole measured sonic compressional DTco *(blue curve)*, DTco calculated from measured shear DTsm *(red curve)*, and measured shear DTsm *(green curve)*.

shows that the increase in compressional transit time (slowdown in compressional velocity) can reach up to 20 μs/ft. This gas effect corrected compressional transit time was used in the log-based pore pressure prediction. (Note the reverse effect in the Knowles formation is due to lithology changes.)

Overburden Stress Gradient

Wireline density logs are available in several wells in the studied area. However, the density data are not very reliable, particularly in the shallow section (<10,000 ft), because of wellbore washouts. We analyzed reliable density log data and obtained the following generalized bulk density-depth relationship:

$$\rho_b = 1.52Z^{0.061} \tag{9.43}$$

where ρ_b is the bulk density in g/cm^3; Z is the TVD in feet. By integrating the bulk density (Eq. 9.43) the overburden stress can be obtained as follows:

$$\sigma_v = 0.6186Z^{1.061} \tag{9.44}$$

where σ_v is the overburden pressure (or *OBP*) in psi; and Z is the TVD in feet. It should be noted that the overburden pressure gradients calculated from bulk density logs in some wells are lower than the one calculated from Eq. (9.44). This is due to the density logs in some wells being affected by wellbore washout. Hence, bulk density logs often need to be corrected, if one uses them for calculating overburden stress.

Fluid Kicks and Influx While Drilling in the Studied Area

Drilling activities show that Bossier and Haynesville shales are gas-bearing and highly overpressured. Because of drilling underbalanced, fluid kicks and influx occurred while drilling in some wells. By analyzing this kick data, we were able to obtain the pore pressure profile in this area, as shown in Figs. 9.20 and 9.21.

The log shows hydrostatic pore pressure (normal pressure) from the surface to the Rodessa formation; then it shows a mild overpressure (10–11 ppg) and follows the hydrostatic pressure gradient till the top of Hosston shale. The pressure profile in this section suggests that aquifers with lateral conductivity may exist. From the Hosston shale and Knowles limestone to the top of Bossier shale, the pore pressure increases further 11–12.5 ppg. From top of Bossier to Lower Bossier there is a sharp increase in pore pressure (12.5–17 ppg). This is caused by gas-bearing formations, and the pore pressure transitions from the mild overpressure to highly overpressured, “hard overpressure.” Therefore, it is critical to choose a casing shoe location in the top of Bossier before the pore pressure reaches the hard overpressure. Otherwise, the high mud weight used, to balance the high pressure, is likely to fracture the formations and cause lost circulation due to the low fracture gradient in the shallow depth. From Mid-Bossier to Lower Bossier the

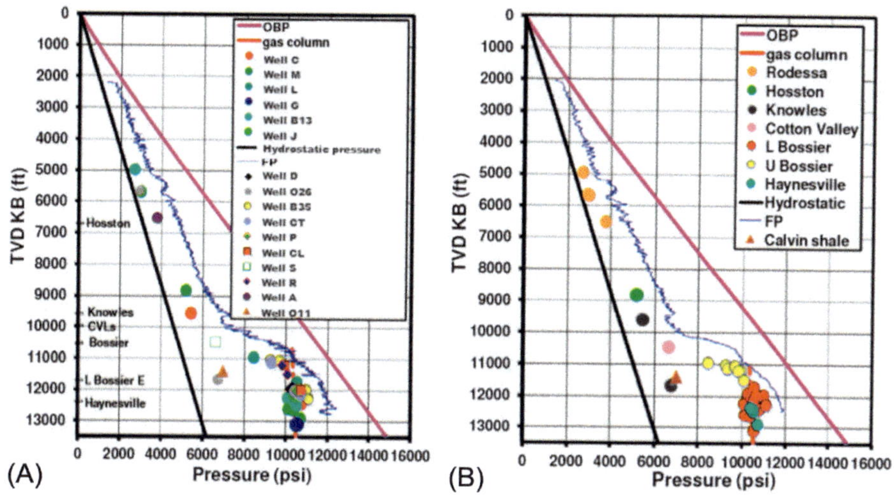

FIG. 9.20

Pore pressure profile obtained from fluid kicks and influx in the offset wells, where the reliable pressure data are available. (A) Plotted based on the wells with a typical formation tops; (B) Plotted on the same formation tops.

pore pressure ramps sharply (up to 17.8 ppg), narrowing the range between the pore pressure and the minimum horizontal stress, or fracture gradient. This makes the drilling mud weight window narrow and increases the problems associated with fluid kicks and mud losses. The high overpressure is a major characteristic of the shale gas formations in this area. Given this pressure profile across the region, one can infer pressure seals where large changes in pressure occur. For instance, a good seal exists between the lower Bossier and the upper Bossier, allowing the pore pressure in the lower Bossier and Haynesville shales to be very different from the upper Bossier shale.

Pore Pressure Prediction From Kicks

Several pressure gradients were identified from the kick data, as shown in Fig. 9.21. There is a normal pressure gradient from surface to the top of Rodessa limestone, then pore pressure becomes slightly overpressured, but follows hydrostatic gradient from Rodessa formation to the top of Hosston (Travis peak) shale. The pore pressure then increases gradually from Hosston shale to Knowles limestone, following the lithostatic gradient. At the upper Bossier shale, the pore pressure increases significantly and transitions zone to hard overpressure in lower Bossier. Hard overpressure exists from Lower Bossier to Haynesville. Based on this pore pressure gradient profile, the pore pressure can be predicted when one knows the predicted formation tops in the predrill wells. A program was written to predict pore pressure based on this model.

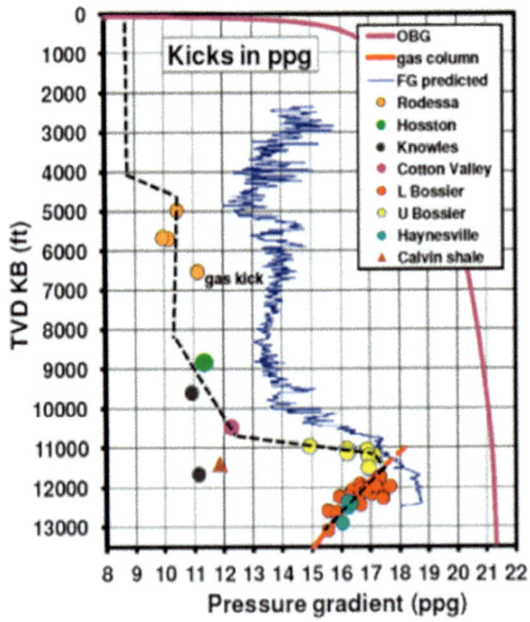

FIG. 9.21

Pore pressure gradient from all kicks in different wells plotted on the same formation tops. Several pressure gradients from different depths are identified in the kick data.

Fig. 9.22 shows an example of predrill pore pressure and fracture gradient predictions in a well. Using the proposed method, predrill pore pressures and fracture gradients were predicted for several dozen wells in this project and drilling risks were reduced greatly.

Pore Pressure Prediction From Well-Log Data

Pore pressure gradient can be estimated from well-logging data when the overpressured pore pressure is mainly generated from compaction disequilibrium. In this theory the pore pressure is predicted for shale lithologies. For other lithologies, such as sandstones, the pore pressure can be obtained using pressure equilibrium or centroid theory. The pore pressure prediction in shales assumes that the rocks' petrophysical properties and pore pressure have similar correlations in an area. Therefore, the pore pressure can be obtained from sonic and resistivity data and calibrated to drilling events. A crucial step in performing a pore pressure analysis from well-logging data is to select clean shales for use in the analysis. Therefore, wireline gamma ray data are used to differentiate shales (with high gamma ray) from other lithologies. The shale base lines in the model are used to identify and separate shale from other rocks. The points where gamma ray values are less than

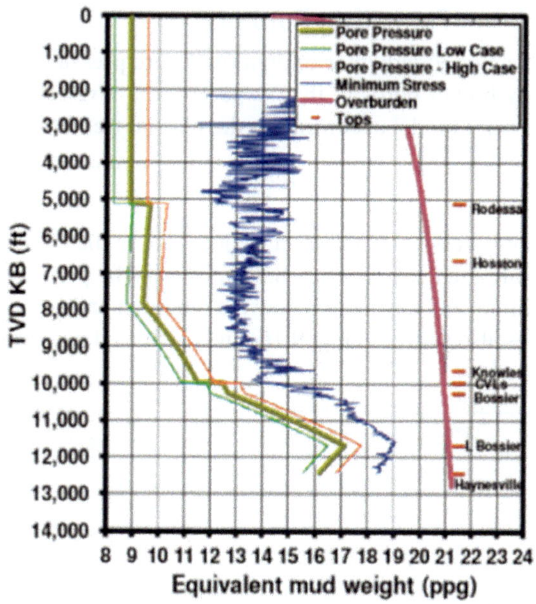

FIG. 9.22

Predrill pressure prediction based on the kick pressure profile and formation tops in a predrill well.

the shale baseline are not used for analysis. Shale points defined on the gamma ray log are transferred to the corresponding sonic and other logging data used for pore pressure analysis.

Pore Pressure From Sonic Velocity

Shell's Tau method (Lopez et al., 2004) was used to predict pore pressure from sonic logging data. The Tau method uses the following equation to calculate pore pressure:

$$P_\tau = OBP - A\left(\frac{C - DT}{DT - D}\right)^{B_s} \tag{9.45}$$

where P_τ is the pore pressure in shale in psi; *OBP* is the overburden pressure in psi; *DT* is the sonic transit time in shale, in μs/ft and *A*, *B*, *C*, *D* are the constants, where *C*, *D* are related to mudline and matrix transit time, respectively. The following parameters were used for pore pressure analysis: $A=1150$, $B=0.98$, $C=207$, and $D=47$, which were obtained based on the pore pressure calibration to fluid kicks and influx in offset wells. The pore pressure prediction from sonic log has a standard error of about 0.84 ppg, which is too big to allow an accurate prediction in the shale gas formations.

Pore Pressure From Formation Porosity

Porosity in shales is a good indicator of abnormal pore pressure, and overpressure is often recognized by higher porosity than expected at a given depth. We derived the following pore pressure model according to the normal trend of porosity:

$$P_{pg} = OBG - (OBG - P_n)\frac{In\varphi_0 - In\varphi}{cZ} \tag{9.46}$$

where P_{pg} is the pore pressure gradient in shale; OBG is the overburden gradient; P_n is the normal pressure gradient (P_n=8.5–8.75 ppg); φ is the porosity in shale, φ_0 is the porosity at the mudline; Z is the depth below the mudline; and c is a constant. Fig. 9.23 presents porosity (as also shown in Fig. 9.16) and calculated pore pressures with depth.

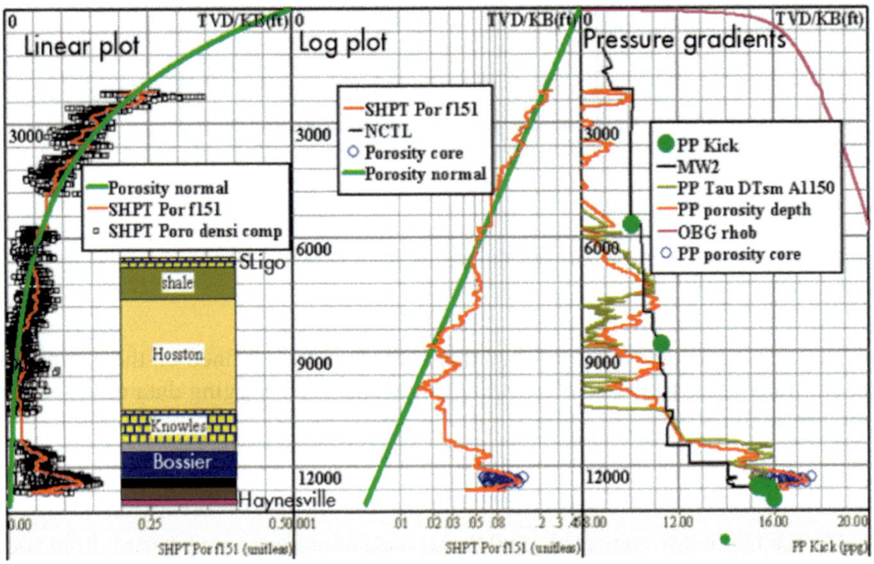

FIG. 9.23

Pore pressure estimations from porosity and sonic log with comparison to kick data showing that porosity model *(red curve in the right track)* can give a better prediction in some formations. The porosity data refers to Fig. 9.16. The right track presents mud weight *(black curve)*, fluid kicks *(green dots)*, overburden gradient *(pink curve)*, pore pressure profile calculated from the porosity by Eq. (9.6) *(red curve)*, and pore pressure profile calculated from the sonic log by Eq. (9.45) *(brown curve)*.

In the preceding figure, Eq. (9.41) ($\varphi_n = \varphi_0 e^{-cZ}$) is used for the normal compaction porosity trend line with the parameters of φ_0=0.5 and c=0.00036. The figure shows that pore pressure increases from Sligo limestone to Hosston shale associated with a higher porosity than the normal trend. In the gas plays at Bossier and Haynesville, the porosity is significantly higher than the normal porosity trend and hard overpressure exists in Bossier and Haynesville formations. The porosity method and Tau model give similar predictions of pore pressure. The log-based method is important for the pore pressure prediction in the pressure transit zone, such as from the Upper Bossier to the Lower Bossier shale. It should be noted that porosity and pore pressures are very different even in Bossier shale, as shown in Fig. 9.23. This implies that several pressure compartments and seals may exist in this shale.

Pore Pressure and Formation Unloading

Unloading associated with the Sabine uplift in the studied area affects log-based pore pressure prediction. Bowers (1995) studied the unloading impact on pore pressure prediction. The sonic velocity and effective vertical stress have a power relation in the loading stage (Bowers, 1995):

$$V_p = 5000 + A\sigma_e^B \tag{9.47}$$

where V_p is the compressional velocity at a given depth (ft/s); σ_e is the vertical effective stress or VES in psi, $\sigma_e = \sigma_e - \sigma_v - P_p$; σ_v is the overburden stress; A and B are parameters calibrated with offset velocity versus effective stress data. Due to formation uplift or unloading, the effective stress and the velocity relation do not follow the loading curve. As demonstrated by the laboratory compaction data in Fig. 9.24, unloading happens and the unloading defines the rebound curves in the original compaction/loading curve.

The unloading occurs along a flatter effective stress path than the initial compaction curve, as shown in Fig. 9.24. In the unloading case a higher pore pressure exists for the same velocity value than the one without unloading. Bowers (1995) proposed the following empirical relation for the unloading curve:

$$V_p = 5000 + A\left[\sigma_{max}\left(\sigma_e / \sigma_{max}\right)^{1/U}\right]^B \tag{9.48}$$

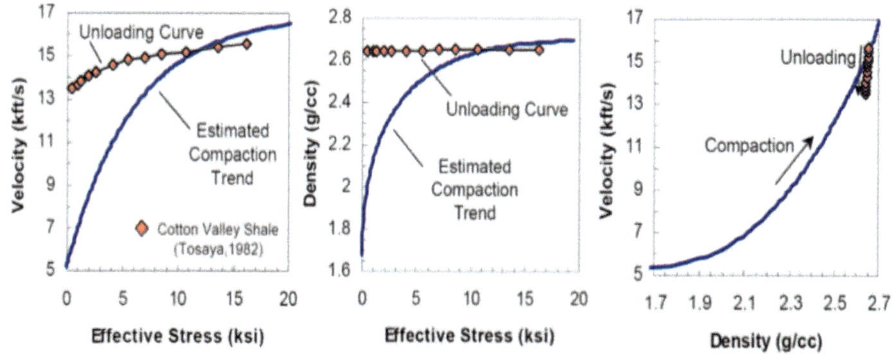

FIG. 9.24

Typical unloading responses from velocity-effective stress, density-effective stress, and velocity-density plots (Bowers, 2001).

where U is a parameter; and σ_{max} is the maximum effective stress when unloading occurs. We plotted the pore pressure (calculated from the fluid kicks) and corresponding sonic velocities in the studied area, in Fig. 9.25.

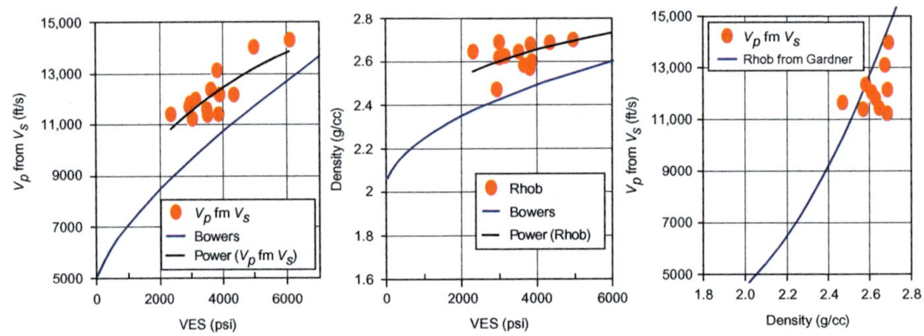

FIG. 9.25

Velocity-effective stress, density-effective stress, and velocity-density plots in some wells showing unloading behavior exists in Bossier and Haynesville. The plots were based on the fluid kicks that have available sonic logging data and VES is the vertical effective stress.

Compared to Bowers's plots, this shows that unloading exists in the area. Geologic evidence also shows that the formation uplift existed in this area, which caused formation unloading. This makes pore pressure prediction from sonic log data more complicated. Analyzing the kick data in this project leads to the following sonic velocity-effective stress correlation (Fig. 9.26), which describes the velocity and vertical effective stress relationship:

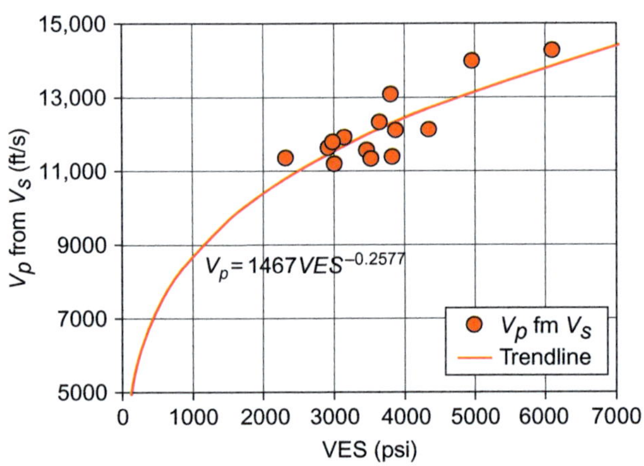

FIG. 9.26

Sonic compressional velocity calculated from shear velocity versus vertical effective stress calculated from kick data.

$$V_p = 1467\sigma_e^{0.2577} \tag{9.49}$$

where V_p is the compressional velocity, in ft/s; and σ_e is the effective stress, in psi.

Conclusions on the Bossier and Haynesville Shale Formation

The sonic compressional velocity in wireline logging is slowed down by the gas presence in Bossier and Haynesville shale formations. This slowdown needs to be corrected in order to calculate pore pressure and fracture gradient. And the shear velocity can be used for this purpose. The Sabine uplift (unloading) was identified, which affects log-based pore pressure prediction, particularly in Bossier and Haynesville shale gas formations.

We investigated 54 kicks across the field to analyze the abnormal pressures in shale gas formations. Using the kick-derived pressure model and improved log-based effective stress methods, we developed a comprehensive pore pressure prediction method. Using the proposed method, predrill work was done on several dozens of wells and drilling risks were reduced greatly.

QUESTION 147

How can we estimate pore pressure in a tight gas reservoir, using a conventional log suite? Is there any workflow?

Carlos Brun

Reservoir Engineer Associate at QEP Resources

ANSWERS PROVIDED BY INDUSTRY EXPERTS

Sam Green

Geopressure Technical Lead (EAME) at Ikon Science

Carlos, There are many reasons that you cannot estimate pore pressure from logs. For example, high TOC (total organic content) which mimics a high pressure response in the logs and/or the presence of gas which does likewise. For an overview on these issues and what is out there in the literature on how various groups have gone about solving the issue as well as key references, please check out the link below to a paper I co-authored on this topic a few months ago in The Leading Edge.

I would say that I have been able to run standard techniques on conventional logs with success in some unconventional plays without manipulating the logs at all, something I put down to a strong compressive stress regime that is maintaining classic pore pressure generating mechanisms in much the same way that vertical stress does in offshore regimes. This is speculation right now, and does not explain why the uplift is not playing more of a role in PP evolution but the fact that I have been able to run standard techniques on conventional logs still holds. In other regimes that are more isotropic and with a weaker overall stress magnitude, the work is more difficult for the reasons our paper outlines.

http://www.ikonscience.com/library/2014/01/23/what-is-the-impact-of-geopressure-on-unconventional-plays

Carlos Brun

Reservoir Engineer Associate at QEP Resources

Sam, realistically, how long would it take to accurately come up with a pore pressure estimate for an unconventional tight gas trap? I need to manage my time to finish the project.

Sam Green

Geopressure Technical Lead (EAME) at Ikon Science

Carlos, the time frame required is an open-ended question. It depends on how many offset wells you have to build the model from, how much seismic you have (or don't have) and what specific objectives the client has. I would say it would be appropriate to allow for an average of 1-2 weeks per well to achieve a typical set of objectives for a full pore pressure model from surface to TD.

Carlos Brun

Reservoir Engineer Associate at QEP Resources

Sam, I do not have seismic data and I am building a hydraulic fracture model which requires a thorough estimation of the pore pressure. I am working on approximately 45 wells.

Sam Green

Geopressure Technical Lead (EAME) at Ikon Science

Carlos, If the 45 wells represent a single geological group of wells then the time frame won't be 45 weeks, it will be less. But it is still a significant undertaking to do a thorough job

Allan Robinson

Geophysicist

Carlos, check your offset logs for MDT or RCI logs. These should contain pressure data points that will allow depth adjustments for correlation purposes in your reservoir. They won't however provide data for the pressure in the overburden.

Kehinde Adesina

Snr. Global Technical Advisor –Drilling & Geomechanics (unconventional & Deepwater), Halliburton

Carlos, you can first try to calibrate initial estimates of PP profiles to MW and look for agreement based on what you get from RES and SONIC and using DXC when you compare profiles from any of these three methods depending on what methods you use. I suggest you use Eaton with unique exponents in each method that align the three profiles together and then you can adjust your trend line to match all the three profiles. This will bring you closer to a good estimate and if you can achieve a +/-0.5ppg from what you think true pressure gradient may be, I would say you have done well.

APPENDIX

DERIVATION OF PORE PRESSURE PREDICTION FROM POROSITY

A

It is well known that the formation porosity and effective stress have the following relationship:

$$\varphi = \varphi_0 e^{-\alpha\sigma_e} \quad \text{(A1)}$$

Therefore, the effective stress can be obtained from Eq. (A1):

$$\sigma_e = \frac{1}{a} In \frac{\varphi_0}{\varphi} \quad \text{(A2)}$$

The effective stress at normal pressure conditions can also be obtained from Eq. (A1), when the porosity is the normal porosity, a condition in which formations are normally compacted, i.e.,

$$\varphi_n = \varphi_0 e^{-\alpha\sigma_n} \quad \text{(A3)}$$

$$\sigma_n = \frac{1}{a} In \frac{\varphi_0}{\varphi_n} \quad \text{(A4)}$$

Combining Eqs. (A2), (A4), one has:

$$\frac{\sigma_e}{\sigma_n} = \frac{In\varphi_0 - In\varphi}{In\varphi_0 - In\varphi_n} \quad \text{(A5)}$$

If we assume the Biot coefficient $\alpha = 1$ in Eq. (A1), then the effective stress and pore pressure have the following relationship:

$$P = \sigma_v - \sigma_e \quad \text{(A6)}$$

Substituting Eq. (A5) into Eq. (A6) and noticing $\sigma_n = \sigma_v - p_n$, we can obtain the pore pressure, overburden stress and porosity relationship:

$$P = \sigma_v - (\sigma_v - P_n)\frac{In\varphi_0 - In\varphi}{In\varphi_0 - In\varphi_n} \tag{A7}$$

where φ_n is the porosity in the condition of the formation with a normal compaction. The normal compaction porosity can be obtained from the following equation:

$$\varphi_n = \varphi_0 e^{-cZ} \tag{A8}$$

Substituting Eq. (A8) into Eq. (A7), we obtain the pore pressure and porosity relationship:

$$P = \sigma_v - (\sigma_v - P_n)\frac{In\varphi_0 - In\varphi}{cZ} \tag{A9}$$

where P is the pore pressure; σ_v is the overburden stress; P_n is the normal pore pressure; φ is porosity in shale, φ_0 is the porosity in the mudline (the ground surface or sea floor); Z is the depth below the mudline; and c is the constant. The pore pressure gradient can be easily obtained from Eq. (A9):

$$P_{pg} = OBG - (OBG - P_{ng})\frac{In\varphi_0 - In\varphi}{cZ} \tag{A10}$$

where P_{pg} is the pore pressure gradient; OBG is the overburden stress gradient; P_{ng} is the normal pressure gradient.

APPENDIX

DERIVATION OF SONIC NORMAL COMPACTION EQUATION

B

From the Wyllie equation, the porosity at the normal compaction condition φ_n can be expressed as:

$$\varphi_n = \frac{\Delta t_n - \Delta t_m}{\Delta t_f - \Delta t_m} \quad \text{(B1)}$$

and the porosity in the mudline φ_{ml} can be written as follows:

$$\varphi_{ml} = \frac{\Delta t_{ml} - \Delta t_m}{\Delta t_f - \Delta t_m} \quad \text{(B2)}$$

From Eq. (A8), we have the following relationship between the normal porosity and mudline porosity:

$$\varphi_n = \varphi_{ml} e^{-cZ} \quad \text{(B3)}$$

Substituting Eqs. (B1), (B2) into Eq. (B3), we can obtain the transit time in the normal compaction trend, i.e.,

$$\Delta t_n = \Delta t_m + \left(\Delta t_{ml} - \Delta t_m\right) e^{-cZ} \quad \text{(B4)}$$

References

Athy, L.F., 1930. Density, porosity, and compaction of sedimentary rocks. AAPG Bull. 14 (1), 1–24.

Bakulin, A., Zdraveva, O., 2010. Building geologically plausible anisotropic depth models using borehole data and horizon-guided interpolation. In: SEG Technical Program Expanded Abstracts 2010. pp. 4118–4122. http://dx.doi.org/10.1190/1.3513721.

Berryman and Milton (1991) for valid fluid substitution equations in a two-mineral rock in that case.

Biot, M.A., 1941. General theory of three-dimensional consolidation. J. Appl. Phys. 12 (1), 155–164.

Bowers, G.L., 1995. Pore pressure estimation from velocity data; accounting for overpressure mechanisms besides undercompaction. SPE Drill. Complet. 10, 89–95.

Bowers, G.L., 2001. Determining an appropriate pore-pressure estimation strategy. Paper OTC 13042.

Burrus, J., 1998. Overpressure models for clastic rocks, their relation to hydrocarbon expulsion: a critical reevaluation. In: Law, B.E., Ulmishek, G.F., Slavin, V.I. (Eds.), Abnormal Pressures in Hydrocarbon Environments: AAPG Memoir, vol. 70. pp. 35–63.

Castagna, J.P., Swan, H.W., 1997. Principles of AVO crossplotting. Lead. Edge 16 (4), 337–344. http://dx.doi.org/10.1190/1.1437626.

Castagna, J.P., Batzle, M.L., Eastwood, R.L., 1985. Relationships between compressional wave and shear-wave velocities in clastic silicate rocks. Geophysics 50, 571–581.

Chapman, R.E., 1983. Petroleum Geology. Elsevier, Amsterdam.

Ciz, R., Shapiro, S.A., 2007. Generalization of Gassmann equations for porous mediasaturated with a solid material. Geophysics 72 (6), A75–A79. http://dx.doi.org/10.1190/1.2772400.

Daines, S.R., 1982. Prediction of fracture pressures for wildcat wells. JPT 34, 863–872.

Davies, R.J., Swarbrick, R.E., Evans, R.J., Huuse, M., 2007. Birth of a mud volcano: East Java, 29 May 2006. GSA Today 17 (2), 4–9. http://dx.doi.org/10.1130/GSAT01702A.1.

Dickinson, G., 1953. Geological aspects of abnormal reservoir pressures in Gulf Coast Louisiana. AAPG Bull. 37 (2), 410–432.

Dodson, J.K., 2004. Gulf of Mexico 'trouble time' creates major drilling expenses. Offshore 64 (1).

Dutta, N.C., 2002. Geopressure prediction using seismic data: current status and the road ahead. Geophysics 67 (6), 2012–2041.

Eaton, B.A., 1968. Fracture gradient prediction and its application in oilfield operations. JPT 21, 25–32. Paper SPE2163.

Eaton, B.A., 1972. The effect of overburden stress on geopressure prediction from well logs. JPT 24, 929–934. Paper SPE3719.

Eaton, B.A., 1975. The Equation for Geopressure Prediction From Well Logs. Society of Petroleum Engineers of AIME. Paper SPE 5544.

Flemings, P.B., Stump, B.B., Finkbeiner, T., Zoback, M., 2002. Flow focusing in overpressured sandstones: theory, observations, and applications. Am. J. Sci. 302, 827–855.

Foster, M.S., 1997. Geophysics 62 (1), 252–261.

Gardner, G.H.F., Gardner, L.W., Gregory, A.R., 1974. Formation velocity and density—the diagnostic basis for stratigraphic traps. Geophysics 39 (6), 2085–2095.

Gartrell, A., et al., 2006. A new model for assessing trap integrity and oil preservation risks associated with postrift fault reactivation in Timor Sea. AAPG Bull. 90 (12), 1921–1944.

Grochau, M., Gurevich, B., 2009. Testing Gassmann fluid substitution: sonic logs versus ultrasonic core measurements. Geophys. Prospect. 57 (1), 75–79. http://dx.doi.org/10.1111/j.1365-2478.2008.00726.x.

Gutierrez, M.A., Braunsdore, N.R., Couzens, B.A., 2006. Calibration and ranking of pore-pressure prediction models. Lead. Edge 25, 1516–1523.

Haimson, B.C., Fairhurst, C., 1970. In situ stress determination at great depth by means of hydraulic fracturing. In: Somerton, W.H. (Ed.), Rock Mechanics—Theory and Practice. Am. Inst. Mining Eng., New York, NY, pp. 559–584.

Hardwick, A., Travis, T., Stokes, S., Hart, M., 2010. Lows and highs: using low frequencies and improved velocity tools to image complex ridges and basement highs in the Faroe-Shetland Basin. First Break 28 (9), 61–67.

Hardwick, A.J., Whittaker, J., Woodburn, N., 2012a. Combining bandwidth extension techniques and pre-stack Q inversion for enhanced imaging of intra-basalt plays. In: EAGE 74th Conference and Exhibition.

Hardwick, A., Woodburn, N., Whittaker, J., 2012b. Broadband verses conventional marine seismic: the importance of compensating for the earth filter. Workshop 11—broadband seismic—technical value to reservoir characterization workflows? In: EAGE 74th Conference and Exhibition.

Heppard, P.D., Cander, H.S., Eggertson, E.B., 1998. Abnormal pressure and the occurrence of hydrocarbons in offshore eastern Trinidad, West Indies. In: Law, B.E., Ulmishek, G.F., Slavin, V.I. (Eds.), Abnormal Pressures in Hydrocarbon Environments: AAPG Memoir, vol. 70. pp. 215–246.

Holand, P., Skalle, P., 2001. Deepwater kicks and BOP performance. SINTEF report for U.S. Minerals Management Service.

Holbrook, P.W., Maggiori, D.A., Hensley, R., 2005. Real-time pore pressure and fracture gradient evaluation in all sedimentary lithologies. SPE Form. Eval. 10 (4), 215–222.

Hottmann, C.E., Johnson, R.K., 1965. Estimation of formation pressures from log-derived shale properties. JPT 17, 717–722. Paper SPE1110.

http://goo.gl/jrTijL.

http://gsa.seg.org/pdf_forms/SEG%20DL%20Handout%20v1.2.pdf.

http://library.seg.org/doi/abs/10.1190/segam2013-1080.1.

http://mpge.ou.edu/research/documents/2011%20thesis/Namrita%20Sondhi_thesis.pdf.

http://subsurfwiki.org/wiki/Phase_determination.

http://www.akamaiuniversity.us/PJST11_1_563.pdf.

http://www.docstoc.com/docs/123302119/PowerPoint-Presentation---DHI---Rose-_amp-Associates.

http://www.exploratoryvision.com/#!velocity-suit/crbm.

http://www.headwave.com/article/articleview/191.

http://www.headwave.com/article/articleview/160.

http://www.iongeo.com/content/documents/ResourceCenter/TechnicalPapers/EAGE_Chalk_tutorial_Jones_130601.pdf.

http://www.petromod.com/.

http://www.roseassoc.com/SAAM.html.

http://www.searchanddiscovery.com/documents/2010/110139forrest/ndx_forrest.pdf.

http://www.spe.org/glossary/wiki/doku.php/terms:velocity_gradient_seismic.

http://www.touchbriefings.com/pdf/1736/ACF1A23.pdf.

https://www.flickr.com/photos/hamedamini/sets/72157645696780941/.

Hubbert, M.K., Willis, D.G., 1957. Mechanics of hydraulic fracturing. Pet. Trans. AIME 210, 153–163.

Issler, D.R., 1992. A new approach to shale compaction and stratigraphic restoration, Beaufort-Mackenzie Basin and Mackenzie Corridor, Northern Canada. AAPG Bull. 76, 1170–1189.

Keys, R.G., Xu, S., 2002. An approximation for the Xu-White velocity model. Geophysics 67, 1406. http://dx.doi.org/10.1190/1.1512786.

KSI, 2001. Best practice procedures for predicting pre-drill geopressures in deep water Gulf of Mexico. DEA project 119 report. Knowledge Systems Inc.

Lang, J., Li, S., Zhang, J., 2011. Wellbore stability modeling and real-time surveillance for deepwater drilling to weak bedding planes and depleted reservoirs. In: SPE/IADC 139708 Presented at the SPE/IADC Drilling Conference and Exhibition Held in Amsterdam, The Netherlands, 1–3 March 2011.

Law, B.E., Spencer, C.W., 1998. Abnormal pressures in hydrocarbon environments. In: Law, B.E., Ulmishek, G.F., Slavin, V.I. (Eds.), Abnormal Pressures in Hydrocarbon Environments: AAPG Memoir, vol. 70. pp. 1–11.

Lin, R., Thomsen, L., 2013. Extracting polar anisotropy parameters from seismic data and well logs. In: SEG Technical Program Expanded Abstracts 2013. pp. 310–314. http://dx.doi.org/10.1190/segam2013-1150.1.

Lopez, J.L., Rappold, P.M., Ugueto, G.A., Wieseneck, J.B., Vu, K., 2004. Integrated shared earth model: 3D pore-pressure prediction and uncertainty analysis. Lead. Edge, 52–59.

Matthews, W.R., Kelly, J., 1967. How to predict formation pressure and fracture gradient. Oil Gas J. 65, 92–106.

Mavko's anisotropic Gassmann implementation, 2009.

Meng, Z., Zhang, J., Wang, R., 2011. In-situ stress, pore pressure and stress-dependent permeability in the Southern Qinshui Basin. Int. J. Rock Mech. Min. Sci. 48 (1), 122–131.

Morley, C.K., King, R., Hillis, R., Tingay, M., Backe, G., 2011. Deepwater fold and thrust belt classification, tectonics, structure and hydrocarbon prospectivity: a review. Earth-Sci. Rev. 104, 41–91.

Mouchet, J.-C., Mitchell, A., 1989. Abnormal Pressures While Drilling. Editions Technip, Paris.

Nelson, H.N., Bird, K.J., 2005. Porosity-depth trends and regional uplift calculated from sonic logs, National Reserve in Alaska. Scientific Investigation Report 2005–5051. U.S. Dept. of the Interior and USGS.

Parra, M., et al., 2010. The Monagas fold–thrust belt of eastern Venezuela. Part I: structural and thermal modeling. Mar. Pet. Geol. 28, 40–69.

Peng, S., Zhang, J., 2007. Engineering Geology for Underground Rocks. Springer, Berlin.

Powley, D.E., 1990. Pressures and hydrogeology in petroleum basins. Earth-Sci. Rev. 29, 215–226.

Raiga-Clemenceau, J., Martin, J.P., Nicoletis, S., 1988. The concept of acoustic formation factor for more accurate porosity determination from sonic transit time data. Log Anal. 29 (1), 54–60.

Ruiz, F., Dvorkin, J., 2010. Predicting elasticity in nonclastic rocks with a differential effective medium model. Geophysics 75 (1), E41–E53. http://dx.doi.org/10.1190/1.3267854.

Rutherford, S.R., Williams, R.H., 1989. Amplitude-versus-offset variations in gas sands. Geophysics 54 (6), 680–688. http://dx.doi.org/10.1190/1.1442696.

Saxena, N., Mavko, G., Mukerji, T., 2015. Fluid substitution in multimineralic rocks with large mineral stiffness contrast. Geophysics 80 (3), L11–L33.

Sayers, C.M., Johnson, G.M., Denyer, G., 2002. Predrill pore pressure prediction using seismic data. Geophysics 67, 1286–1292.

Schneider, J., Flemings, P.B., Dugan, B., Long, H., Germaine, J.T., 2009. Overpressure and consolidation near the seafloor of Brazos-Trinity Basin IV, northwest deepwater Gulf of Mexico. J. Geophys. Res. 114, B05102.

Simm, R., 2007. Practical Gassmann fluid substitution in sand/shale sequences. First Break 25 (12), 61–68.

Slotnick, M.M., 1936. On seismic computation with applications. Geophysics 1, 9–22.

Snedden, Sarg, Ying, 2003. Exploration play analysis from a sequence stratigraphic perspective.

Swarbrick, R.E., Osborne, M.J., 1998. Mechanisms that generate abnormal pressures: an overview. In: Law, B.E., Ulmishek, G.F., Slavin, V.I. (Eds.), Abnormal Pressures in Hydrocarbon Environments: AAPG Memoir, vol. 70. pp. 13–34.

Swarbrick, R.E., Osborne, M.J., Yardley, G.S., 2002. Comparison of overpressure magnitude resulting from the main generating mechanisms. In: Huffman, A.R., Bowers, G.L. (Eds.), Pressure Regimes in Sedimentary Basins and Their Prediction: AAPG Memoir, vol. 76. pp. 1–12.

Terzaghi, K., Peck, R.B., Mesri, G., 1996. Soil Mechanics in Engineering Practice, third ed. John Wiley & Sons, New York, NY.

Thomsen, L., 2012. On the fluid dependence of the parameters of anisotropy. In: Society of Exploration Geophysics Convention Expanded Abstracts, 82.

Tingay, M.R.P., Hillis, R.R., Swarbrick, R.E., Morley, C.K., Damit, A.R., 2009. Origin of overpressure and pore-pressure prediction in the Baram province, Brunei. AAPG Bull. 93 (1), 51–74.

Tosaya, C.A., 1982. Acoustical Properties of Clay Bearing Rocks (Ph.D. dissertation). Stanford University.

Traugott, M., 1997. Pore pressure and fracture gradient determinations in deepwater. World Oil 218, 68–70.

Tsvankin, L., Grechka, V., 2009. Seismology of Azimuthally Anisotropic Media and Seismic Fracture Characterization. Geophysics References Series No. 17.

Van Ruth, P., Hillis, R., Tingate, P., 2004. The origin of overpressure in the Carnarvon Basin, Western Australia: implications for pore pressure prediction. Pet. Geosci. 10, 247–257.

Verm, R., Hilterman, F., 1995. Lithology color-coded seismic sections: the calibration of AVO cross plotting to rock properties. Lead. Edge 14, 847–853.

Wang, Q., et al., 2001. Present-day crustal deformation in China constrained by global positioning system measurements. Science 294, 574. http://dx.doi.org/10.1126/science.1063647.

Woerpel, C., 2010. Anisotropic velocity modeling for microseismic processing: part 2—fast and accurate model calibration with a cross-well source. In: SEG Technical Program Expanded Abstracts 2010. pp. 2135–2139. http://dx.doi.org/10.1190/1.3513268.

Woodburn, N., Hardwick, A., Travis, T., 2011. Enhanced low frequency signal processing for sub-basalt imaging. In: Extended Abstracts, EAGE 73rd Conference and Exhibition.

Woodburn, N., Hardwick, A., Travis, T., 2012. Enhanced low frequency signal processing for sub-basalt imaging on the Faroese continental shelf. In: 4th Faroe Islands Exploration Conference.

www.openflow.com.

Wyllie, M.R.J., Gregory, A.R., Gardner, L.W., 1956. Elastic wave velocities in heterogeneous and porous media. Geophysics 21, 41–70.

Xu, S., White, R.E., 1995. A new velocity model for clay-sand mixtures. Geophys. Prospect. 43 (1), 91–118. http://dx.doi.org/10.1111/j.1365-2478.1995.tb00126.x.

Yang, Y., et al., 2013. QTL analysis of high thermotolerance with superior and downgraded parental yeast strains reveals new minor QTLs and converges on novel causative alleles involved in RNA processing. PLoS Genet. 9 (8), e1003693.

Yardley, G.S., Swarbrick, R.E., 2000. Lateral transfer: a source of additional overpressure? Mar. Pet. Geol. 17, 523–537.

York, P., Prithard, D., Dodson, J.K., Dodson, T., Rosenberg, S., Gala, D., Utama, B., 2009. Eliminating non-productive time associated drilling trouble zone. In: OTC20220 Presented at the 2009 Offshore Technology Conference Held in Houston.

Zhang, J., 2008. Unpublished reports and training notes.

Zhang, J., Roegiers, J.-C., 2005. Double porosity finite element method for borehole modeling. Rock Mech. Rock. Eng. 38, 217–242.

Zhang, J., Roegiers, J.-C., 2010. Discussion on "Integrating borehole-breakout dimensions, strength criteria, and leak-off test results, to constrain the state of stress across the Chelungpu Fault, Taiwan". Tectonophysics 492, 295–298.

Zhang, J., Standifird, W.B., Lenamond, C., 2008. Casing ultradeep, ultralong salt sections in deep water: a case study for failure diagnosis and risk mitigation in record-depth well. In: Paper SPE 114273 Presented at SPE Annual Technical Conference and Exhibition, 21–24 September 2008, Denver, Colorado, USA.

Zhang, R., Sen, M.K., Srinivasan, S., 2013. A pre-stack seismic basis pursuit inversion. Geophysics 78 (1), R1–R11.

Index

Note: Page numbers followed by *f* indicate figures and *t* indicate tables.

T

Edwards Brothers Malloy
Ann Arbor MI. USA
July 31, 2017